Control System Analysis and Identification with MATLAB®

Block Pulse and Related Orthogonal Functions

T0315220

Control System Analysis and Identification with MATLAB®

Block Pulse and Related Orthogonal Functions

Anish Deb
Srimanti Roychoudhury

CRC Press
Taylor & Francis Group
Boca Raton London New York

CRC Press is an imprint of the
Taylor & Francis Group, an **informa** business

CRC Press
Taylor & Francis Group
6000 Broken Sound Parkway NW, Suite 300
Boca Raton, FL 33487-2742

First issued in paperback 2020

© 2019 by Taylor & Francis Group, LLC
CRC Press is an imprint of Taylor & Francis Group, an Informa business

No claim to original U.S. Government works

ISBN 13: 978-0-367-57123-8 (pbk)
ISBN 13: 978-1-138-30322-5 (hbk)

Library of Congress Cataloging-in-Publication Data

Names: Deb, Anish, author. | Roychoudhury, Srimanti, 1984- author.
Title: Control system analysis and identification with MATLAB : block pulse and related orthogonal functions / Anish Deb and Srimanti Roychoudhury.
Description: First edition. | Boca Raton, FL : Taylor & Francis Group, 2018.
| "A CRC title, part of the Taylor & Francis imprint, a member of the Taylor & Francis Group, the academic division of T&F Informa plc." |
Includes bibliographical references and index.
Identifiers: LCCN 2018004696| ISBN 9781138303225 (hardback : acid-free paper)
| ISBN 9780203731291 (ebook)
Subjects: LCSH: Automatic control--Mathematics. | Discrete-time systems--Mathematical models. | Functions, Orthogonal. | MATLAB.
Classification: LCC TJ213 .D3536 2018 | DDC 629.80285/53--dc23
LC record available at https://lccn.loc.gov/2018004696

**Visit the Taylor & Francis Web site at
http://www.taylorandfrancis.com**

**and the CRC Press Web site at
http://www.crcpress.com**

To my grandson Aryarup Basak who now is in class seven, but

wants to be a physicist and a writer when he grows up.

Anish Deb

To my mother Dipshikha Roychoudhury and father Sujit Roychoudhury

who always helped and guided me along the journey of my life

and whatever I have achieved in life is a gift from them.

Srimanti Roychoudhury

Contents

List of Principal Symbols

δ	Increasing or decreasing width of generalised block pulse function set
$\delta_i(t)$	$(i+1)$-th component of the delta function set
δ_{pq}	Kronecker delta
$\Delta_{(m)}(t)$	Delta function vector of dimension m, having m component functions
∇	Backward difference operator
φ_n	$(n+1)$th component function of a Walsh function set
Φ	Walsh function vector
ψ_n	$(n+1)$th component function of a block pulse function set
ψ'_n	$(n+1)$th component function of the non-optimal block pulse function set
ψ_{ng}	$(n+1)$th component function of a generalised block pulse function set
ψ_{ngp}	$(n+1)$th component function of a positive pulse width modulated (PPWM) generalised block pulse function set
ψ_{ngn}	$(n+1)$th component function of a negative pulse width modulated (NPWM) generalised block pulse function set
$\Psi_{(m)}$	Block pulse function vector of dimension m
$\Psi'_{(m)}$	Non-optimal block pulse function vector of dimension m
$\Psi_{g(m)}$	Generalised block pulse function vector of dimension m
$\Psi_{gp(m)}$	Positive pulse-width modulated (PPWM) generalised block pulse function vector of dimension m
$\Psi_{gn(m)}$	Negative pulse-width modulated (NPWM) generalised block pulse function vector of dimension m
μ_i	A point in the $(i+1)$th interval
μ_{max}	Maximum value of μ_i
$\otimes$	Kronecker product operation
$\mathbf{A}$	System matrix in state model
AMP error	Average of Mod of Percentage Error
$\mathbf{B}$	Input matrix in state model
$\mathbf{BPOTF}_{(m)}$	Block pulse operational transfer function of order m
c_j	Coefficient or weight connected to the $(j+1)$th member of the orthogonal set
$C(s)$	Laplace transformed output
$C(z)$	z-transform of the output $c(t)$
$\mathbf{C_c}$	Convolution matrix of order m, in block pulse function domain
$\mathbf{C'_c}$	Convolution matrix of order m, in non-optimal block pulse function domain

$\mathbf{C}_e$ Convolution matrix of order m, formed by the elements of the block pulse function vector of the error signal $e(t)$

$\mathbf{C}'_e$ Convolution matrix of order m, formed by the elements of the non-optimal block pulse function vector of error signal $e(t)$

$\hat{c}_i$ $(i + 1)$th coefficient of output, obtained through convolution operation in BPF domain

$CV_{12}(t)$ Convolution of two time functions $f_1(t)$ and $f_2(t)$

$\mathbf{D}_{(m)}$ Operational matrix for differentiation of order m, in block pulse function domain

$\mathbf{D}_G$ Diagonal matrix formed by the elements of the block pulse function vector $\mathbf{G}$

$\mathbf{D}_{g(m)}$ Operational matrix for differentiation of dimension m, in generalised block pulse function domain

$\mathbf{D}(n)_{(m)}$ One-shot operational matrix of order m for n times repeated differentiation, in block pulse function domain

$\mathbf{D}_{W(m)}$ Operational matrix for differentiation of order m in Walsh domain

$\mathbf{DOTF}_{(m)}$ Delta operational transfer function of order m

$f(t)$ Time function

$f(t - \tau)$ Function $f(t)$ delayed by τ seconds

$f(t/\lambda)$ Scaled function

$f^*(t)$ Sampled function

$\hat{f}(t)$ First order Taylor series approximation of the function $f(t)$

$\dot{f}_{max}$ Maximum value of first order derivative of $f(t)$ over a time period T

$G(s)$ Forward path transfer function

$\hat{\mathbf{G}}$ Block pulse vector of the impulse response of the system $g(t)$, obtained through 'deconvolution' operation in BPF domain

$\mathbf{GCVM}_{(m)}$ Generalised convolution matrix of order m, in linearly pulse-width modulated generalised block pulse function (LPWM-GBPF) domain

$\overline{\mathbf{GCVM}}_{(m)}$ Generalised convolution matrix of order m, in conventional block pulse function domain

$\mathbf{GCVMC}_{(m)}$ Generalised convolution matrix of order m, formed by the elements of the linearly pulse-width modulated generalised block pulse function (LPWM-GBPF) vector of output signal $c(t)$

$\mathbf{GCVME}_{(m)}$ Generalised convolution matrix of order m, formed by the elements of the linearly pulse-width modulated generalised block pulse function (LPWM- GBPF) vector of the error signal $e(t)$

h Width of a segment or sub-interval

h_n Width of $(n + 1)$th component of the generalised block pulse function set

$\mathbf{har}_{(m)}$ Haar function set of order m

$\mathbf{H}_{(m)}$ Haar matrix of dimension $(m \times m)$

$\mathbf{I}_{(m)}$	Identity matrix of dimension $(m \times m)$
$\mathbf{J}_{(m)}$	Relational matrix between the Haar function set of order m and the Walsh function set of dimension m
m	Number of sub-intervals considered in a time period T
$\mathbf{MBPOTF}_{(m)}$	Modified block pulse operational transfer function of order m
MISE	Mean integral square error
$\mathbf{N}_{(m)}$	Transformation matrix of order m, relating the delayed unit step function set and the block pulse function set
$\mathbf{P}_{(m)}, \mathbf{P}_{B(m)}$	First order integration operational matrix of dimension m, in block pulse function domain
$\mathbf{P}_{g(m)}$	Operational matrix for integration of dimension m, in generalised block pulse function domain
$\mathbf{P1}_{(m)}$	Improved integration operational matrix of dimension m, in block pulse function domain
$\mathbf{P1}_{DEL(m)}$	Operational matrix for integration of order m, in delta function domain
$\mathbf{P1}_{n(m)}$	Operational matrix for integration of order m, in non-optimal block pulse function domain
$\mathbf{P1}_{s(m)}$	Operational matrix for integration of order m, in sample-and-hold function domain
$\mathbf{P2}_{(m)}$	Integration Operational matrix of dimension m, further improved than $\mathbf{P1}_{(m)}$, in block pulse function domain
$\mathbf{P}(n)_{(m)}$	One-shot operational matrix of order m for n times repeated integration in block pulse function domain
$\mathbf{P1}(n)_{DEL(m)}$	One-shot operational matrix of order m for n times repeated integration of the delta function set
$\mathbf{P1}(n)_{S(m)}$	One-shot operational matrix for n times repeated integration of order m, in sample-and-hold function domain
$\mathbf{P}_{(m)}^{n}$	n times repeated integration matrix of order m, using $\mathbf{P}_{(m)}$ in block pulse function domain
$\mathbf{P1}_{DEL(m)}^{n}$	n times repeated integration, using the first order operational matrix for integration of order m, in delta function domain
$\mathbf{P1}_{S(m)}^{n}$	n times repeated integration, using the first order integration operational matrix of order m, in sample-and-hold function domain
$\mathbf{P}_{D(m)}$	Operational matrix for integration of dimension m, in delayed unit step function domain
$\mathbf{Q}_{(m)}$	Delay matrix of dimension m
$\mathbf{r}(\phi)$	Walsh vector of the time function $r(t)$
$R(s)$	Laplace transformed input
$\mathbb{R}^{m \times m}$	Any real matrix of dimension $(m \times m)$
rad_n	$(n + 1)$th component function of a Rademacher function set
$\mathbf{S}$	Sample-and-hold matrix
S_n	$(n + 1)$th component function of a sample-and-hold function set

$\mathbf{S}_{(m)}$ Sample-and-hold function vector of dimension m, having m
            component functions
$\mathbf{S}_B$ Stretch matrix of order m, in block pulse function domain
$\mathbf{S}_D$ Stretch matrix of order m, in delayed unit step function
            domain
$\mathbf{S}_W$ Stretch matrix of order m, in Walsh function domain
$\mathbf{SHOTF}_{(m)}$ Sample-and-hold operational transfer function of order m
t Time in seconds
T Time period
T_S Sampling period
$TR(z)$ z-transfer function
u_n $(n + 1)$th component function of the delayed unit step function
            (DUSF) set
$\mathbf{U}_{(m)}$ Delayed unit step function vector of dimension m
$\mathbf{W}_{(m)}$ Walsh matrix of dimension $(m \times m)$
$\mathbf{WOTF}_{(m)}$ Walsh operational transfer function of order m
$\mathbf{x}_{(0)}$ Initial value vector
$\mathbf{x}(t)$ State vector
$\mathbf{x}(t/\lambda)$ Scaled state vector
$\mathbf{ZOH}$ Zero order hold

Preface

The first author encountered Walsh functions way back in 1982, when he came across a book entitled *Transmission of Information by Orthogonal Functions* by Henning F. Harmuth. At that time he was working in the area of power electronics. So, the very look of Walsh functions attracted him because of the similarities of some of the power electronic waveforms with Walsh functions. So he thought of bridging these two waveforms through mathematics. That is how he entered the wonderful world of piecewise constant orthogonal basis functions and pursued his research. The task was not easy because then only two books were available across the globe: one book by Harmuth, mentioned above, and the other book *Walsh Functions and Their Applications* by K. G. Beauchamp. Also, the internet was a thing of the future. But in spite of these hurdles the first author stuck to Walsh functions out of sheer passion.

So his journey started and what an enjoyable journey it was! He traveled from Walsh functions to orthogonal hybrid functions spanning a period of three and a half decades. On his way, he met, apart from Walsh functions, block pulse functions, pulse-width modulated generalized block pulse functions, delayed unit step functions, a set of Dirac delta functions, sample-and-hold functions, triangular functions, "non-optimal" block pulse functions and finally orthogonal hybrid functions—a combination of sample-and-hold functions and triangular functions. Out of these nine kinds of orthogonal functions, five were invented by him and the same were applied to the study of control system analysis and identification.

The second author joined the first author in 2010. Since she already had the basic exposure to Walsh and block pulse functions, she became a companion of my research-journey and we two started working together in the above domain of non-sinusoidal orthogonal functions. We were enjoying our association with Walsh functions, block pulse functions and the like, and never felt the toil while we waded to and fro in the sea of our research domain and at the same time worked fanatically on the manuscript of this book.

Application of non-sinusoidal piecewise constant orthogonal functions was initiated by Walsh functions which were introduced by J. L. Walsh in 1922 in a conference and a year later his paper was published in the American Journal of Mathematics. The very look of the Walsh function set was very different from the set of sine-cosine functions in the basic sense that it did not contain any curved lines at all!

For more than four decades, the Walsh function set remained dormant by way of its applications. It became attractive to a few researchers only during mid-1960s. But in the next 10 to 15 years, the Walsh function set found its application in many areas of electrical engineering like communication, solution of differential as well as integral equations, control system analysis, control system identification and in various other fields. But from the beginning of the 1980s, the spotlight shifted to block pulse functions (BPF). The BPF set was also orthogonal and piecewise constant. Further, it was related to Walsh functions and Haar functions by similarity transformation. This function set was the most fundamental and simplest of all piecewise constant basis functions (PCBF). So it is no wonder that the BPF set has been enjoying moderate popularity till date.

Now, let's say a few words about the book. This book revolves around block pulse functions. The focus of the book is on control theory and mainly it deals with analysis and identification of control systems by the use of block pulse and related orthogonal functions. After a brief introduction of the function sets in Chapter 1, the following nine aspects have been addressed in the remaining nine chapters:

Chapter 2 The technique of function approximation

Chapter 3 Block pulse operational matrices for integration and differentiation

Chapter 4 One-shot operational matrices for repeated integration

Chapter 5 Operational transfer function for system analysis

Chapter 6 Convolution and "deconvolution" for system analysis and identification

Chapter 7 Sample-and-hold functions for the analysis of sample-and-hold systems

Chapter 8 Discrete time system analysis through a set of Dirac delta functions

Chapter 9 "Non-optimal" block pulse functions for system analysis and identification

Chapter 10 System analysis and identification using linearly pulse-width modulated generalized block pulse functions

Prerequisite for using this book is a primary knowledge of control theory and linear algebra (an introduction to linear algebra is provided in Appendix A). The material of the book may be used as a special paper at post graduate level of engineering courses. The ideas presented in the book may also be used in designing systems in academic or industrial arena. Further, the book may also help researchers to pursue in a new direction where the space is populated only by piecewise orthogonal functions like the kind discussed in the book.

All the topics in the book are supported with relevant numerical examples. And to make the book user friendly, many MATLAB® program are appended at the end of the book in Appendix B.

Finally, we acknowledge the support of our institute Budge Budge Institute of Technology, Kolkata, India, and especially the support of Mr. Jagannath Gupta, Chairman, Mr. K. K. Gupta, Vice-Chairman and Dr. Shubhangi Gupta, Executive Director of our institute for their moral support and keen interest in our book writing project. The support of our *alma mater*, the Department of Applied Physics, University of Calcutta, is also most gratefully acknowledged.

Anish Deb
Srimanti Roychoudhury

MATLAB® is a registered trademark of The MathWorks, Inc. For product information, please contact:

The MathWorks, Inc.
3 Apple Hill Drive
Natick, MA 01760-2098 USA
Tel: 508 647 7000
Fax: 508-647-7001
E-mail: info@mathworks.com
Web: www.mathworks.com

Authors

Anish Deb (b.1951) did his B. Tech. (1974), M. Tech. (1976), and Ph.D. (Tech.) (1990) from the Department of Applied Physics, University of Calcutta. He started his career as a design engineer (1978) in industry and joined the Department of Applied Physics, University of Calcutta as a Lecturer in 1983. In 1990, he became a Reader and later became a Professor (1998) in the same Department.

He has retired from the University of Calcutta in November 2016 and is presently a Professor and Head of the Department in the Department of Electrical Engineering, Budge Budge Institute of Technology, Kolkata.

His research interest includes automatic control in general and application of "alternative" orthogonal functions like Walsh functions, block pulse functions, triangular functions, etc., in systems and control. He has published more than 70 research papers in different national and international journals and conferences. He is the principal author of the books *Triangular Orthogonal Functions for the Analysis of Continuous Time Systems* published by Elsevier (India) in 2007 and Anthem Press (UK) in 2011, *Power Electronic Systems: Walsh Analysis with MATLAB* published by CRC Press (USA) in 2014, and *Analysis and Identification of Time-Invariant Systems, Time-Varying Systems and Multi-Delay Systems using Orthogonal Hybrid Functions: Theory and Algorithms with MATLAB®* published by Springer (Switzerland) in 2016.

Srimanti Roychoudhury (b.1984) did her B. Tech. (2006) from Jalpaiguri Government Engineering College, under West Bengal University of Technology, M. Tech. (2010) and Ph.D. (Tech.) (March 2018) from the Department of Applied Physics, University of Calcutta. The topic of her doctoral research was "Application of hybrid functions for control system analysis and identification."

During 2006–2007, she worked in the Department of Electrical Engineering of Jalpaiguri Government Engineering College as a part-time Faculty. She also acted as a visiting Faculty during 2012–2013 in the Department of Polymer Science & Technology and during 2015–2016 in the Department of Applied Physics, University of Calcutta. She started her teaching career in 2010 and joined the Department of Electrical Engineering, Budge Budge Institute of Technology, Kolkata as an Assistant Professor. In 2018, she became an Associate Professor in the same institute.

Her research area includes control theory in general and application of "alternative" orthogonal functions like Walsh functions, block pulse functions, triangular functions, hybrid functions, etc., in different areas of systems and control.

She has published eight research papers in different national and international journals and conferences. She is the second author of the book *Analysis and Identification of Time-Invariant Systems, Time-Varying Systems and Multi-Delay Systems using Orthogonal Hybrid Functions: Theory and Algorithms with MATLAB®* published by Springer (Switzerland) in 2016.

1

Block Pulse and Related Basis Functions

1.1 Block Pulse and Related Basis Functions

Orthogonal properties [1] of familiar sine-cosine functions have been known for more than two centuries, but the use of such functions to solve complex analytical problems was initiated by the work of the famous mathematician Baron Jean-Baptiste-Joseph Fourier [2]. Fourier introduced the idea that an arbitrary function, even the one defined by different equations in adjacent segments of its range, could nevertheless be represented by a single analytic expression. Although this idea encountered resistance at the time, it proved to be pivotal to many later developments in many areas of mathematics, science, and engineering.

In many spheres of electrical engineering, the basis for any analysis is usually a system of sine-cosine functions. This is mainly due to the desirable properties of frequency domain representation of a large class of functions encountered in engineering design and also immense popularity of sinusoidal voltage in most engineering applications. In the fields of circuit analysis, control theory, communication, and analysis of stochastic problems, ample examples are found where the *completeness* and *orthogonal properties* [1] of such a system of sine-cosine functions lead to attractive solutions. But with the application of digital techniques in these areas, awareness for other more general complete systems of orthogonal functions has developed. This "new" class of functions, though not possessing some of the desirable properties of sine-cosine functions, has other advantages to be useful in many applications in the context of digital technology. Many members of this class are piecewise constant binary valued, and therefore indicate their possible suitability and applicability in the analysis and synthesis of systems leading to piecewise constant solutions.

1.2 Orthogonal Functions and Their Properties

Any continuous time function can be synthesized completely to a reasonable degree of accuracy by using a set of orthogonal functions. For such accurate representation of a time function, the orthogonal set should be *complete* [1].

Let a time function $f(t)$, defined over a semi-open time interval $[0, T)$, be represented by an orthogonal function set $\mathbf{S}(t)$. Then

$$f(t) = \sum_{j=0}^{\infty} c_j s_j(t) \tag{1.1}$$

where c_j is the coefficient or weight connected to the $(j + 1)$-th member of the orthogonal set.

The members of the set $\mathbf{S}(t)$ are said to be orthogonal in the interval $0 \leq t < T$ if for any positive integral values of p and q, we have

$$\int_0^T s_p(t) s_q(t) \, dt = \delta_{pq} \ (\text{a constant}) \tag{1.2}$$

where δ_{pq} is the Kronecker delta and

$$\delta_{pq} = \begin{cases} 0 & \text{for } p \neq q \\ \text{constant} & \text{for } p = q. \end{cases}$$

When $\delta_{pq} = 1$, for $p = q$, the set is said to be *orthonormal*.

An orthogonal set is said to be *complete* or *closed* if no other function can be found which is normal to each member of the defined set, satisfying equation (1.2).

Since, only a finite number of terms of the series $\mathbf{S}(t)$ can be considered for practical realization of any time function $f(t)$, right-hand side (RHS) of equation (1.1) has to be truncated and we write

$$f(t) \approx \sum_{j=0}^{N} c_j s_j(t) \tag{1.3}$$

where N is a large integer. A point to note is, N is required to be large enough to come up with a solution of the problem with desired accuracy for all practical purposes. Also, it is necessary to choose the coefficients c_j's in such a manner that the mean integral square error (MISE) [3] is minimized.

The MISE is defined as

$$\text{MISE} \triangleq \frac{1}{T} \int_0^T \left[f(t) - \sum_{j=0}^N c_j s_j(t) \right]^2 dt \qquad (1.4)$$

and its minimization is achieved by making

$$c_j = \frac{1}{\delta} \int_0^T f(t) s_j(t) \, dt \qquad (1.5)$$

where δ is considered to be the non-zero value of Kronecker delta.

For a *complete* orthogonal function set, the MISE in equation (1.4) decreases monotonically to zero as N tends to infinity.

The proof of the statement that MISE is minimized if c_j is chosen as per equation (1.5) is presented in the following section.

1.2.1 Minimization of Mean Integral Square Error (MISE)

Let a time function $f(t)$, defined over a semi-open time interval $[0, T)$, be represented by an orthogonal function set $\Phi_{(m)}(t)$ having component functions $\phi_0(t)$, $\phi_1(t)$, $\phi_2(t)$, ..., $\phi_{(m-1)}(t)$.

Then

$$f(t) \approx \sum_{n=0}^{m-1} c_n \phi_n(t) = c_0 \phi_0(t) + c_1 \phi_1(t) + \ldots + c_{(m-1)} \phi_{(m-1)}(t)$$

where c_n's are the coefficients or weights connected to respective members of the orthogonal set.

Let us define

$$\alpha_n \triangleq \frac{1}{\delta} \int_0^T f(t) \phi_n(t) \, dt \qquad (1.6)$$

where δ is the non-zero value of Kronecker delta for the particular orthogonal function set.

Over the period T, the MISE, following (1.4), is given by

$$\text{MISE} = \frac{1}{T} \int_0^T \left[f(t) - \sum_{n=0}^{m-1} c_n \phi_n(t) \right]^2 dt$$

$$= \frac{1}{T} \int_0^T \{ f(t) \}^2 \, dt - \frac{1}{T} \int_0^T 2 f(t) \sum_{n=0}^{m-1} c_n \phi_n(t) dt + \frac{1}{T} \int_0^T \left\{ \sum_{n=0}^{m-1} c_n \phi_n(t) \right\}^2 dt$$

$$(1.7)$$

The third term in equation (1.7) reduces to $\dfrac{1}{T}\displaystyle\sum_{n=0}^{m-1} c_n^2 \delta$, because when the term is expanded, integrations of the cross terms vanish due to property (1.2) of an orthogonal function set.

Using (1.6), the second term of (1.7) reduces to $\dfrac{2}{T}\displaystyle\sum_{n=0}^{m-1} c_n \alpha_n \delta$. Hence, we may write

$$\text{MISE} = \frac{1}{T}\int_0^T \{f(t)\}^2\, dt - \frac{2}{T}\sum_{n=0}^{m-1} c_n \alpha_n \delta + \frac{1}{T}\sum_{n=0}^{m-1} c_n^2 \delta \tag{1.8}$$

Now, we add and subtract the term $\dfrac{1}{T}\delta\displaystyle\sum_{n=0}^{m-1}\alpha_n^2$ to the RHS of (1.8) to get

$$\text{MISE} = \frac{1}{T}\int_0^T \{f(t)\}^2\, dt - \frac{2}{T}\delta\sum_{n=0}^{m-1} c_n \alpha_n + \frac{1}{T}\delta\sum_{n=0}^{m-1} c_n^2 - \frac{1}{T}\delta\sum_{n=0}^{m-1}\alpha_n^2 + \frac{1}{T}\delta\sum_{n=0}^{m-1}\alpha_n^2$$

or,

$$\text{MISE} = \frac{1}{T}\int_0^T \{f(t)\}^2\, dt - \frac{1}{T}\delta\sum_{n=0}^{m-1}\alpha_n^2 + \frac{1}{T}\delta\sum_{n=0}^{m-1}(\alpha_n - c_n)^2$$

From the above equation, we see that the MISE will be minimized if the third term in the RHS could be made zero. This is possible *only* if $\alpha_n = c_n$. That is

$$c_n = \frac{1}{\delta_{mn}}\int_0^T f(t)\phi_n(t)\, dt$$

Thus,

$$\text{MISE} = \frac{1}{T}\left[\int_0^T \{f(t)\}^2 dt - \delta\sum_{n=0}^{m-1} c_n^2\right] \tag{1.9}$$

This is the minimized expression for MISE.

1.2.2 Haar Functions

In 1910, Hungarian mathematician Alfred Haar [4] proposed a *complete* set of piecewise constant binary-valued orthogonal functions. Figure 1.1 shows the first eight members of the set.

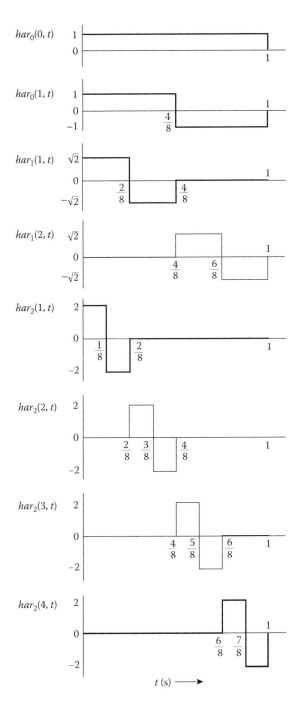

FIGURE 1.1
A set of first eight Haar functions.

Haar functions have three possible states 0 and $\pm A$, where A is a function of $\sqrt{2}$. Thus, the amplitude of the component functions varies with their place in the series.

The component functions of the Haar function set have both scaling and shifting properties. These properties are a necessity for any wavelet [5]. That is why it is now recognized as the first known wavelet basis and at the same time, it is the simplest possible wavelet.

An m-set of Haar functions may be defined mathematically in the semi-open interval $t \in [0, 1)$ as given below:

The first member of the set is

$$har_0(0, t) = 1, \ t \in [0, 1)$$

while the general term for other members is given by

$$har_j(n, t) = \begin{cases} 2^{j/2}, & (n-1)/2^j \leq t < \left(n - \frac{1}{2}\right)/2^j \\ -2^{j/2}, & \left(n - \frac{1}{2}\right)/2^j \leq t < n/2^j \\ 0, & \text{elsewhere} \end{cases} \tag{1.10}$$

where j, n, and m are integers governed by the relations $0 \leq j < \log_2(m)$ and $1 \leq n \leq 2^j$. The number of members in the set is of the form $m = 2^k$, k being a positive integer.

Following (1.10), the members of the set of Haar functions can be obtained in a sequential manner. In Figure 1.1, k is taken to be 3, thus giving $m = 8$.

Haar's set is such that formal expansion of a given continuous function in terms of these new functions converges uniformly to the given function as k tends to infinity.

1.2.3 Rademacher Functions

In 1922, inspired by Haar, German mathematician H. Rademacher presented another set of two-valued orthonormal functions [6] that are shown in Figure 1.2. The set of Rademacher functions is orthonormal but *incomplete*. As seen from Figure 1.2, the first member of the set rad_0 has a constant value of unity throughout the interval, while the 5th member rad_4 of the set is given by a square wave of unit amplitude and 2^{4-1} cycles in the semi-open interval $[0, 1)$. Thus, the $(n + 1)$th member will have 2^{n-1} cycles in the interval.

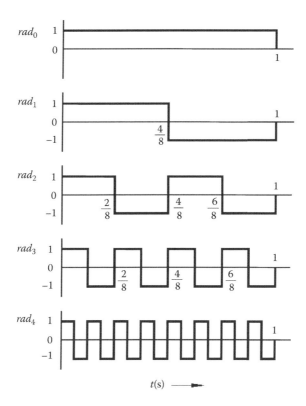

FIGURE 1.2
A set of first five Rademacher functions.

1.2.4 Walsh Functions

After the Rademacher functions were introduced in 1922, around the same time, American mathematician J. L. Walsh independently proposed yet another binary-valued *complete* set of normal orthogonal functions **Φ**, later named Walsh functions [7,8], which is shown in Figure 1.3.

As indicated by Walsh, there are many possible orthogonal function sets of this kind and several researchers, in later years, have suggested orthogonal sets [9,10] formed with the help of combinations of the well-known piecewise constant orthogonal functions.

In his original paper, Walsh pointed out that, "... *Haar's set is, however, merely one of an infinity of sets which can be constructed of functions of this same character.*" While proposing his new set of orthonormal functions **Φ**, Walsh wrote "... *each function φ takes only the values +1 and –1, except at a finite number of points of discontinuity, where it takes the value zero.*"

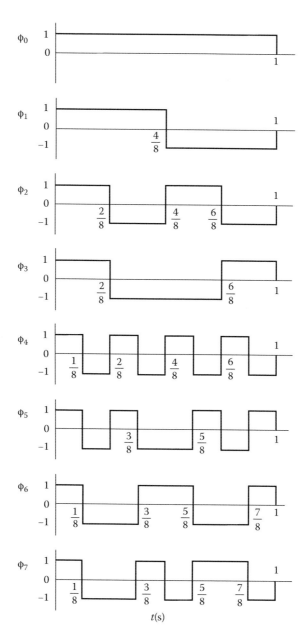

FIGURE 1.3
A set of first eight Walsh functions arranged in dyadic order.

However, the Rademacher functions were found to be a true *subset* of the Walsh function set.

The Walsh function set possesses the following properties all of which are not shared by other orthogonal functions belonging to the same class. These are:

i. Its members are all two-valued functions.

ii. It is a *complete* orthonormal set.

iii. It has striking similarity with the sine-cosine functions, primarily with respect to their zero-crossing patterns.

Like the Haar functions, a set of Walsh functions, comprised of m members, also follows the rule $m = 2^k$, k being an integer.

1.2.4.1 Relation between Walsh Functions and Rademacher Functions [11]

It is *not* at all straightforward to draw the components of Walsh functions in proper order. But it is easy to determine the component Walsh functions using Rademacher functions. This is shown in the following, where, for simplicity, we express $\phi_n(t)$ as ϕ_n.

The relationships between Rademacher functions and Walsh functions are as follows:

$$
\begin{aligned}
\phi_0 &= rad_0 \\
\phi_1 &= rad_1 \\
\phi_2 &= (rad_2)^1 (rad_1)^0 \\
\phi_3 &= (rad_2)^1 (rad_1)^1 \\
\phi_4 &= (rad_3)^1 (rad_2)^0 (rad_1)^0 \\
\phi_5 &= (rad_3)^1 (rad_2)^0 (rad_1)^1 \\
\phi_6 &= (rad_3)^1 (rad_2)^1 (rad_1)^0 \\
\phi_7 &= (rad_3)^1 (rad_2)^1 (rad_1)^1 \\
&\ \ \vdots \\
\phi_n &= (rad_p)^{b_1} (rad_{p-1})^{b_2} (rad_{p-2})^{b_3} \cdots
\end{aligned}
\tag{1.11}
$$

where $p = [\log_2 n] + 1$ in which $[\log_2 n]$ means the greatest integer in $\log_2 n$, and $b_1, b_2, b_3, \ldots$ are binary digits obtained by converting the decimal number n to its binary equivalent.

That is

$$
n = b_1 2^{p-1} + b_2 2^{p-2} + b_3 2^{p-3} + \cdots
$$

1.2.4.2 Numerical Example

Example 1.1

Determine the Walsh function $\phi_6(t)$ with the help of Rademacher functions.

Here, $n = 6$.

So, following relation (1.11), we have

$$p = [\log_2 n] + 1 = 3$$

Expressing n in binary form, we write

$$6 = 1 \cdot 2^2 + 1 \cdot 2^1 + 0 \cdot 2^0$$
$$\uparrow \qquad \uparrow \qquad \uparrow$$
$$b_1 \qquad b_2 \qquad b_3$$

Substituting the values of b_1, b_2, and b_3 in the general expression (1.11), we obtain

$$\phi_6 = (rad_3)^1 (rad_2)^1 (rad_1)^0$$

which means that Walsh function $\phi_6(t)$ is obtained as a product of Rademacher functions rad_3 and rad_2, as shown in Figure 1.4.

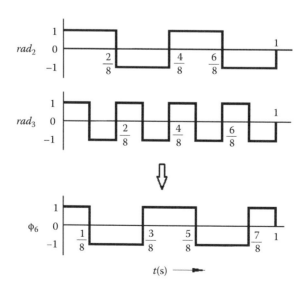

FIGURE 1.4

Walsh function ϕ_6 shown as a product of two Rademacher functions, rad_2 and rad_3.

1.2.5 Slant Functions

A special orthogonal function set, known as the slant function set, was introduced by Enomoto and Shibata [12] for image transmission analysis. These functions are also applied successfully to image processing problems [13,14].

Slant functions have a finite but a large number of possible states as can be seen from Figure 1.5. The superiority of the slant function set lies in its transform characteristics, which permit a compaction of the image energy to only a few transformed samples. Thus, the efficiency of image data transmission in this form is improved.

However, the slant function set enjoyed very limited applications.

1.2.6 Block Pulse Functions (BPF)

During the 19th century, voltage and current pulses, such as Morse code signals, were generated by mechanical switches, and finally detected by different magneto-mechanical devices. These pulses were nothing but block pulses—the most important function set used for communication.

However, until the 1980s of the last century, the set of block pulses received less attention from the mathematicians as well as application engineers possibly due to their apparent *incompleteness*. But disjoint and orthogonal properties of such a function set were well known. However, the convergence properties of block pulse series were established in 1981 by Kwong and Chen [15].

The BPF's were first mentioned in Reference [9] and later formally introduced in the literature by Chen et al. [16,17].

A set of first eight block pulse functions [16–19] in the semi-open interval $t \in [0, T)$ is shown in Figure 1.6.

An m-set BPF is defined as

$$\psi_i(t) = \begin{cases} 1 & \text{for } ih \le t < (i+1)h \\ 0 & \text{elsewhere} \end{cases} \tag{1.12}$$

where $i = 0, 1, 2, \ldots, (m-1)$ and $h = \dfrac{T}{m}$, the width of each component block pulse.

The BPF set is a complete [15] orthogonal function set and can easily be normalized by defining the component functions in the interval [0, T) as

$$\psi_i(t) = \begin{cases} \dfrac{1}{\sqrt{h}} & \text{for } ih \le t < (i+1)h \\ 0 & \text{elsewhere} \end{cases}$$

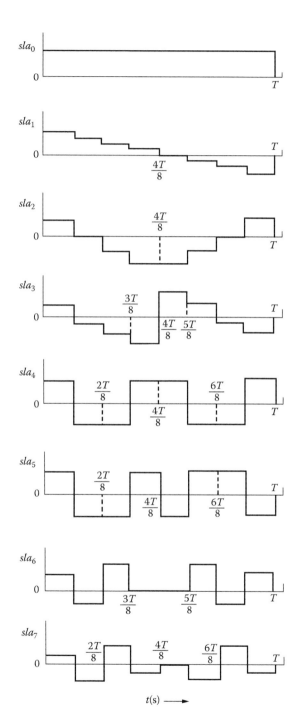

FIGURE 1.5
A set of first eight slant functions.

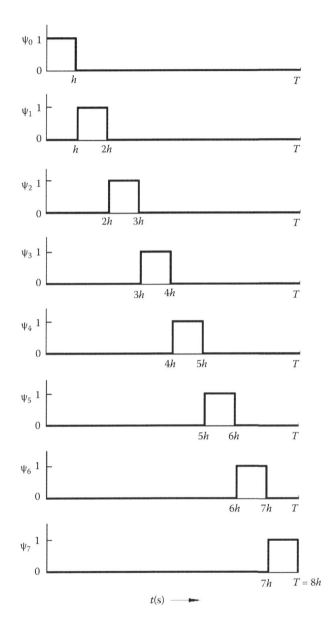

FIGURE 1.6
A set of first eight block pulse functions.

It is well known [18] that an absolutely integrable real-valued function $f(t)$ of Lebesgue measure, defined in the semi-open interval $[0, T)$, can be expanded in an m-term equal width BPF series, as

$$f(t) \approx \sum_{i=0}^{m-1} f_i \psi_i = \begin{bmatrix} f_0 & f_1 & \cdots & f_{(m-1)} \end{bmatrix} \Psi_{(m)}(t) \triangleq \mathbf{F}^{\mathrm{T}} \Psi_{(m)}(t) \qquad (1.13)$$

where $\Psi_{(m)}(t)$ is the BPF vector of order m given by

$$\Psi_{(m)}(t) = \begin{bmatrix} \psi_0(t) & \psi_1(t) & \cdots & \psi_{(m-1)}(t) \end{bmatrix}$$

and the coefficients f_i, $i = 0, 1, 2, \ldots, m - 1$, associated with respective BPF members are [18]

$$f_i = \frac{1}{h} \int_{ih}^{(i+1)h} f(t)\mathrm{d}t, \quad i = 0, 1, 2, \ldots, m - 1 \qquad (1.14)$$

That is, f_i is the average value of the function $f(t)$ in the $(i + 1)$-th interval with a time span of T/m.

1.2.7 Relation among Haar, Walsh, and Block Pulse Functions

Haar, Walsh, and BPF are related through *similarity transformation*. In the following, the relations are shown.

The matrix relating the Walsh function set with the BPF set is known as *Walsh matrix* [16,17] and the relation is given by

$$\Phi_{(m)} = \mathbf{W}_{(m)} \Psi_{(m)} \qquad (1.15)$$

where $\Phi_{(m)}$ is the Walsh function set of order m, $\Psi_{(m)}$ is the BPF set of order m, and $\mathbf{W}_{(m)}$ is the Walsh matrix of dimension $(m \times m)$. When $m = 4$, $\mathbf{W}_{(4)}$ is given by

$$\mathbf{W}_{(4)} = \begin{bmatrix} 1 & 1 & 1 & 1 \\ 1 & 1 & -1 & -1 \\ 1 & -1 & 1 & -1 \\ 1 & -1 & -1 & 1 \end{bmatrix} \qquad (1.16)$$

The Walsh matrix has the following properties:

$$\mathbf{W}_{(m)}^T = \mathbf{W}_{(m)} \text{ and } \mathbf{W}_{(m)}\mathbf{W}_{(m)}^T = \mathbf{W}_{(m)}^2 = m\mathbf{I}_{(m)} \tag{1.17}$$

where $\mathbf{I}_{(m)}$ is the identity matrix of order m.

Similarly, the matrix that relates the Haar function set and the BPF set is known as the *Haar matrix* [20] and the relation is given by

$$\mathbf{har}_{(m)} = \mathbf{H}_{(m)}\mathbf{\Psi}_{(m)} \tag{1.18}$$

where $\mathbf{har}_{(m)}$ is the Haar function set of order m, and
$\mathbf{H}_{(m)}$ is the Haar matrix of dimension $(m \times m)$.
When $m = 4$, $\mathbf{H}_{(4)}$ is given by

$$\mathbf{H}_{(4)} = \begin{bmatrix} 1 & 1 & 1 & 1 \\ 1 & 1 & -1 & -1 \\ \sqrt{2} & -\sqrt{2} & 0 & 0 \\ 0 & 0 & \sqrt{2} & -\sqrt{2} \end{bmatrix} \tag{1.19}$$

The Haar matrix has the following property:

$$\mathbf{H}_{(m)}\mathbf{H}_{(m)}^T = m\mathbf{I}_{(m)} \tag{1.20}$$

From equations (1.15) and (1.18), we can determine the relationship between the Haar function set and the Walsh function set.
From equation (1.15), we can use (1.17) to write

$$\mathbf{\Psi}_{(m)} = \mathbf{W}_{(m)}^{-1}\mathbf{\Phi}_{(m)} = \frac{1}{m}\mathbf{W}_{(m)}\mathbf{\Phi}_{(m)} \tag{1.21}$$

Using (1.21) in (1.18), we have

$$\mathbf{har}_{(m)} = \frac{1}{m}\mathbf{H}_{(m)}\mathbf{W}_{(m)}\mathbf{\Phi}_{(m)} \triangleq \mathbf{J}_{(m)}\mathbf{\Phi}_{(m)} \tag{1.22}$$

where the $\mathbf{J}_{(m)}$ is the relational matrix of dimension $(m \times m)$.
Equation (1.22) provides us the relationship between the Haar function set of order m and the Walsh function set of order m.

For $m = 4$, $\mathbf{J}_{(4)}$ is given by

$$\mathbf{J}_{(4)} = \begin{bmatrix} 1 & 0 & 0 & 0 \\ 0 & 1 & 0 & 0 \\ 0 & 0 & \dfrac{1}{\sqrt{2}} & \dfrac{1}{\sqrt{2}} \\ 0 & 0 & \dfrac{1}{\sqrt{2}} & -\dfrac{1}{\sqrt{2}} \end{bmatrix} \tag{1.23}$$

This relational matrix $\mathbf{J}_{(m)}$ has the following property:

$$\mathbf{J}_{(m)}\mathbf{J}_{(m)}^{\mathrm{T}} = \mathbf{I}_{(m)} \tag{1.24}$$

Inspection of equations (1.17), (1.20), and (1.24) reveals that when we multiply a relational matrix with its transpose, the result is $m\mathbf{I}_{(m)}$ for Walsh matrix and Haar matrix. These matrices relate BPF with Walsh functions and Haar functions respectively. Since the BPF sets are not *orthonormal*, the constant m appears in the RHS of (1.17) and (1.20). Had the BPF set been normalized, the above result would have been $\mathbf{I}_{(m)}$ only.

This is so for equation (1.24) for the $\mathbf{J}$ matrix relating two orthonormal function sets like Haar and Walsh.

1.2.8 Generalized Block Pulse Functions (GBPF)

If the subinterval h in (1.12) is of different width for different components of BPF, like h_0, h_1, h_2, ..., $h_{(m-1)}$ etc., the set of BPF is known as a *generalized* BPF set.

The generalized BPFs are defined in the literature [21,22] as

$$\psi_{ig}(t) = \begin{cases} 1, & t_i \leq t < t_{i+1} \\ 0, & \text{otherwise} \end{cases}$$

Figure 1.7 represents a set of m GBPFs with unit amplitude and different widths h_i ($i = 0, 1, 2, ..., m - 1$). The nomenclature is such that it will be able to characterize a particular member of the GBPF set in a unique manner. From Figure 1.7 it is observed that there are three characterizing properties of any member of a GBPF set. These are:

a. The distance of the member from the origin,
b. The width of the member,
c. The total time interval under consideration.

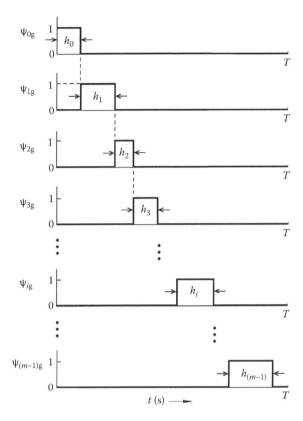

FIGURE 1.7
A set of *m*-generalized block pulse functions (GBPF).

These three properties can represent any member of the GBPF set uniquely. Thus, the $(i + 1)$-th member of the GBPF set can be represented as:

$$\psi_{ig}\left[\sum_{j=0}^{i-1}h_j, h_i, \sum_{j=0}^{m-1}h_j\right] = \begin{cases} 1, & \sum_{j=0}^{i-1}h_j \leq t < \sum_{j=0}^{i}h_j \\ 0, & \text{otherwise} \end{cases} \tag{1.25}$$

where the arguments of ψ_{ig} represent

a. the distance of ψ_{ig} from the origin, i.e.

$$\sum_{j=0}^{i-1}h_j$$

and when $j = 0$, the summation is assumed to be zero, indicating a
zero delay of ψ_{0g} from the origin,

b. the width of ψ_{ig} $(= h_i)$,

c. the total time interval T under consideration, i.e.

$$\sum_{i=0}^{m-1} h_i$$

It is evident that (1.25) does not have an aesthetic form, mainly because we
have considered a set of GBPF with no restriction on the widths of its members. Also, it is obvious that using a set of GBPF for system analysis will meet
with unnecessary hindrance to form the required algorithm only because of
its so-called generality.

However, if we make all the h_i's equal, then this set of GBPF becomes "well-
behaved" and is reduced to the well-known set of conventional BPF.

If $h_0 = h_1 = h_2 = \cdots = h_{(m-1)} = h$, we have $h = T/m$. Then from (1.25) we write

$$\psi_{ig}(ih, h, T) = \begin{cases} 1, & ih \le t < (i+1)h \\ 0, & \text{elsewhere} \end{cases} \tag{1.26}$$

Since h and T are fixed for a defined set, the main characteristic of any
member of a BPF set is the delay time ih.

The stress on the delay may seem irrelevant, but we would like to point
out one particular characteristic of BPFs, which distinguishes them from
the sine-cosine family as well as other sequency functions [9]. In the case of
BPFs, it is not possible to define a frequency (as for sine-cosine functions) or
sequency (as for Walsh functions) because these functions, though fundamental in nature, do not have any zero-crossing at all. Therefore, in contrast,
while denoting ψ_{ig} we can use ih as an indicator similar to the frequency or
sequency of other orthogonal functions. Since T/m is constant in the delay ih,
i may be used to define the order of the BPF ψ_{ig}. Thus, (1.26) can be further
simplified to

$$\psi_{ig}(h, T) = \begin{cases} 1, & ih \le t \le (i+1)h \\ 0, & \text{elsewhere} \end{cases} \tag{1.27}$$

where i serves as a double indicator: the order of the function, and its delay
ih from the origin. Since the order of any member of the BPF set depends
entirely upon its delay, the subscript i in ψ_{ig} is consistent in indicating the
specific function in a qualitative manner.

1.2.8.1 Advantages of Using Generalized BPF over Conventional BPF

Generalized BPF's are able to reconstruct many functions with less number of components within the same error bounds compared to its equal-width conventional counterpart.

Also, in case of staircase functions, if we approximate the function by equal-width BPFs, we find that the total error introduced may be much more, because, the widths of the BPF components may not match exactly with the jumps of the staircase function (which is what happens most of the time). Thus, the error is evidently increased. But if we use generalized BPFs and the widths of different BPF components are matched with the shape of the staircase function, then error is drastically reduced. The only disadvantage being the *a priori* knowledge required about the jump points of the staircase function for a decent reconstruction. A typical example of such a function with equal-width BPF approximation is shown in Figure 1.8.

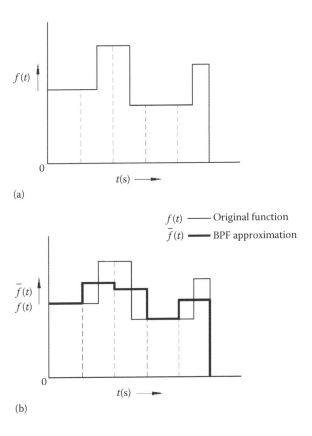

FIGURE 1.8
(a) A staircase function and (b) its equal-width block pulse function representation. The original function becomes unrecognizable.

From Figure 1.8, it is observed that for equal-width approximation, not only the error is much more, but the shape of the approximated function greatly deviates from the original as well. Such approximation makes the BPF represented function "unrecognizable."

1.2.9 Pulse-Width Modulated Generalized Block Pulse Functions (PWM-GBPF) [23,24]

1.2.9.1 Conversion of a GBPF Set to a Pulse-Width Modulated (PWM) GBPF Set

It is evident from Section 1.2.8 that to approximate a function $f(t)$ by a set of generalized BPF is not simple. Also, the selection of the widths $h_i = (i = 0, 1, 2, ..., m - 1)$ is difficult since it demands *a priori* knowledge of the nature of variation of $f(t)$. That is, where the variation of $f(t)$ is larger, we need smaller widths of the BPF for better approximation, and *vice versa*. This selection will reduce the number of BPF components in the set while keeping the accuracy of approximation at the desired level. However, in this case, the computational algorithm becomes complicated.

To simplify the processing steps, we present a GBPF set with members of monotonically *increasing* or *decreasing* width, where the magnitude of increase or decrease δ (say) is constant.

For instance, if the first member of a monotonically increasing set has a width of h_0, then the next members of the set will be of widths $(h_0 + \delta)$, $(h_0 + 2\delta)$,

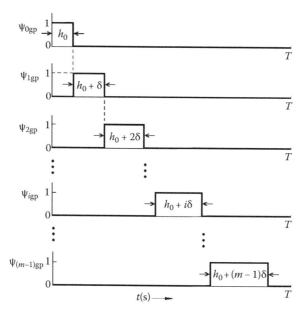

FIGURE 1.9

A set of *m-positive* pulse width modulated (PPWM) generalized block pulse function. The constant positive increment of the set being δ.

$(h_0 + 3\delta)$, ..., $[h_0 + (m - 1)\,\delta]$ for an m-set of GBPF. This is shown in Figure 1.9. If we consider this to be positive pulse-width modulation, then we can call this set a positive pulse-width modulated (PPWM) generalized BPF set. Hence, for a negative pulse-width modulated (NPWM) set of m-members, the widths of the functions will be h_0, $(h_0 - \delta)$, $(h_0 - 2\delta)$, ... , $[h_0 - (m - 1)\delta]$ as shown in Figure 1.10.

Any function $f(t)$ may be approximated by a set of PWM-GBPF by investigating the nature of $f(t)$ through computing the derivatives $\dot{f}(0)$ and $\dot{f}(\infty)$, and comparing their values. That is, if $\left|\dot{f}(0)\right| > \left|\dot{f}(\infty)\right|$, we select the PPWM-GBPF set, and if $\left|\dot{f}(0)\right| < \left|\dot{f}(\infty)\right|$, then the NPWM-GBPF set is chosen. It is noted that for such cases, $\left|\ddot{f}(t)\right| \neq 0$.

Obviously, when $\dot{f}(0) = \dot{f}(\infty)$, we put $\delta = 0$ and the analysis is carried out with a conventional set. However, by this method neither is the assessment of $f(t)$ at intermediate points possible, nor is it computationally attractive. For oscillatory functions, the best choice will be either to put $\delta = 0$ or to represent such a function as a sum of piecewise monotonically increasing or decreasing functions. Then it is possible to follow the above-mentioned strategy.

The total period of a PWM-GBPF set is given by

$$T = mh_0 + \frac{m(m-1)}{2}\delta \qquad (1.28)$$

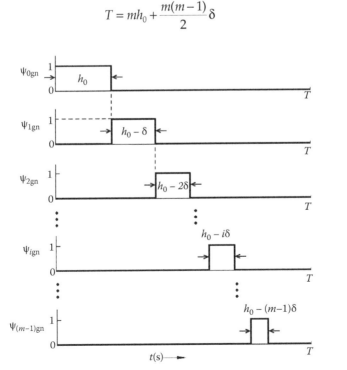

FIGURE 1.10
A set of *m-negative* pulse width modulated (NPWM) generalized block pulse functions. The constant negative decrement of the set being δ.

where δ will assume a positive value for the PPWM-GBPF set and a negative one for the NPWM-GBPF set.

However, for the sake of convenience in analyses using PWM-GBPF, the following restriction on the choice of δ is suggested:

$$(m-1)|\delta| < h_0 \tag{1.29}$$

To summarize, the algorithm to choose a PWM-GBPF set for approximation of a time function $f(t)$ will be as follows:

Step 1: Select m, the number of generalized BPF in the set.

Step 2: Check whether $\ddot{f}(t) \neq 0$.

Step 3: Select T.

Step 4: Compute $|\dot{f}(0)|$ and $|\dot{f}(\infty)|$.

Step 5: If $|\dot{f}(0)| > |\dot{f}(\infty)|$, then δ is positive.

Step 6: If $|\dot{f}(0)| < |\dot{f}(\infty)|$, then δ is negative.

Step 7: If $|\dot{f}(0)| = |\dot{f}(\infty)|$, then δ = 0.

Step 8: Compute h_0 from (1.28), following constraint (1.29).

Step 9: Form the PWM-GBPF set for generalized block pulse domain analysis.

1.2.9.2 Principle of Representation of a Time Function via a Pulse-Width Modulated (PWM) GBPF Set

Let us consider a PWM-GBPF set as defined above and as shown in Figure 1.9. If $f(t)$ is expanded in terms of such a set with m members, then $f(t)$ is given by

$$f(t) \triangleq \mathbf{C}^{\mathrm{T}} \mathbf{\Psi}_{gp(m)}(t) \tag{1.30}$$

where $\mathbf{C} \triangleq \begin{bmatrix} c_0 & c_1 & \cdots & c_{(m-1)} \end{bmatrix}^{\mathrm{T}}$ (where [...]^{T} denotes transpose) and $\mathbf{\Psi}_{gp(m)}(t)$ is an m-term set of PWM-GBPF series. Obviously, the width of the $(i+1)$-th member of the set, namely $\psi_{igp}(t)$, is

$$\text{width} [\psi_{igp}(t)] = h_0 + i\delta, \quad i = 0, 1, 2, \dots, m-1$$

and the coefficients c_i, $i = 0, 1, 2, \dots, m - 1$, associated with respective PWM-GBPF members, are given by

$$c_i = \frac{1}{(h_0 + i\delta)} \int_{ih_0 + i(i-1)\delta/2}^{(i+1)h_0 + i(i+1)\delta/2} f(t)\mathrm{d}t \tag{1.31}$$

That is, c_i is the average value of the function $f(t)$ in the $(i + 1)$-th interval having a time span of $(h_0 + i\delta)$.

When $\delta = 0$, $h = h_0$, and we have a BPF set of members with equal widths. Thus, (1.31) reverts back to (1.14), giving $c_i = f_i$.

1.2.10 Non-Optimal Block Pulse Functions (NOBPF)

The "non-optimal" BPFs were suggested by Deb et. al. [25,26]. For coefficient computation of NOBPF expansion, Deb et. al. employed trapezoidal [27] integration instead of the exact integration of (1.14).

In equation (1.14), the BPF coefficients f_i' 's were computed using the traditional integration formula. These coefficients may be termed "optimal" coefficients because they imply minimum MISE with respect to function approximation.

The "non-optimal" expansion procedure for computation of coefficients used the trapezoidal rule for integration where only the terminal samples of the function to be approximated are needed in any particular time interval. Thus, the computational burden was much reduced in lieu of guaranteed MISE.

Let us employ the well-known trapezoidal rule to compute the NOBPF coefficients of a time function f_t. Calling these coefficients f_i' 's, we get

$$f_i' \approx \frac{\frac{1}{2}\Big[f(ih) + f\{(i+1)h\}\Big]h}{h} = \frac{\Big[f(ih) + f\{(i+1)h\}\Big]}{2} \qquad (1.32)$$

It is observed that f_i' 's are, in effect, the average values of two consecutive samples of the function f_t, and this is again a significant deviation from the traditional formula (1.14).

The process of function approximation in NOBPF domain is shown in Figure 1.11.

A time function f_t can be approximated in NOBPF domain as

$$f(t) \approx \begin{bmatrix} f_0' & f_1' & f_2' & \cdots & f_i' & \cdots & f_{(m-1)}' \end{bmatrix} \Psi'_{(m)}(t) \triangleq \mathbf{F'}^{\mathrm{T}}\Psi'_{(m)}(t) \qquad (1.33)$$

where, $\Psi'_{(m)}$ is the non-optimal block pulse vector of order m.

It is evident that f_i' 's in (1.32) are different from f_i' 's of equation (1.14) and f_t in (1.33) is different from that of (1.13). Hence, it will not be approximated with guaranteed MISE.

1.2.11 Delayed Unit Step Functions (DUSF)

Delayed unit step functions, shown in Figure 1.12, were suggested by Hwang [28] in 1983. Though not of much use due to its dependence on BPFs,

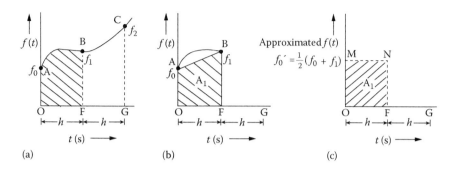

(a)　　　　　　　　　　　(b)　　　　　　　　　　　(c)

FIGURE 1.11
(a) The function $f(t)$ and its three equidistant samples A, B, and C; (b) piecewise linear approximation of $f(t)$ in the first interval by joining the samples A and B with a straight line; and (c) non-optimal block pulse function (NOBPF) approximation of $f(t)$ in the first interval through area preserving transformation, where the area of the trapezium AOFB is represented via the equivalent rectangle MOFN.

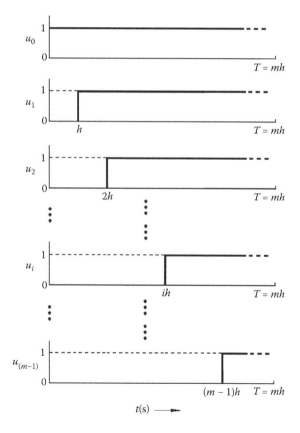

FIGURE 1.12
A set of DUSF for m-component functions.

proved by Deb et al. [19], it deserves to be included in the historical record of piecewise constant basis functions as a new variant.

The $(i + 1)$-th member of this function set is defined as

$$u_i(t) = \begin{cases} 1 & t \geq ih \\ 0 & t < ih \end{cases}$$

where $i = 0, 1, 2, \ldots, (m - 1)$ and h is the amount of delay between consecutive members of the DUSF set.

As in the case of BPF, any square integrable function of Lebesgue measure in the interval $[0, T)$ can be approximated by

$$f(t) \approx \sum_{i=0}^{m-1} g_i u_i(t) \triangleq \mathbf{G}^{\mathrm{T}} \mathbf{U}_{(m)}(t) \tag{1.34}$$

where the DUSF coefficient vector $\mathbf{G}$ is

$$\mathbf{G} = \begin{bmatrix} g_0 & g_1 & \cdots & g_{(m-1)} \end{bmatrix}^{\mathrm{T}} \tag{1.35}$$

and the DUSF vector of order m is

$$\mathbf{U}_{(m)}(t) \triangleq \begin{bmatrix} u_0(t) & u_1(t) & \cdots & u_{(m-1)}(t) \end{bmatrix}^{\mathrm{T}}$$

In (1.34), g_i is given by [8]

$$g_i = \frac{1}{h} \left[\int_{ih}^{(i+1)h} f(t)\,\mathrm{d}t - \int_{(i-1)h}^{ih} f(t)\,\mathrm{d}t \right] \tag{1.36}$$

Table 1.1 compares the basic properties of BPF and delayed unit step functions. The qualitative comparison fairly indicates the superiority of BPFs over DUSFs. This superiority can be established analytically and this has been shown later in Chapter 6.

From equations (1.34) and (1.36), we can write

$$g_0 = f_0 \text{ and } g_i = f_i - f_{(i-1)} \tag{1.37}$$

Hence,

$$\mathbf{G} = \begin{bmatrix} f_0 & (f_1 - f_0) & (f_2 - f_1) & \cdots & (f_{(m-1)} - f_{(m-2)}) \end{bmatrix}^{\mathrm{T}}$$

TABLE 1.1

Basic Properties of BPFs and DUSFs

	BPF	DUSF
Piecewise constant	Yes	Yes
Orthogonal	Yes	No
Finite	Yes	No
Disjoint	Yes	No
Implementation	Easily implementable	Has to be truncated for implementation
Orthonormal	Can easily be normalized	Cannot be normalized
Accuracy of analysis	Same as with DUSF	Same as with BPF

Substituting **G** in (1.34), we obtain

$$f(t) \approx \begin{bmatrix} f_0 & (f_1 - f_0) & (f_2 - f_1) & \cdots & (f_{(m-1)} - f_{(m-2)}) \end{bmatrix} \begin{bmatrix} u_0(t) \\ u_1(t) \\ u_2(t) \\ \vdots \\ u_{(m-1)}(t) \end{bmatrix}$$

or, $f(t) \approx f_0 u_0(t) + (f_1 - f_0)u_1(t) + (f_2 - f_1)u_2(t) + \cdots + (f_{(m-1)} - f_{(m-2)})u_{(m-1)}(t)$

Rearranging, we have [19]

$$f(t) \approx f_0[u_0(t) - u_1(t)] + f_1[u_1(t) - u_2(t)] + \cdots$$

$$+ f_{(m-2)}\Big[u_{(m-2)}(t) - u_{(m-1)}(t)\Big] + f_{(m-1)}u_{(m-1)}(t)$$

or, $f(t) \approx f_0\psi_0(t) + f_1\psi_1(t) + \cdots + f_{(m-2)}\psi_{(m-2)}(t) + f_{(m-1)}\psi_{(m-1)}(t)$ (1.38)

From equation (1.38), it is clear that the expansion coefficients of a function $f(t)$ in DUSF domain is not different from equation (1.14), obtained in BPF domain.

Dependence of DUSFs on BPFs is clear from the following general equation

$$\psi_i(t) = u_i(t) - u_{(i+1)}(t)$$ (1.39)

and Figures 1.6 and 1.12.

1.2.12 Sample-and-Hold Functions (SHF)

Any square integrable function $f(t)$ may be represented by a sample-and-hold function set [29] in the semi-open interval $[0, T)$ by considering the $(i + 1)$-th component of the expansion to be

$$f_i(t) \approx f(ih), \ ih \le t(i+1)h, \ i = 0,1,2,\ldots,(m-1)$$

where h is the sampling period $(= T/m)$, $f_i(t)$ is the part of the function $f(t)$ in the $(i + 1)$-th interval and $f(ih)$ is the first term of the Taylor series expansion of the function $f(t)$ around the point $t = ih$. Thus $f_i(t) = f_i S_i(t)$.
   A set of SHF, comprised of m component functions, is defined as

$$S_i(t) = \begin{cases} 1 & \text{for } ih \le t < (i+1)h \\ 0 & \text{elsewhere} \end{cases} \tag{1.40}$$

where $i = 0, 1, 2, \ldots, (m - 1)$.
   The basic functions of the SHF set are look-alikes of the members of the BPF set shown in Figure 1.6. Only the method of computation of the coefficients differs in the two cases. That is, the expansion coefficients in SHF domain do not depend upon the traditional integration formula (1.14).
   As far as MISE is concerned, SHF approximations incur more error compared to its BPF equivalent for most of the cases. When we deal with systems having sample-and-hold, the SHF function set is the most useful set as an analysis tool because it incurs least error [9].
   Figure 1.13 shows a time scale history of all the functions discussed earlier.

1.3 BPF in Systems and Control

The BPF set started gaining ground over Walsh function set from the early 1980s of the last century. Its accelerating popularity was mainly due to the following attributes:

   a. Compatibility with digital technology with respect to 1 and 0, as is apparent from Figure 1.6.

   b. Its simplicity which added to its potential.

   c. The operational matrix for integration and differentiation (discussed in Chapter 3) in block pulse domain are patterned matrices and also

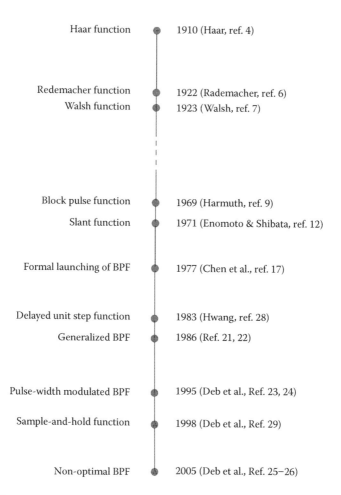

FIGURE 1.13
Time scale history of piecewise constant and related basis function family.

are of upper triangular nature. This immensely helps in mathematical manipulation and related computational effort is drastically reduced.

Though such advantages are strong enough, the apparent *incompleteness* of the BPF set prohibited its use amongst researchers during the early stage.

The earliest work concerning completeness and suitability of BPF for use in place of Walsh functions, is a small technical note of Rao and Srinivasan [30]. Later Kwong and Chen [15], Chen and Lee [31] discussed convergence properties of BPF series and the BPF solution of a linear time invariant system.

Chen et al. [16,17] introduced the BPF set formally in 1977 and they used it along with Walsh functions through similarity transformation to solve fractional calculus problems for distributed parameter systems. At about the same time, Sannuti's paper [32] on the analysis and synthesis of dynamical systems in state space was a significant step toward BPF applications. Shieh et al. [33] dealt with the same problems. Rao and Srinivasan proposed methods of analysis and synthesis for delay systems [34] where an operational matrix for delay via BPF was proposed.

Chen and Jeng [35] considered systems with piecewise constant delays. Differential equations, related to the dynamics of current collection mechanism of electric locomotives, contain terms with a stretched argument. Such equations have been treated in Ref. [36] using BPF. Chen [37] also dealt with scaled systems. BPFs have been used in obtaining discrete-time solutions of continuous-time systems with sample-and-hold. The results obtained via BPF method has been compared with other techniques [38].

The higher powers of the operational matrix for integration accomplished the task of repeated integration. However, the use of higher powers led to accumulation of error at each stage of integration. This has been recognized by Rao and Palanisamy [39] who gave one-shot operational matrices for repeated integration via BPF and Walsh functions. Later Deb et al. presented such matrices in different forms [40]. Chi-Hsu [41] deals with the same aspect suggesting improvements in operational matrices for fractional and operational calculus. Palanisamy reveals certain interesting aspects of the operational matrix for integration [42].

Replacement of Walsh function by block pulse took place in system identification algorithms for computational advantage. Shih and Chia [43] used BPF in identifying delay systems. Jan and Wong [44] and Cheng and Hsu [45] identified bilinear models. Multidimensional BPFs have been proposed by Rao and Srinivasan [46]. These were used in solving partial differential equations. Nath and Lee [47] discussed multidimensional extensions of block pulse with applications.

Identification of nonlinear distributed systems and linear feedback systems via BPF were done by Hsu and Cheng [48] and Kwong and Chen [49]. Palanisamy and Bhattacharya also used BPF in system identification [50] and in analyzing stiff systems [51]. Solution of multipoint boundary value problems and integral equations were obtained using a set of BPF [52,53]. In parameter estimation of bilinear systems, Cheng and Hsu [54] applied BPF. Still many more applications of BPF remain to be mentioned.

Thus BPF continued to reign over other piecewise constant orthogonal functions with its simple but powerful attributes. But numerical instability is observed when deconvolution operation [55] in BPF domain is executed for system identification. Also, oscillations where observed [56] for system analysis in BPF domain.

In the year 2000, Razzaghi and Marzban [57] proposed a direct method for solving variational problems using a hybrid of BPF and orthogonal Chebyshev polynomials. However, they followed the same philosophy of transforming the variational problem into a problem of solving an algebraic equation, as was followed by earlier researchers like Chen [11] and Rao [3].

Razzaghi and Marzban's work was supplemented by Maleknejad and Kajani [58] and Marzban and Razzaghi [59] who used BPF in conjunction with Legendre polynomials. Maleknejad and Mahmoudi [60] used Taylor series for numerical solution of integral equations. Maleknejad and Rahimi [61] used a modification of BPF to solve Volterra integral equation of the first kind. Maleknejad et al. [62] proposed a stochastic operational matrix for integration in the BPF domain to solve a stochastic Volterra integral equation by reducing it to a linear lower triangular system and using forward substitution. In 2012, Park [63] used BPF and subsequent transformations to solve two point boundary value problems for optimal tracking of large scale systems.

It has been more than 35 years that researchers have been working with the BPF set and applying it successfully in many areas of engineering and technology. However, the presented overview focuses on only a few of the applications of BPF in the broad area of systems and control. The set of BPF thus seems to be getting more and more useful to control engineering researchers day by day and it possibly assumes the position of uncrowned leader amongst all piecewise constant basis function sets.

References

1. G. Sansone, *Orthogonal functions*, Interscience Publishers Inc., New York, 1959.
2. J. B. Fourier, *Theorie analytique de la Chaleur*, 1828. *The analytic theory of heat*, 1878, (English edition) Reprinted by Dover Pub. Co, New York, 1955.
3. G. P. Rao, *Piecewise constant orthogonal functions and their application to systems and control*, LNC1S, vol. **55**, Springer-Verlag, Berlin, 1983.
4. Alfred Haar, Zur theorie der orthogonalen funktionen systeme, Math. Annalen, vol. **69**, pp. 331–371, 1910.
5. F. Mix Dwight and Kraig J. Olejniczak, *Elements of Wavelets for Engineers and Scientists*, Wiley Interscience, Hoboken, 2003.
6. H. Rademacher, Einige saitze von allegemeinen orthogonal funktionen, Math. Annalen, vol. **87**, pp. 122–138, 1922.
7. J. L. Walsh, A closed set of normal orthogonal functions, Amer. J. Math., vol. **45**, pp. 5–24, 1923.
8. K. G. Beauchamp, *Walsh functions and their applications*, Academic Press, London, 1975.
9. H. F. Harmuth, *Transmission of information by orthogonal functions*, Springer-Verlag, Berlin, 1969.
10. D. M. Huang, Walsh-Hadamard-Haar hybrid transform, IEEE Proc. 5th Int. Conf. on Pattern Recognition, pp. 180–182, 1980.
11. C. F. Chen and C. H. Hsiao, A state space approach to Walsh series solution of linear systems, Int. J. Syst. Sci., vol. **6**, no. 9, pp. 833–858, 1975.

12. H. Enomoto and K. Shibata, Orthogonal transform coding system for television signals, Proc. Symp. Appl. of Walsh functions, Washington DC, USA, pp. 11–17, 1971.

13. W. K. Pratt, L. R. Welch, and W. Chen, Slant transform for image coding, Proc. Symp. Appl. of Walsh functions, Washington DC, USA, pp. 229–234, March 1972.

14. W. K. Pratt, *Digital image processing*, John Wiley & Sons, New York, 1978.

15. C. P. Kwong and C. F. Chen, The convergence properties of block pulse series, Int. J. Syst. Sci., vol. **12**, no. 6, pp. 745–751, 1981.

16. T. T. Wu, C. F. Chen, and Y. T. Tsay, Walsh operational matrices for fractional calculus and their application to distributed systems, IEEE Symposium on Circuits and Systems, Munich, Germany, April, 1976.

17. C. F. Chen, Y. T. Tsay, and T. T. Wu, Walsh operational matrices for fractional calculus and their application to distributed systems, J. Franklin Inst., vol. **303**, no. 3, pp. 267–284, 1977.

18. Z. H. Jiang and W. Schaufelberger, *Block pulse functions and their applications in control systems*, LNCIS, vol. **179**, Springer-Verlag, Berlin, 1992.

19. Anish Deb, Gautam Sarkar, and Sunit K. Sen, Block pulse functions, the most fundamental of all piecewise constant basis functions, Int. J. Syst. Sci., vol. **25**, no. 2, pp. 351–363, 1994.

20. Anish Deb and Suchismita Ghosh, *Power electronic systems: Walsh analysis with MATLAB®*, CRC Press, Boca Raton, 2014.

21. C. H. Wang and R. S. Marleau, Application of a generalized block pulse operational matrices for the approximation of continuous-time systems, Int. J. Syst. Sci., vol. **17**, no. 9, pp. 1269–1278, 1986.

22. M. L. Wang, S. Y. Yang, and R. Y. Chang, Application of a generalized block pulse function to a scaled system, Int. J. Syst. Sci., vol. **18**, no. 8, pp. 1495–1503, 1987.

23. Anish Deb, Gautam Sarkar, and Sunit K. Sen, A new set of pulse width modulated generalized block pulse functions (PWM-GBPF) and their application to cross/auto-correlation of time varying functions, Int. J. Syst. Sci., vol. **26**, no. 1, pp. 65–89, 1995.

24. Anish Deb, Gautam Sarkar, and Sunit K. Sen, Linearly pulse-width modulated block pulse functions and their application to linear SISO feedback control system identification, Proc. IEE, Part – D, Control Theory and Appl., vol. **142**, no. 1, pp. 44–50, 1995.

25. Anish Deb, Gautam Sarkar, Priyaranjan Mandal, Amitava Biswas, and Anindita Sengupta, Optimal block pulse function (OBPF) vs. Non-optimal block Pulse function (NOBPF), Proceedings of International Conference of IEE (PEITSICON) 2005, pp. 195–199, Kolkata, 28–29th Jan., 2005.

26. Anish Deb, Gautam Sarkar, Priyaranjan Mandal, Amitava Biswas, and Anindita Sengupta, Non-optimal block pulse function (NOBPF) based analysis and identification of SISO control systems, Journal of Automation & Systems Engineering, vol. **7**, no. 1, pp. 24–58, 2013.

27. S. C. Chapra and R. P. Canale, *Numerical Methods for Engineers*, (6th Ed.), McGraw-Hill Higher Education, Boston, 2012.

28. Chyi Hwang, Solution of functional differential equation via delayed unit step functions, Int. J. Syst. Sci., vol. **14**, no. 9, pp. 1065–1073, 1983.

29. Anish Deb, Gautam Sarkar, Manabrata Bhattacharjee, and Sunit K. Sen, A new set of piecewise constant orthogonal functions for the analysis of linear SISO systems with sample-and-hold, J. Franklin Inst., vol. **335B**, no. 2, pp. 333–358, 1998.

30. G. P. Rao and T. Srinivasan, Remarks on "Author's reply" to "Comments on design of piecewise constant gains for optimal control via Walsh functions," IEEE Trans. Automatic Control, vol. **AC-23**, no. 4, pp. 762–763, 1978.

31. W. L. Chen and C. L. Lee, On the convergence of the block-pulse series solution of a linear time-invariant system, Int. J. Syst, Sci., vol. **13**, no. 5, pp. 491–498, 1982.

32. P. Sannuti, Analysis and synthesis of dynamic systems via block pulse functions, Proc. IEE, vol. **124**, no. 6, pp. 569–571, 1977.

33. L. A. Shieh, R. E. Yates, and J. M. Navarro, Representation of continuous time state equations by discrete-time state equations, IEEE Trans. Signal, Man and Cybernatics, vol. **SMC-8**, no. 6, pp. 485–492, 1978.

34. G. P. Rao and T. Srinivasan, Analysis and synthesis of dynamic systems containing time delays via block-pulse functions, Proc. IEE, vol. **125**, no. 9, pp. 1064–1068, 1978.

35. W. L. Chen and B. S. Jeng, Analysis of piecewise constant delay systems via block-pulse functions, Int. J. Syst. Sci., vol. **12**, no. 5, pp. 625–633, 1981.

36. G. P. Rao and T. Srinivasan, An optimal method of solving differential equations characterizing the dynamic of a current collection system for an electric locomotive, J. Inst. Maths. Appl., vol. **25**, no. 4, pp. 329–342, 1980.

37. W. L. Chen, Block-pulse series analysis of scaled systems, Int. J. Syst. Sci., vol. **12**, no. 7, pp. 885–891, 1981.

38. A. Deb, G. Sarkar, M. Bhattacharjee and S. K. Sen, Analysis of linear SISO (single input single output) control systems with sample-and-hold using block pulse function (BPF) operational technique, Indian J. Engg. & Materials Sc., vol. **6**, pp. 5–8, 1999.

39. G. P. Rao and K. R. Palanisamy, Improved algorithms for parameter identifications in lumped continuous systems via Walsh functions, Proc. IEE, Part-D, CTA, vol. **130**, no. 1, pp. 9–16, 1983.

40. Anish Deb, G. Sarkar, M. Bhattacharjee, and S. K. Sen, All-integrator approach to linear SISO control system analysis using block pulse function (BPF), J. Franklin Inst., vol. **334B**, no. 2, pp. 319–335, 1997.

41. W. Chi-Hsu, On the generalization of block pulse operational matrices for fractional and operational calculus, J. Franklin Inst., vol. **315**, no. 2, pp. 91–102, 1983.

42. K. R. Palanisamy, A note on block-pulse function operational matrix for integration, Int. J. Syst. Sci., vol. **14**, no. 11, pp. 1287–1290, 1983.

43. Y. P. Shih and W. K. Chia, Parameter estimation of delay systems via block-pulse functions, J. Dynamic Syst., Measurement and Control, vol. **102**, no. 3, pp. 159–162, 1980.

44. Y. G. Jan and K. M. Wong, Bilinear system identification by block pulse functions, J. Franklin Inst., vol. **512**, no. 5, pp. 349–359, 1981.

45. Bing Cheng and Ning-Show Hsu, Analysis and parameter estimation of bilinear systems via block-pulse functions, Int. J. Control, vol. **36**, no. 1, pp. 53–65, 1982.

46. G. P. Rao and T. Srinivasan, Multidimensional block pulse functions and their use in the study of distributed parameter systems, Int. J. Syst. Sci., vol. **11**, no. 6, pp. 689–708, 1980.

47. A. K. Nath and T. T. Lee, On the multidimensional extension of block-pulse functions and their applications, Int. J. Syst. Sci., vol. **14**, no. 2, pp. 201–208, 1983.

48. Ning-Show Hsu and Bing Cheng, Identification of nonlinear distributed systems via block pulse functions, Int. J. Control, vol. **36**, no. 2, pp. 281–291, 1982.

49. C. P. Kwong and C. F. Chen, Linear feedback systems identification via block pulse functions, Int. J. Syst. Sci., vol. **12**, no. 5, pp. 635–642, 1981.

50. K. R. Palanisamy and D. K. Bhattacharya, System identification via block-pulse functions, Int. J. Syst. Sci., vol. **12**, no. 5, pp. 643–647, 1981.

51. K. R. Palanisamy and D. K. Bhattacharya, Analysis of stiff systems via single step method of block-pulse functions, Int. J. Syst. Sci., vol. **13**, no. 9, pp. 961–968, 1982.

52. J. Kalat and P. N. Paraskevopoulos, Solution of multipoint boundary value problems via block pulse functions, J. Franklin Inst., vol. **324**, no. 1, pp. 73–81, 1987.

53. F. C. Kung and S. Y. Chen, Solution of integral equations using a set of block pulse functions, J. Franklin Inst., vol. **306**, no. 4, pp. 283–291, 1978.

54. B. Cheng and N. S. Hsu, Analysis and parameter estimation of bilinear systems via block pulse functions, Int. J. Control, vol. **36**, no. 1, pp. 53–65, 1982.

55. Anish Deb, Gautam Sarkar, Amitava Biswas, and Priyaranjan Mandal, Numerical instability of deconvolution operation via block pulse functions, J. Franklin Inst., vol. **345**, pp. 319–327, 2008.

56. Anish Deb and D. W. Fountain, A note on oscillations in Walsh domain analysis of first order systems, IEEE Trans. Circuits and Syst., vol. **CAS-38**, no. 8, pp. 945–948, 1991.

57. Mohsen Razzaghi and Hamid-Reza Marzban, Direct method for variational problems via hybrid of block-pulse and Chebyshev functions, Mathematical Problems in Engineering, vol. **6**, no. 1, pp. 85–97, 2000.

58. K. Maleknejad and M. Tavassoli Kajani, Solving second kind integral equations by Galerkin methods with hybrid Legendre and Block-Pulse functions, Applied Mathematics and Computation, vol. **145**, no. 2–3, pp. 623–629, 2003.

59. H. R. Marzban and M. Razzaghi, Optimal control of linear delay systems via hybrid of block-pulse and Legendre polynomials, J. Franklin Inst., vol. **341**, no. 3, pp. 279–293, 2004.

60. K. Maleknejad and Y. Mahmoudi, Numerical solution of linear Fredholm integral equation by using hybrid Taylor and Block-Pulse functions, Applied Mathematics and Computation, vol. **149**, no. 3, pp. 799–806, 2004.

61. K. Maleknejad and B. Rahimi, Modification of Block Pulse Functions and their application to solve numerically Volterra integral equation of the first kind, Communications in Nonlinear Science and Numerical Simulation, vol. **16**, no. 6, pp. 2469–2477, 2011.

62. K. Maleknejad, M. Khodabin, and M. Rostami, Numerical solution of stochastic Volterra integral equations by a stochastic operational matrix based on block pulse functions, Mathematical and Computer Modelling, vol. **55**, no. 3–4, pp. 791–800, 2012.

63. J. H. Park, Trajectory optimization for large scale systems via block pulse functions and transformations, International J. Control & Automation, vol. **5**, no, 4, pp. 39–48, 2012.

Study Problems

1.1 What are piecewise constant orthogonal functions? What are the advantages of using these functions instead of sine-cosine functions?

1.2 State the properties of an orthogonal function set.

1.3 Write down the Haar matrix for $m = 4$.

1.4 Write down the Walsh matrix for $m = 4$.

1.5 Define MISE. Show that the selection of c_j as

$$c_j = \frac{1}{T} \int_0^T f(t)\varphi_j(t)\,dt$$

reduces MISE to a minimum.

1.6 Show how Rademacher functions are related to Walsh functions. Then, find φ_{17} of a Walsh function set using Rademacher functions.

1.7 Show with a neat sketch a block pulse function (BPF) set of order 6. Show mathematically, how a BPF set is normalized.

1.8 Compare the major piecewise constant basis orthogonal functions, e.g., Walsh, Haar, block pulse function (BPF) and sample-and-hold function (SHF) with respect to their attributes and properties.

1.9 What are the basic differences between the optimal block pulse function set and the non-optimal block pulse function set?

2

Function Approximation via Block Pulse Function and Related Functions

In this chapter, first of all, a few elementary properties of the block pulse function (BPF) [1–3] set are discussed. Then some square integrable time functions of Lebesgue measure are approximated using the BPF set and other related piecewise constant basis functions like generalized block pulse function (GBPF) [4,5], pulse-width modulated generalized block pulse function (PWM-GBPF) [6,7], non-optimal block pulse function (NOBPF) [8,9], delayed unit step function (DUSF) [3,10], and sample-and-hold function (SHF) [11]. All these approximations are compared with the conventional equal-width BPF domain approximation qualitatively as well as quantitatively.

2.1 Block Pulse Functions: Properties [2]

Basic qualitative properties of the BPF set are tabulated in Table 2.1 to provide a qualitative appraisal. Of many properties of the BPFs four important properties are presented here.

2.1.1 Disjointedness

The BPFs are disjointed with each other in the interval $t \in [0, T)$. That is,

$$\psi_i(t)\psi_j(t) = \begin{cases} \psi_i(t) & \text{for } i = j \\ 0 & \text{for } i \neq j \end{cases} \tag{2.1}$$

where $i, j = 0, 1, 2, \ldots, (m - 1)$.

This property can directly be obtained from the definition of BPFs.

For definition of two block pulses $\psi_i(t)$ and $\psi_j(t)$ we have

$$\frac{\psi_i(t)}{\psi_j(t)} = \frac{\psi_i(t)\, \psi_j(t)}{\psi_j(t)\, \psi_j(t)} \qquad \text{[multiplying the numerator and denominator by } \psi_j(t)\text{]}$$

TABLE 2.1

Basic Properties of Block Pulse Function (BPF)

Properties	BPF
Piecewise constant	Yes
Orthogonal	Yes
Finite	Yes
Disjoint	Yes
Orthonormal	Can easily be normalized
Implementation	Easily implementable
Coefficient determination of $f(t)$	Involves integration of $f(t)$ and scaling
Accuracy of analysis	Provides staircase solution

By virtue of (2.1) above

$$\frac{\psi_i(t)}{\psi_j(t)} = \begin{cases} 1 & \text{for } i = j \\ 0 & \text{for } i \neq j \end{cases} \tag{2.2}$$

2.1.2 Orthogonality

The BPFs are orthogonal to each other in the interval $t \in [0, T)$. That is,

$$\int_0^T \psi_i(t)\psi_j(t)\, dt = \begin{cases} h & \text{for } i = j \\ 0 & \text{for } i \neq j \end{cases} \tag{2.3}$$

where $i, j = 0, 1, 2, \ldots, (m - 1)$.

This property can directly be obtained from the disjointedness of BPFs.

2.1.3 Addition

For addition of two time functions $f(t)$ and $g(t)$, let us consider

$$a(t) \triangleq f(t) + g(t)$$

where $a(t)$ is the resulting function.

First we expand the continuous functions $f(t)$ and $g(t)$ directly into BPF domain following equation (1.13) and then add them to find the resultant function in BPF domain.

Thus,

$$
\left.
\begin{aligned}
f(t) \approx \overline{f}(t) &= \sum_{i=0}^{m-1} f_i \psi_i = \begin{bmatrix} f_0 & f_1 & \cdots & f_{(m-1)} \end{bmatrix} \mathbf{\Psi}_{(m)}(t) \triangleq \mathbf{F}^{\mathrm{T}} \mathbf{\Psi}_{(m)}(t) \\
g(t) \approx \overline{g}(t) &= \sum_{i=0}^{m-1} g_i \psi_i = \begin{bmatrix} g_0 & g_1 & \cdots & g_{(m-1)} \end{bmatrix} \mathbf{\Psi}_{(m)}(t) \triangleq \mathbf{G}^{\mathrm{T}} \mathbf{\Psi}_{(m)}(t)
\end{aligned}
\right\}
\tag{2.4}
$$

The resultant function [$f(t) + g(t)$] may now be obtained from (2.4).
Now we expand $a(t)$, the sum of $f(t)$ and $g(t)$, in BPF domain to obtain

$$
a(t) \approx \overline{a}(t) = \sum_{i=0}^{m-1} a_i \psi_i = \begin{bmatrix} a_0 & a_1 & \cdots & a_{(m-1)} \end{bmatrix} \mathbf{\Psi}_{(m)}(t) \triangleq \mathbf{A}^{\mathrm{T}} \mathbf{\Psi}_{(m)}(t)
\tag{2.5}
$$

The BPF domain expanded forms of $f(t)$, $g(t)$, and $a(t)$ are respectively called $\overline{f}(t)$, $\overline{g}(t)$, and $\overline{a}(t)$. It should be noted that these $\overline{f}(t)$, $\overline{g}(t)$, and $\overline{a}(t)$ are piecewise constant in nature.

Let us now write down the $(i + 1)$th BPF coefficient of the function $\overline{a}(t)$

$$
\begin{aligned}
a_i &= \frac{1}{h} \int_{ih}^{(i+1)h} [f(t) + g(t)] \mathrm{d}t, \quad i = 0, 1, 2, \ldots, m-1 \\
&= \frac{1}{h} \int_{ih}^{(i+1)h} f(t) \mathrm{d}t + \frac{1}{h} \int_{ih}^{(i+1)h} g(t) \mathrm{d}t \\
&= f_i + g_i
\end{aligned}
\tag{2.6}
$$

That means,

$$
\sum_{i=0}^{m-1} a_i \psi_i = \sum_{i=0}^{m-1} f_i \psi_i + \sum_{i=0}^{m-1} g_i \psi_i
$$

or, $\mathbf{A}^{\mathrm{T}} = \mathbf{F}^{\mathrm{T}} + \mathbf{G}^{\mathrm{T}}$
\tag{2.7}

From equations (2.6) and (2.7), it is seen that both the results of addition are identical.

2.1.4 Subtraction

For subtraction of two time functions $f(t)$ and $g(t)$, let us consider

$$s(t) = f(t) - g(t)$$

where $s(t)$ is the resulting function.

First the continuous functions $f(t)$ and $g(t)$ are expanded directly into BPF domain following equation (1.13) and then subtracted to find the resultant function in BPF domain.

Thus,

$$\left.\begin{aligned}
f(t) \approx \bar{f}(t) = \sum_{i=0}^{m-1} f_i \psi_i = \begin{bmatrix} f_0 & f_1 & \cdots & f_{(m-1)} \end{bmatrix} \Psi_{(m)}(t) \triangleq \mathbf{F}^{\mathrm{T}} \Psi_{(m)}(t) \\
g(t) \approx \bar{g}(t) = \sum_{i=0}^{m-1} g_i \psi_i = \begin{bmatrix} g_0 & g_1 & \cdots & g_{(m-1)} \end{bmatrix} \Psi_{(m)}(t) \triangleq \mathbf{G}^{\mathrm{T}} \Psi_{(m)}(t)
\end{aligned}\right\} \tag{2.8}$$

Like the previous section, the resultant function $[f(t) - g(t)]$ may now be obtained from (2.8).

Now we expand $s(t)$, the difference of $f(t)$ and $g(t)$, in BPF domain to obtain

$$s(t) \approx \bar{s}(t) = \sum_{i=0}^{m-1} s_i \psi_i = \begin{bmatrix} s_0 & s_1 & \cdots & s_{(m-1)} \end{bmatrix} \Psi_{(m)}(t) \triangleq \mathbf{S}^{\mathrm{T}} \Psi_{(m)}(t) \tag{2.9}$$

The BPF domain expanded forms of $f(t)$, $g(t)$, and $s(t)$ are respectively called $\bar{f}(t)$, $\bar{g}(t)$, and $\bar{s}(t)$. It should be noted that these $\bar{f}(t)$, $\bar{g}(t)$, and $\bar{s}(t)$ are piecewise constant in nature.

Let us write down the $(i + 1)$th BPF coefficient of the function $\bar{s}(t)$

$$\begin{aligned}
s_i &= \frac{1}{h} \int_{ih}^{(i+1)h} [f(t) - g(t)] \mathrm{d}t, \quad i = 0, 1, 2, \ldots, m-1 \\
&= \frac{1}{h} \int_{ih}^{(i+1)h} f(t) \mathrm{d}t - \frac{1}{h} \int_{ih}^{(i+1)h} g(t) \mathrm{d}t \\
&= f_i - g_i
\end{aligned} \tag{2.10}$$

That means,

$$\sum_{i=0}^{m-1} s_i \psi_i = \sum_{i=0}^{m-1} f_i \psi_i - \sum_{i=0}^{m-1} g_i \psi_i$$

$$\text{or,} \quad \mathbf{S}^{\mathrm{T}} = \mathbf{F}^{\mathrm{T}} - \mathbf{G}^{\mathrm{T}} \tag{2.11}$$

From equations (2.10) and (2.11), it is seen that both the results of subtraction are identical.

2.1.5 Multiplication

For two functions $f(t)$ and $g(t)$, we determine the block pulse series of their product using the BPF expansion of each function.

That is,

$$\begin{aligned} f(t)g(t) &\approx \left(\sum_{i=0}^{m-1} f_i \psi_i(t) \right) \left(\sum_{j=0}^{m-1} g_j \psi_j(t) \right) \\ &= \sum_{i=0}^{m-1} f_i g_i \psi_i(t) \end{aligned} \tag{2.12}$$

where either of $f(t)$ or $g(t)$ is not equal to zero.

In deriving (2.12), the cross products will be zero, because of the disjointedness property of the BPFs.

Expressing in vector forms, the product of (2.12) may be expressed as under.

$$\begin{aligned} f(t)g(t) &\approx \left[f_0 g_0 \quad f_1 g_1 \quad \cdots \quad f_{(m-1)} g_{(m-1)} \right] \mathbf{\Psi}(t) \\ &= \mathbf{F}^{\mathrm{T}} \mathbf{D}_{\mathrm{G}} \mathbf{\Psi}(t) \\ &= \mathbf{G}^{\mathrm{T}} \mathbf{D}_{\mathrm{F}} \mathbf{\Psi}(t) \end{aligned} \tag{2.13}$$

where $\mathbf{D}_{\mathrm{F}}$ and $\mathbf{D}_{\mathrm{G}}$ are diagonal matrices formed by the elements of the BPF vectors $\mathbf{F}$ and $\mathbf{G}$, respectively, where

$$\mathbf{F} \triangleq \left[f_0 \quad f_1 \quad \cdots \quad f_{(m-1)} \right]^{\mathrm{T}}$$

and

$$\mathbf{G} \triangleq \left[g_0 \quad g_1 \quad \cdots \quad g_{(m-1)} \right]^{\mathrm{T}}$$

Equation (2.13) shows that the block pulse coefficients of the product are only the products of respective block pulse coefficients of individual functions in the same subintervals.

2.1.6 Division

For functions $f(t)$ and $g(t)$, we have the block pulse series of their quotient:

$$f(t)/g(t) \approx \left(\sum_{i=0}^{m-1} f_i \psi_i(t) \right) \Bigg/ \left(\sum_{j=0}^{m-1} g_j \psi_j(t) \right); \quad \text{for } g(t) \neq 0$$

$$= \sum_{i=0}^{m-1} (f_i/g_i) \psi_i(t) \tag{2.14}$$

Expressing in vector forms, these equations are:

$$f(t)/g(t) \approx \left[f_0/g_0 \quad f_1/g_1 \quad \cdots \quad f_{(m-1)}/g_{(m-1)} \right] \mathbf{\Psi}(t)$$

$$= \mathbf{F}^\mathrm{T} \mathbf{D}_\mathrm{G}^{-1} \mathbf{\Psi}(t) \tag{2.15}$$

where $\mathbf{D}_\mathrm{G}$ is the diagonal matrix formed by the elements of block pulse coefficient vector $\mathbf{G}$.

Here, all the joint terms in the division of block pulse series disappear owing to the disjointedness of BPFs.

Thus, while working with block pulse series expansions, above rules simplifies calculations significantly.

2.2 Function Approximation

Earlier, in Chapter 1, we presented the definitions for the BPF set and related piecewise constant orthogonal basis functions (PCOBF). Also, we discussed the principles of function approximation using all these functions. Using those principles, we now approximate different types of time functions using the conventional BPFs and other related PCOBFs. For different types of functions, we compare the approximations quantitatively with respect to MISEs.

2.2.1 Using Block Pulse Functions

At first we determine the coefficients of a time function $f(t)$ in BPF domain [1–3] using equation (1.14). We consider the following two examples:

2.2.1.1 Numerical Examples

Example 2.1

Let us expand the function $f_1(t) = t$ in BPF domain taking $m = 8$ and $T = 1$ s. Following the method mentioned above, the result is,

$$f_1(t) \approx \Big[0.06250000 \quad 0.18750000 \quad 0.31250000 \quad 0.43750000$$
$$0.56250000 \quad 0.68750000 \quad 0.81250000 \quad 0.93750000 \Big] \Psi_{(8)}(t)$$

$$(2.16)$$

Figure 2.1 shows the exact function along with its BPF approximation.

Example 2.2

Now we take up the function $f_2(t) = \sin(\pi t)$ and express it in BPF domain for $m = 8$ and $T = 1$ s. The result is,

$$f_2(t) \approx \Big[0.19383918 \quad 0.55200729 \quad 0.82613728 \quad 0.97449537$$
$$0.97449537 \quad 0.82613728 \quad 0.55200729 \quad 0.19383918 \Big] \Psi_{(8)}(t)$$

$$(2.17)$$

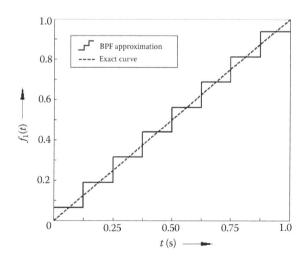

FIGURE 2.1

Exact curve for $f_1(t) = t$ of Example 2.1 and its block pulse function approximation for $m = 8$ and $T = 1$ s.

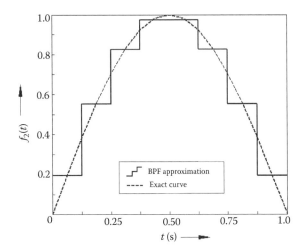

FIGURE 2.2
Exact curve for $f_2(t) = \sin(\pi t)$ of Example 2.2 and its block pulse function approximation for $m = 8$ and $T = 1$ s (vide Appendix B, Program no. 2.1).

Figure 2.2 shows the exact function along with its BPF domain approximation.

2.2.2 Using Generalized Block Pulse Functions (GBPF)

We can determine the coefficients of a time function $f(t)$ in GBPF domain [4,5] using equation (1.25), as discussed in Section 1.2.8. We consider the function $f_2(t)$ for GBPF domain approximation.

2.2.2.1 Numerical Example

Example 2.3

We consider the typical function $f_2(t) = \sin(\pi t)$ and express it in GBPF domain for $m = 8$ and $T = 1$ s.

For better approximation of the function in GBPF domain with respect to MISE, compared to BPF domain, the subintervals chosen are:

$h_0 = 0.1$ s, $h_1 = 0.1$ s, $h_2 = 0.12$ s, $h_3 = 0.18$ s, $h_4 = 0.18$ s, $h_5 = 0.12$ s, $h_6 = 0.1$ s, $h_7 = 0.1$ s.

The result of approximation in GBPF domain is,

$$f_2(t) \approx \begin{bmatrix} 0.15579195 & 0.45212584 & 0.72465951 & 0.94754981 \\ 0.94754981 & 0.72465951 & 0.45212584 & 0.15579195 \end{bmatrix} \Psi_{g(8)}(t)$$

$$(2.18)$$

Figure 2.3 shows the qualitative comparison of GBPF based approximation along with its BPF domain approximation. In case of GBPF approximation, we can choose the subintervals to obtain minimum approximation error, whereas in conventional BPF domain, we do not have such choice.

Apart from this qualitative approximation via GBPF, we compute MISE for Example 2.3 to judge the effectiveness of GBPF approximation quantitatively. Table 2.2 shows this quantitative comparison between these two approaches.

From Table 2.2 it is noted that by adjusting the width of each subinterval for approximating the function of Example 2.3 in GBPF domain, we can approximate the function more accurately compared to the BPF approach.

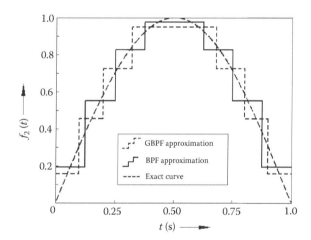

FIGURE 2.3

Exact curve for $f_2(t) = \sin(\pi t)$ of Example 2.3 and its graphical comparison of the GBPF domain approximation with the BPF based approximation, for $m = 8$ and $T = 1$ s (vide Appendix B, Program no. 2.2).

TABLE 2.2

MISEs in Approximating the Function of Example 2.3, in GBPF Domain and Conventional BPF Domain, for $m = 8$ and $T = 1$ s (Vide Appendix B, Program No. 2.2)

	BPF Approximation	**GBPF Approximation**
MISEs	0.00639258	0.00500445

2.2.3 Using Pulse-Width Modulated Generalized Block Pulse Functions (PWM-GBPF)

Similar to the GBPF set, the selection of widths $h_i(i = 0, 1, 2, …, m − 1)$ in PWM-GBPF domain [6,7], will always depend upon the variation of time function $f(t)$, as discussed in Section 1.2.9. Here we consider a function $f_3(t)$ for PWM-GBPF based approximation.

2.2.3.1 Numerical Example

Example 2.4

Now we take up the function $f_3(t) = \exp(−t)$ and approximate it in PWM-GBPF domain for $m = 8$ and $T = 4$ s. The δ in equation (1.28) is chosen as 0.1 and using (1.28) we determine h_0 for approximating the function in PWM-GBPF domain.

For PWM-GBPF approximation, the subintervals chosen are $h_0 = 0.15$ s, $h_1 = 0.25$ s, $h_2 = 0.35$ s, $h_3 = 0.45$ s, $h_4 = 0.55$ s, $h_5 = 0.65$ s, $h_6 = 0.75$ s, $h_7 = 0.85$ s.

Figure 2.4 compares graphically the PWM-GBPF based approximation of the function with its BPF equivalent. Choosing a proper δ considering the constraint of equation (1.29), it is possible to minimize the error in PWM-GBPF based approximation compared to conventional BPF domain.

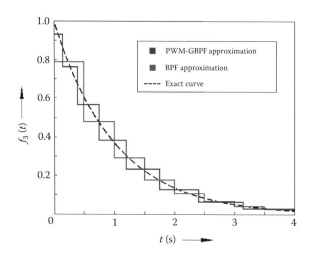

FIGURE 2.4

Exact curve for $f_3(t) = \exp(−t)$ of Example 2.4 and its graphical comparison between the PWM-GBPF approximation and the BPF approximation, for $m = 8$ and $T = 4$ s (vide Appendix B, Program no. 2.3).

TABLE 2.3

MISEs in Approximating the Function of Example 2.4, Using the
PWM-GBPF Approach and the Conventional BPF Approach, for
$m = 8$ and $T = 4$ s (Vide Appendix B, Program No. 2.3)

	BPF Based Approximation	PWM-GBPF Based Approximation
MISEs	0.00253982	0.00113198

The result of approximation in PWM-GBPF domain is,

$$f_3(t) \approx \left[0.92861349\ 0.76155172\ 0.56558141\ 0.38038298 \right.$$
$$\left. 0.23167322\ 0.12777845\ 0.06382110\ 0.02886646 \right] \mathbf{\Psi}_{gp(8)}(t) \quad (2.19)$$

To verify the effectiveness of the PWM-GBPF based approximation quantitatively, Table 2.3 is presented. The Table provides a comparison between MISEs in approximating the given function in the two domains. The table is self explanatory. That is, for this kind of function, as discussed in Section 1.2.9 the PWM-GBPF based approximation is better compared to the BPF domain.

2.2.4 Using Non-Optimal Block Pulse Functions (NOBPF)

In NOBPF domain [8,9], for computation of expansion coefficients the trapezoidal rule for integration is employed. That means, only the samples of the function to be approximated are needed to compute the expansion coefficients. This reduces the computational burden, as discussed in Section 1.2.10.

Here, we consider the function $f_2(t)$ for NOBPF based approximation, so that we can verify the effectiveness of this approach in comparison to the optimal block pulse function (OBPF)—that is the conventional BPF approach. But in the NOBPF approach, minimized MISE is not guaranteed.

2.2.4.1 Numerical Example

Example 2.5

We consider the function $f_2(t) = \sin(\pi t)$ and approximate it in NOBPF domain for $m = 8$ and $T = 4$ s.

The result of approximation in NOBPF domain is,

$$f_2(t) \approx \left[0.19134172\ 0.54489511\ 0.81549316\ 0.96193977 \right.$$
$$\left. 0.96193977\ 0.81549316\ 0.54489511\ 0.19134172 \right] \mathbf{\Psi}'_{(8)}(t) \quad (2.20)$$

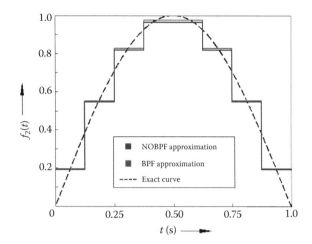

FIGURE 2.5
Exact curve for $f_2(t) = \sin(\pi t)$ of Example 2.5 and its graphical comparison of the NOBPF based approximation with the BPF approximation, for $m = 8$ and $T = 1$ s (vide Appendix B, Program no. 2.4).

TABLE 2.4

MISEs in Approximating the Function of Example 2.5, Using the NOBPF Based Approach and the Conventional BPF Domain, for $m = 8$ and $T = 1$ s (Vide Appendix B, Program No. 2.4)

	BPF Approximation	NOBPF Approximation
MISEs	0.00253982	0.00647452

Figure 2.5 shows the graphical comparison of function approximation using the NOBPF approach and the OBPF approach. For a quantitative comparison, we determine the MISEs for these two methods and present them in Table 2.4.

2.2.5 Using Delayed Unit Step Functions (DUSF)

When a time function is approximated in DUSF domain [3,10], it has been established by Deb et al. [3] that the DUSF set gives rise to an approximation identical to that obtained via BPF domain. This is because the expansion coefficients in DUSF domain, though seemingly different from that of BPF domain, are related to BPF expansion coefficients via simple linear equations, as shown in Section 1.2.11.

As discussed in Section 1.2.11, Reference [3] proved that the DUSF set is derived from the BPF set, and the BPF set is the most fundamental of all piecewise constant orthogonal functions.

To prove the dependence of DUSF on BPF, we take up the same function treated in Example 2.2, for approximation in DUSF domain.

2.2.5.1 Numerical Example

Example 2.6 (vide Appendix B, Program no. 2.5)

Consider the function $f_2(t) = \sin(\pi t)$. We expand it in DUSF domain for $m = 8$ and $T = 1$ s. The result is,

$$f_2(t) \approx \big[0.19383918 \;\; 0.35816810 \;\; 0.274129996 \;\; 0.14835808$$
$$0.00000000 \;\; -0.14835808 \;\; -0.27413000 \;\; -0.35816809 \big] \mathbf{U}_{(8)}(t)$$

$$(2.21)$$

This is equivalent to,

$$\big[0.19383918 \;\; 0.55200729 \;\; 0.82613728 \;\; 0.97449537$$
$$0.97449537 \;\; 0.82613728 \;\; 0.55200729 \;\; 0.19383918 \big] \mathbf{\Psi}_{(8)}(t)$$

$$(2.22)$$

After comparing the equation (2.21) with (2.16) of Example 2.2, we find that the approximation results are identical, as indicated by equation (1.38) of Section 1.2.11.

2.2.6 Using Sample-and-Hold Functions (SHF)

Any square integrable function $f(t)$ may be expanded in sample-and-hold function (SHF) domain [11] in the semi-open interval $[0, T)$ by computing the expansion coefficients from the samples of the function to be approximated. As discussed in Section 1.2.12, the component functions of SHF set are look-alikes of the members of the BPF set. But the computation of the expansion coefficients does not depend on the traditional integration formula (1.5).

We take up one function $f_4(t)$ for SHF based approximation, and compare the same qualitatively and quantitatively with the conventional BPF set.

2.2.6.1 Numerical Example

Example 2.7

We consider the function $f_4(t) = 1 - \exp(-t)$ and approximate it in SHF domain for $m = 8$ and $T = 1$ s.

The result of approximation in SHF domain is,

$$f_4(t) \approx \Big[0.00000000 \quad 0.11750310 \quad 0.22119922 \quad 0.31271072$$
$$0.39346934 \quad 0.46473857 \quad 0.52763345 \quad 0.58313798 \Big] \mathbf{S}_\theta(t) \qquad (2.23)$$

Figure 2.6 graphically compares the approximation in SHF domain with that of in BPF domain. From the figure, it is apparent that the SHF approximation introduces much more error compared with its BPF equivalent. For a quantitative comparison, we determine the MISEs for these two methods and present them in Table 2.5.

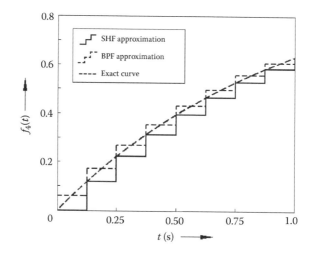

FIGURE 2.6
Exact curve for $f_4(t) = 1 - \exp(-t)$ of Example 2.7 and its graphical comparison between the SHF approximation with the BPF approximation, for $m = 8$ and $T = 1$ s (vide Appendix B, Program no. 2.6).

TABLE 2.5

MISEs in Approximating the Function of Example 2.7, in the SHF Domain and the Conventional BPF Domain, for $m = 8$ and $T = 1$ s (Vide Appendix B, Program No. 2.6)

	BPF Domain Approximation	SHF Domain Approximation
MISEs	0.00056205	0.00231965

2.3 Error Analysis for Function Approximation in BPF Domain

The representational error for conventional equal-width BPF expansion of any square integrable function $f(t)$ of Lebesgue measure was formally investigated by Rao and Srinivasan [12,13]. For m blocks of equal width h over an interval $[0, T)$ such that $T = mh$, the representational error in the $(i + 1)$-th interval is,

$$e(t) = f(t) - c_i \psi_i(t)$$

The mean integral square error (MISE) in the $(i + 1)$th interval is,

$$
\begin{aligned}
\left[\varepsilon_i^2 \right] &= \int_{ih}^{(i+1)h} [f(t) - c_i \psi_i(t)]^2 dt \\
&= \int_{ih}^{(i+1)h} [f(t)]^2 dt - \frac{1}{h} \left[\int_{ih}^{(i+1)h} f(t)\, dt \right]^2 \\
&= \int_{ih}^{(i+1)h} [f(t)]^2 dt - hc_i^2
\end{aligned}
\tag{2.24}
$$

The MISE over the interval $[0, T)$ is given by,

$$\left[\in^2 \right] = \frac{1}{T} \int_0^T \left[f(t) - \sum_{i=0}^{m-1} c_i \psi_i(t) \right]^2 dt$$

In equation (2.24), we represent $f(t)$ by its first order Taylor series approximation as the linear function

$$f(t) \approx \hat{f}(t) = f(ih) + h\dot{f}(ih)$$

Therefore, the BPF coefficient of $\hat{f}(t)$ is,

$$c_i = \frac{f(ih) + \{f(ih) + h\dot{f}(ih)\}}{2} = f(ih) + \frac{h}{2}\dot{f}(ih) \tag{2.25}$$

The MISE in the $(i + 1)$th interval, using equations (2.24) and (2.25), is

$$\left[\in^2 \right] = \sum_{i=0}^{m-1} \left[\varepsilon_i^2 \right] = \frac{mh^3}{12T} \dot{f}(ih)^2$$

We expand $f(t)$ in the $(i + 1)$th interval around any point μ_i, by Taylor series, neglecting second and higher derivatives of $f(t)$.

Then the upper bound of MISE is given as:

$$\left[\in^2\right]_u = \sum_{i=0}^{m-1}\left[\varepsilon_i^2\right]_u = \frac{mh^3}{12T}(\dot{f}_{max})^2 \qquad (2.26)$$

where $\dot{f}_{max}$ is largest among all $\dot{f}(\mu_i)$.

2.4 Conclusion

This chapter presents the approximation of different time functions, in conventional BPF domain as well as in other related piecewise constant basis function domains like the GBPF domain, PWM-GBPF domain, NOBPF domain, DUSF domain, and finally the SHF domain.

In the BPF domain, we have approximated two time functions in Examples 2.1 and 2.2. Since we know that the BPF set is the most fundamental [3] of all the piecewise constant basis function sets, we have compared these approximations with the approximations obtained using the rest of the piecewise constant basis function sets. Also we have discussed some basic properties of BPF.

In GBPF domain, we approximated a half sine wave which was already approximated in BPF domain. We compared the results of these two approximations with respect to MISE and the quantitative results are tabulated in Table 2.2. It is observed that judicious selection of the widths of the subintervals of the GBPF components has paid off and the approximation of the function was better than that obtained using the BPF set.

Similarly, for a typical class of functions, the PWM-GBPF set is more suitable for function approximation than the BPF set. This has been proved for a typical function with respect to MISE, vide Example 2.4.

The NOBPF set has also been used for approximation of time functions and have been compared with the BPF approach. This NOBPF set, mainly because of the trapezoidal integration, effectively uses only samples of the function to be expanded via BPF and thus reduces computational burden drastically. But NOBPF approximation does not guarantee the minimum MISE.

Similarly, we have approximated the function of Example 2.2, using the DUSF set. Since, DUSF is an alternative representation of the BPF set, we obtain the same approximation results as in the case of BPF domain.

Finally, we have approximated a typical time function using the sample-and-hold function (SHF) set and have compared the results with that obtained via the BPF set with respect to MISE. This SHF set was designed for analyzing discrete time systems with a sample-and-hold device. Thus, though it belongs to the PCOBF family, it is not a contender to the BPF set for function approximation.

All the function sets discussed have their respective advantages for specific application areas. But the BPF set maintains its popularity because of its simplicity, potential, wide range of application areas and stunning suitability for digital manipulations.

References

1. C. F. Chen, Y. T. Tsay, and T. T. Wu, Walsh operational matrices for fractional calculus and their application to distributed systems, J. Franklin Inst., vol. **303**, no. 3, pp. 267–284, 1977.
2. Z. H. Jiang and W. Schaufelberger, *Block pulse functions and their applications in control systems*, LNCIS, vol. **179**, Springer-Verlag, Berlin, 1992.
3. Anish Deb, Gautam Sarkar, and Sunit K. Sen, Block pulse functions, the most fundamental of all piecewise constant basis functions, Int. J. Syst. Sci., vol. **25**, no. 2, pp. 351–363, 1994.
4. C. H. Wang and R. S. Marleau, Application of a generalized block pulse operational matrices for the approximation of continuous-time systems, Int. J. Syst. Sci., vol. **17**, no. 9, pp. 1269–1278, 1986.
5. M. L. Wang, S. Y. Yang, and R. Y. Chang, Application of a generalized block pulse function to a scaled system, Int. J. Syst. Sci., vol. **18**, no. 8, pp. 1495–1503, 1987.
6. Anish Deb, Gautam Sarkar, and Sunit K. Sen, A new set of pulse width modulated generalised block pulse functions (PWM-GBPF) and their application to cross/auto-correlation of time varying functions, Int. J. Syst. Sci., vol. **26**, no. 1, pp. 65–89, 1995.
7. Anish Deb, Gautam Sarkar, and Sunit K. Sen, Linearly pulse-width modulated block pulse functions and their application to linear SISO feedback control system identification, Proc. IEE, Part – D, Control Theory and Appl., vol. **142**, no. 1, pp. 44–50, 1995.
8. Anish Deb, Gautam Sarkar, Priyaranjan Mandal, Amitava Biswas, and Anindita Sengupta, Optimal block pulse function (OBPF) vs. non-optimal block pulse function (NOBPF), Proceedings of International Conference of IEE (PEITSICON) 2005, pp. 195–199, Kolkata, 28-29th Jan., 2005.
9. Anish Deb, Gautam Sarkar, Priyaranjan Mandal, Amitava Biswas, and Anindita Sengupta, Non-optimal block pulse function (NOBPF) based analysis and identification of SISO control systems, J. Automation & Systems Engineering, vol. **7**, no. 1, pp. 24–58, 2013.

10. Chyi Hwang, Solution of functional differential equation via delayed unit step functions, Int. J. Syst. Sci., vol. **14**, no. 9, pp. 1065–1073, 1983.
11. Anish Deb, Gautam Sarkar, Manabrata Bhattacharjee, and Sunit K. Sen, A new set of piecewise constant orthogonal functions for the analysis of linear SISO systems with sample-and-hold, J. Franklin Inst., vol. **335B**, no. 2, pp. 333–358, 1998.
12. G. P. Rao, *Piecewise constant orthogonal functions and their application to systems and control*, LNC1S, vol. **55**, Springer-Verlag, Berlin, 1983.
13. G. P. Rao and T. Srinivasan, Analysis and synthesis of dynamic systems containing time delays via block-pulse functions, Proc. IEE, vol. **125**, no. 9, pp. 1064–1068, 1978.

Study Problems

2.1 Derive the relation $\mathbf{\Phi} = \mathbf{W\Psi}$

where $\mathbf{W}$ is the Walsh matrix,

$\mathbf{\Phi}$ is the Walsh function vector,

$\mathbf{\Psi}$ is the block pulse function vector of appropriate orders.

2.2 Show that the Walsh matrix $\mathbf{W}$ has the property $\mathbf{W}_{(m)}^2 = m\mathbf{I}_{(m)}$, where each of the matrices is of order m.

2.3 Derive the Walsh coefficients for the following functions for $m = 4$:

$$
\text{a.} \quad f(t) = \begin{cases} 0, & 0 \le t < \dfrac{1}{4}s \\[2mm] 1, & \dfrac{1}{4} \le t < \dfrac{2}{4}s \\[2mm] 0, & \dfrac{2}{4} \le t < \dfrac{4}{4}s \end{cases}
$$

b. $f(t) = \sin t, 0 \le t < 1$ s

c. $f(t) = Kt, 0 \le t < 1$ s, where K is a positive constant.

2.4 Derive the block pulse function (BPF) coefficients for the functions given in Problem 2.3.

2.5 Find the BPF coefficients of the following functions for $m = 6$:

a. $f(t) = \exp(-2t), 0 \le t < 1$ s

b. $f(t) = \begin{cases} t, & 0 \le t < 1 \text{ s} \\ (2 - t), & 1 \le t < 2 \text{ s} \end{cases}$

c. $f(t) = u(t), 0 \le t < 2$ s

d. $f(t) = 1 - \exp(-t), 0 \le t < 1$ s.

2.6   Derive the NOBPF coefficients of the following functions for $m = 7$:

a. $f(t) = t^2, 0 \leq t < 1$ s

b. $f(t) = \exp(-t), 0 \leq t < 2$ s

Draw the exact curve along with the NOBPF domain approximated staircase curve in each case.

2.7   Derive the delayed unit step function (DUSF) coefficients for the functions of Problem 2.6 with $m = 10$. Also show that the results in DUSF domain are same as that obtained using BPF domain approximation.

3

Block Pulse Domain Operational Matrices for Integration and Differentiation

In this chapter, we have discussed operational matrices for integration and differentiation in block pulse function (BPF) domain. These matrices are used to perform numerical integration and differentiation in an operational manner, *i.e.* by way of simple multiplication. Though operational calculus [1,2] is quite old and established, such calculus in connection to piecewise constant orthogonal functions was independently introduced into the literature by Chen and Hsiao [3,4], Le Van et al. [5] and Rao [6] in 1975. From then on such matrices became popular with researchers [7–9] and are used extensively in solving different problems related to control theory.

The basic philosophy of using the integration operational matrix is:

Step 1: Take a complete orthogonal function set $\Phi_{(m)}$ (say) having m component functions. Φ can be a set of Walsh functions, BPFs, Haar functions, etc.

Step 2: Integrate each of the m members of the set Φ and express each of the results in terms of m component functions of the set $\Phi_{(m)}$.

That is, for the first member ϕ_0 of the set $\Phi_{(m)}$, we can write:

$$\int \phi_0 \, dt \approx \mathbf{s}_0^T \Phi_{(m)} \tag{3.1}$$

where $\mathbf{s}_0$ is a matrix having the expansion coefficients (vide Section 1.2) of the function $\int \phi_0 \, dt$ as its elements. Thus

$$\mathbf{s}_0 \triangleq \begin{bmatrix} s_{00} & s_{01} & s_{02} & \cdots & s_{0(m-1)} \end{bmatrix}^T$$

where $[\ldots]^T$ denotes transpose.

Following the above procedure, we can have similar relations for other members of the set as:

$$
\left.
\begin{aligned}
\int \phi_1 \, dt &\approx \mathbf{s}_1^T \mathbf{\Phi}_{(m)} \\
\int \phi_2 \, dt &\approx \mathbf{s}_2^T \mathbf{\Phi}_{(m)} \\
&\vdots \\
\int \phi_{(m-1)} \, dt &\approx \mathbf{s}_{(m-1)}^T \mathbf{\Phi}_{(m)}
\end{aligned}
\right\}
\tag{3.2}
$$

Equations (3.1) and (3.2) may be written in the follow compact form:

$$
\int \mathbf{\Phi}_{(m)} \, dt \approx \mathbf{S}_{(m)} \mathbf{\Phi}_{(m)}
\tag{3.3}
$$

where $\mathbf{S}_{(m)}$ is a square matrix of order m, known as the *operational matrix for integration* in the domain of the set $\mathbf{\Phi}_{(m)}$.

By inverting the square matrix $\mathbf{S}_{(m)}$, we obtain $\mathbf{R}_{(m)}$, the *operational matrix for differentiation*. That is,

$$
\mathbf{S}_{(m)}^{-1} = \mathbf{R}_{(m)}
\tag{3.4}
$$

It is implied that while the square matrix $\mathbf{S}_{(m)}$ operates as an *integrator*, the matrix $\mathbf{R}_{(m)}$ operates as a *differentiator* in the domain of the set $\mathbf{\Phi}_{(m)}$.

3.1 Operational Matrix for Integration [3,4,7]

Let us integrate the $(i + 1)$th BPF component and expand the integrated function back into the block pulse domain.

The equation of the $(i + 1)$th block pulse is given by:

$$
\psi_i(t) = u(t - ih) - u\{t - (i + 1)h\}
$$

Thus

$$
\begin{aligned}
\int \psi_i(t) \, dt &= \int [u(t - ih) - u\{t - (i + 1)h\}] \, dt \\
&= (t - ih)u(t - ih) - \{t - (i + 1)h\}u\{t - (i + 1)h\}
\end{aligned}
\tag{3.5}
$$

Figure 3.1 shows the $(i + 1)$th block pulse and its first integration.

If we expand the above integrated function to its equivalent block pulse series, it will look like the piecewise constant function shown in Figure 3.2.

Hence, the general mathematical expression for equation (3.5) may be written as:

$$\int \psi_i(t)\, dt \approx 0.\psi_0(t) + \ldots + 0.\psi_{(i-1)}(t) + \frac{h}{2}\psi_i(t) + h\psi_{i+1}(t) + \ldots + h\psi_{(m-1)}(t) \quad (3.6)$$

Now, let us consider the first integration of the first four BPFs. These integrated functions and their BPF expansions are shown in Figure 3.3.

Putting proper values of i for the first four BPFs in equation (3.6) we have:

$$\left.\begin{aligned}
\int \psi_0(t)\, dt &= \frac{h}{2}\psi_0(t) + h\psi_1(t) + h\psi_2(t) + h\psi_3(t)\\[4pt]
\int \psi_1(t)\, dt &= 0.\psi_0(t) + \frac{h}{2}\psi_1(t) + h\psi_2(t) + h\psi_3(t)\\[4pt]
\int \psi_2(t)\, dt &= 0.\psi_0(t) + 0.\psi_1(t) + \frac{h}{2}\psi_2(t) + h\psi_3(t)\\[4pt]
\int \psi_3(t)\, dt &= 0.\psi_0(t) + 0.\psi_1(t) + 0.\psi_2(t) + \frac{h}{2}\psi_3(t)
\end{aligned}\right\} \quad (3.7)$$

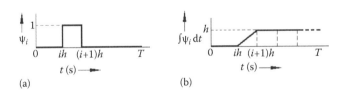

(a) (b)

FIGURE 3.1
(a) The $(i + 1)$th block pulse and (b) its first integration.

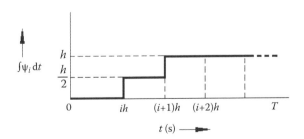

FIGURE 3.2
First integration of $\psi_i(t)$ expressed in block pulse domain.

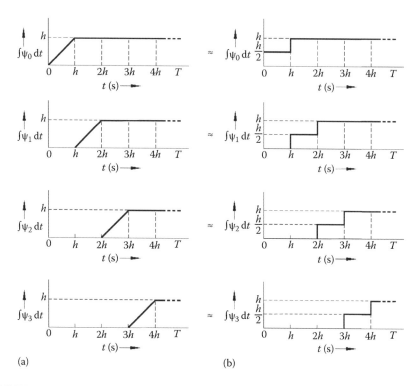

FIGURE 3.3
(a) Integration of first four block pulse functions ($m = 4$) and (b) their block pulse representations.

Writing the above set of equations in matrix form:

$$
\begin{bmatrix}
\int \psi_0(t)\,dt \\[2mm]
\int \psi_1(t)\,dt \\[2mm]
\int \psi_2(t)\,dt \\[2mm]
\int \psi_3(t)\,dt
\end{bmatrix}
\approx h
\begin{bmatrix}
\dfrac{1}{2} & 1 & 1 & 1 \\[2mm]
0 & \dfrac{1}{2} & 1 & 1 \\[2mm]
0 & 0 & \dfrac{1}{2} & 1 \\[2mm]
0 & 0 & 0 & \dfrac{1}{2}
\end{bmatrix}
\begin{bmatrix}
\psi_0(t) \\[2mm]
\psi_1(t) \\[2mm]
\psi_2(t) \\[2mm]
\sqrt{\psi}_3(t)
\end{bmatrix}
\tag{3.8}
$$

Equation (3.8) may be represented in a more compact form like equation (3.3). That is:

$$
\int \boldsymbol{\Psi}_{(4)}(t)\,dt \approx \mathbf{P}_{(4)}\,\boldsymbol{\Psi}_{(4)}(t)
$$

where $\mathbf{P}_{(4)}$ is a constant upper triangular matrix of order 4. This matrix is called the *operational matrix for integration* (of first order) in block pulse domain.

Generalizing this fact for an m-set BPF, we write:

$$\int \mathbf{\Psi}_{(m)}(t)\,dt \approx \mathbf{P}_{(m)}\mathbf{\Psi}_{(m)}(t) \tag{3.9}$$

where the operational matrix for integration of first order is:

$$\mathbf{P}_{(m)} = h \begin{bmatrix} \dfrac{1}{2} & 1 & 1 & \cdots & 1 \\[6pt] & \dfrac{1}{2} & 1 & \cdots & 1 \\[6pt] & & \dfrac{1}{2} & \ddots & \vdots \\[6pt] \mathbf{0} & & & \ddots & 1 \\[6pt] & & & & \dfrac{1}{2} \end{bmatrix}_{(m\times m)} \tag{3.10}$$

A special form of representation of the matrix $\mathbf{P}_{(m)}$ [7,10] is:

$$\mathbf{P}_{(m)} = h\left[\frac{1}{2}\mathbf{I}_{(m)} + \mathbf{Q}_{(m)} + \mathbf{Q}_{(m)}^2 + \ldots + \mathbf{Q}_{(m)}^{(m-1)} \right] \tag{3.11}$$

where $\mathbf{Q}_{(m)}$ is the delay matrix [7] given by

$$\mathbf{Q}_{(m)} = \begin{bmatrix} 0 & 1 & 0 & \cdots & 0 \\ 0 & 1 & \ddots & & 0 \\ & & 0 & \ddots & \vdots \\ \mathbf{0} & & & \ddots & 1 \\ & & & & 0 \end{bmatrix}_{(m\times m)} = \left[\begin{array}{c|ccc} 0 & & & \\ 0 & & \mathbf{I}_{(m-1)} & \\ \vdots & & & \\ 0 & & & \\ \hline 0 & 0 & \cdots & \cdots & 0 \end{array} \right]_{(m\times m)}$$

and it has the property $\mathbf{Q}_{(m)}^m = \mathbf{0}$.
Equation (3.11) may be written as:

$$\begin{aligned}
\mathbf{P}_{(m)} &= h\left[\left\{ \mathbf{I}_{(m)} + \mathbf{Q}_{(m)} + \mathbf{Q}_{(m)}^2 + \ldots \text{ to } \infty \right\} - \frac{1}{2}\mathbf{I}_{(m)} \right] \\
&= h\left[\mathbf{I}_{(m)} + \mathbf{Q}_{(m)} \right]\left[\mathbf{I}_{(m)} - \mathbf{Q}_{(m)} \right]^{-1}
\end{aligned} \tag{3.12}$$

3.1.1 Nature of Integration of a Function in BPF Domain Using the Operational Matrix P [11]

Let $F(t) = \int_0^t f(\tau)\, d\tau$

where

$$f(t) \underset{\approx}{\Delta} \begin{bmatrix} f_0 & f_1 & \cdots & f_{(m-1)} \end{bmatrix} \Psi_{(m)}(t) \triangleq \mathbf{f}^{\mathrm{T}} \Psi_{(m)}(t) \tag{3.13}$$

Using equation (3.9), $F(t)$ can be evaluated approximately in the block pulse domain as:

$$F(t) \approx \mathbf{f}^{\mathrm{T}} \mathbf{P}_{(m)} \Psi_{(m)}(t) \tag{3.14}$$

Substituting $\mathbf{f}$ and $\mathbf{P}_{(m)}$ in equation (3.14) yields:

$$F(t) \approx h\left[\left(\frac{1}{2} f_0 \right) \quad \left(f_0 + \frac{1}{2} f_1 \right) \quad \left(f_0 + f_1 + \frac{1}{2} f_2 \right) \quad \cdots \right.$$

$$\left. \left(f_0 + f_1 + \cdots + f_{(m-2)} + \frac{1}{2} f_{(m-1)} \right) \right] \Psi_{(m)}(t)$$

$$\triangleq \begin{bmatrix} F_0 & F_1 & \cdots & F_i & \cdots & F_{(m-1)} \end{bmatrix} \Psi_{(m)}(t)$$

where

$$F_0 = \frac{1}{2} h f_0$$

and

$$F_i = \frac{1}{2} h f_i + h \sum_{j=0}^{i-1} f_j, \quad j = 0, 1, 2, \ldots, (i-1) \text{ and } i = 0, 1, 2, \ldots, (m-1) \tag{3.15}$$

It was shown in Reference [11] that F_i can be reduced to:

$$F_i = \frac{1}{h}\left[h\left[\frac{F(ih) + F((i+1)h)}{2} \right] \right] \tag{3.16}$$

That is, F_i is the 'average area' (which is nothing but the BPF coefficient) of $F(t)$ in the interval $ih \leq t < (i + 1) h$ with the area obtained by the trapezoidal rule [12,13].

If $F(t)$ is directly expanded in a BPF series using equation (1.13), then

$$F(t) \approx \begin{bmatrix} \bar{F}_0 & \bar{F}_1 & \cdots & \bar{F}_i & \cdots & \bar{F}_{(m-1)} \end{bmatrix} \mathbf{\Psi}_{(m)}(t) \tag{3.17}$$

where the coefficient $\bar{F}_i$ is the 'average area' of $F(t)$ in the interval $ih \leq t < (i + 1) h$, the area being obtained by *exact integration*.

It is obvious that coefficient $\bar{F}_i$ in equation (3.17), obtained using equation (1.14), will be more accurate than F_i obtained using equation (3.16).

3.1.2 Exact Integration and Operational Matrix Based Integration of a BPF Series Expanded Function

We will show in this section that when a function is basically a constant or ramp type then its actual integration converges to that of approximate integration via the trapezoidal rule. Hence, the difference between F_i and $\bar{F}_i$ will no longer exist.

Now, following equation (3.13), we write:

$$f(t) \approx f_0 \psi_0(t) + f_1 \psi_1(t) + \ldots + f_{(m-1)} \psi_{(m-1)}(t)$$

This is a staircase waveform. Let us now find out the exact integration of this BPF series expanded function.

$$\begin{aligned}
F(t) &= \int_0^t f(\tau) \, d\tau \\
&= f_0 \int_0^t \psi_0(\tau) d\tau + f_1 \int_0^t \psi_1(\tau) d\tau + \ldots + f_{(m-1)} \int_0^t \psi_{(m-1)}(\tau) d\tau \\
&= f_0 \int_0^t \big[u(\tau) - u(\tau - h) \big] d\tau + f_1 \int_0^t \big[u(\tau - h) - u(\tau - 2h) \big] d\tau + \ldots \\
&\quad + f_{(m-1)} \int_0^t \big[u\{\tau - (m-1)h\} - u(\tau - T) \big] d\tau \\
&= f_0 t + (f_1 - f_0)(t - h)u(t - h) + (f_2 - f_1)(t - 2h)u(t - 2h) + \ldots \\
&\quad + (f_{(m-1)} - f_{(m-2)})\{t - (m-1)h\}u\{t - (m-1)h\} - f_{(m-1)}(t - T)u(t - T)
\end{aligned} \tag{3.18}$$

where $T = mh$.

Now to expand $F(t)$ in the form of a BPF series we need to evaluate its BPF series coefficients. Following equation (1.14), the coefficient $\bar{\bar{F}}_i$ of equation (3.18) is expressed as:

$$\bar{F}_i = \frac{1}{h} \int_{ih}^{(i+1)h} \Big[f_0 t + (f_1 - f_0)(t - h)u(t - h)$$
$$+ (f_2 - f_1)(t - 2h)u(t - 2h) + \ldots$$
$$+ \big(f_{(m-1)} - f_{(m-2)} \big) \{t - (m-1)h\}u\{t - (m-1)h\} - f_{(m-1)}(t-T)u(t-T) \Big] dt$$

Integrating term by term we have:

$$\bar{F}_i = \left[f_0 \left[\frac{t^2}{2} \right]_{ih}^{(i+1)h} + (f_1 - f_0)\left[\frac{t^2}{2} - ht \right]_{ih}^{(i+1)h} \right.$$
$$\left. + (f_2 - f_1)\left[\frac{t^2}{2} - 2ht \right]_{ih}^{(i+1)h} + \ldots + (f_{i-1} - f_{i-2})\left[\frac{t^2}{2} - (i-1)ht \right]_{ih}^{(i+1)h} \right]$$

where terms having delays greater than ih have been omitted because they are zero.

After simplification, we have:

$$\bar{F}_i = h\Big[f_0 + f_1 + \ldots + f_{(i-1)} \Big] + \frac{1}{2} h f_i = h \sum_{j=0}^{i-1} f_j + \frac{1}{2} h f_i \qquad (3.19)$$

which is same as equation (3.15).

Hence, it is shown that once we consider a function $f(t)$ in the BPF series expanded form, BPF coefficients $\bar{F}_i$ of the integral of $f(t)$, i.e. $F(t)$, determined using the operational matrix **P**, is always as good as the exact integration using (1.14). But if $f(t)$ is considered in its exact form, which may not be a staircase function, then its integration using **P** introduces error. Under such circumstances $F_i \neq \bar{F}_i$.

3.1.3 Numerical Example

Example 3.1

Let us consider the function $f(t) = \dfrac{t^2}{2}$ for $t \geq 0$ and use the operational matrix for integration for $m = 10$ and $T = 1$ s. Results obtained are

TABLE 3.1

Comparison of the Results of Integration Obtained via the Operational Matrix **P** and Direct BPF Expansion along with Percentage Errors for Example 3.1, for $m = 10$ and $T = 1$ s (Vide Appendix B, Program No. 3.1)

t (sec)	BPF Coefficients Using Operational Matrix P	BPF Coefficients Using Direct Expansion of the Integrated Function	Percentage Error
0.0			
	0.00008333	0.00004167	– 100.00000000
0.1			
	0.00075000	0.00062500	– 20.00000000
0.2			
	0.00291667	0.00270833	– 7.69230769
0.3			
	0.00758333	0.00729167	– 4.00000000
0.4			
	0.01575000	0.01537500	– 2.43902439
0.5			
	0.02841667	0.02795833	– 1.63934426
0.6			
	0.04658333	0.04604167	– 1.17647059
0.7			
	0.07125000	0.07062500	– 0.88495575
0.8			
	0.10341667	0.10270833	– 0.68965517
0.9			
	0.14408333	0.14329167	– 0.55248619
1.0			

compared with the direct BPF expansion of the exact integration of $f(t)$. Table 3.1 shows the quantitative comparison of the results along with percentage errors.

3.2 Operational Matrices for Integration in Generalized Block Pulse Function Domain

Let us consider four members of a generalized block pulse function (GBPF) set [9–11] over a time interval $T = (h_0 + h_1 + h_2 + h_3)$ s.

The Figure 3.4(a) shows four generalized block pulses and their first integrations. The integrated functions are then converted to their GBPF equivalents as shown in Figure 3.4(b).

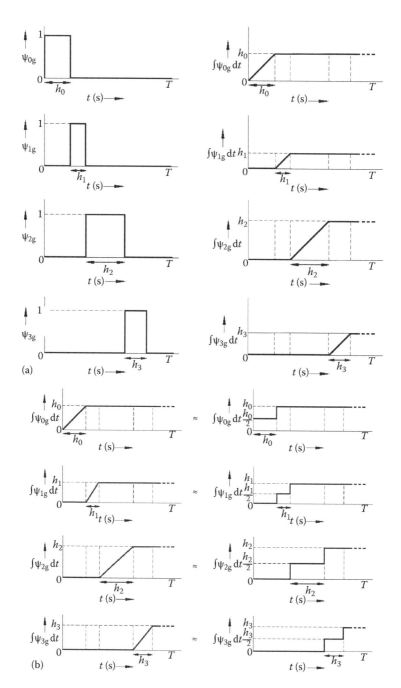

FIGURE 3.4
(a) First four members of the generalized block pulse function (GBPF) set and their first integrations. (b) Integration of first four generalized block pulse functions (GBPFs) ($m = 4$) and their approximate GBPF domain representations.

Thus, a relation similar to equation (3.8) may be written for GBPF domain integration. That is,

$$
\begin{bmatrix}
\int \psi_{0g}(t)\,dt \\
\int \psi_{1g}(t)\,dt \\
\int \psi_{2g}(t)\,dt \\
\int \psi_{3g}(t)\,dt
\end{bmatrix}
\approx
\begin{bmatrix}
\dfrac{h_0}{2} & h_0 & h_0 & h_0 \\
0 & \dfrac{h_1}{2} & h_1 & h_1 \\
0 & 0 & \dfrac{h_2}{2} & h_2 \\
0 & 0 & 0 & \dfrac{h_3}{2}
\end{bmatrix}
\begin{bmatrix}
\psi_{0g}(t) \\
\psi_{1g}(t) \\
\psi_{2g}(t) \\
\psi_{3g}(t)
\end{bmatrix}
\tag{3.20}
$$

Hence, the general form of the operational matrix for integration for m number of GBPFs is:

$$
\mathbf{P}_{g(m)} =
\begin{bmatrix}
\dfrac{h_0}{2} & h_0 & h_0 & \cdots & h_0 \\
 & \dfrac{h_1}{2} & h_1 & \cdots & h_1 \\
 & & \dfrac{h_2}{2} & \ddots & \vdots \\
 & \mathbf{0} & & \ddots & h_{(m-2)} \\
 & & & & \dfrac{h_{(m-1)}}{2}
\end{bmatrix}_{(m \times m)}
\tag{3.21}
$$

where $\mathbf{P}_{g(m)}$ is the operational matrix for integration of order m in the GBPF domain.

3.2.1 Numerical Example

Example 3.2

We consider again the function of Example 3.1. We integrate this function in GBPF domain via the operational matrix $\mathbf{P}_g$, for $m = 10$ and $T = 1$ s. The subintervals chosen are:

$$
h_0 = 0.2 \text{ s}, \; h_1 = 0.17 \text{ s}, \; h_2 = 0.15 \text{ s}, \; h_3 = 0.12 \text{ s}, \; h_4 = 0.1 \text{ s}, \; h_5 = 0.08 \text{ s},
$$
$$
h_6 = 0.07 \text{ s}, \; h_7 = 0.05 \text{ s}, \; h_8 = 0.04 \text{ s}, \; h_9 = 0.02 \text{ s}.
$$

Results of integration are compared with the direct GBPF expansion of the exact integration of $f(t)$. Table 3.2 shows the quantitative comparison of the results along with percentage errors.

In Figure 3.5(a), we show the BPF domain result along with the exact curve. And in Figure 3.5(b), we present the GBPF domain result with the exact curve. By comparing the mean integral square errors (MISEs) for both the cases, we find that, the MISE for equal width BPF integration is 0.00004167, whereas it is 0.00002362 for GBPF domain. Thus, integration using the operational matrix in GBPF domain is advantageous if we have *a priori* knowledge about the function to be integrated, so that the subintervals can be chosen judiciously.

TABLE 3.2

Comparison of Results of Integration Obtained via the Operational Matrix $\mathbf{P}_g$ and Direct Generalized Block Pulse Function (GBPF) Expansion along with Percentage Errors for Example 3.2, for $m = 10$ and $T = 1$ s (Vide Appendix B, Program No. 3.2)

t (sec)	GBPF Coefficients Using Operational Matrix $\mathbf{P}_g$	GBPF Coefficients Using Direct Expansion of the Integrated Function	Percentage Error
0.00			
	0.00066667	0.00033333	− 100.00000000
0.20			
	0.00488775	0.00420138	− 16.33691351
0.37			
	0.01593842	0.01510404	− 5.52418365
0.52			
	0.03356267	0.03286667	− 2.11764706
0.64			
	0.05561400	0.05503900	− 1.04471375
0.74			
	0.07971600	0.07930000	− 0.52459016
0.82			
	0.10469475	0.10434563	− 0.33458518
0.89			
	0.12796275	0.12777213	− 0.14919138
0.94			
	0.14764800	0.14752000	− 0.08676790
0.98			
	0.16176600	0.16173300	− 0.02040400
1.00			

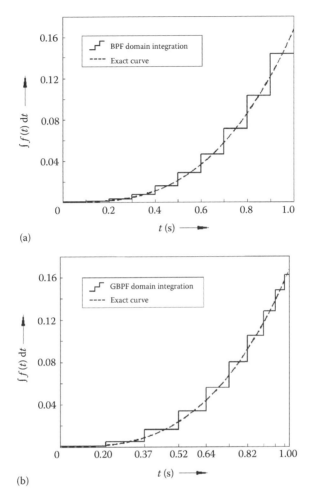

(a)

(b)

FIGURE 3.5
Integration of the function $f(t)$ of Example 3.2 using (a) the operational matrix **P** in BPF domain and (b) the operational matrix $\mathbf{P}_g$ in generalized block pulse function (GBPF) domain, for $m = 10$ and $T = 1$ s.

3.3 Improvement of the Integration Operational Matrix of First Order

In block pulse domain analysis, the block pulse operational matrix [7] for integration, **P**, has been utilized quite frequently by many researchers. This matrix was first derived by Chen et al. [7] and later utilized by other researchers [6,10]. This matrix has led to generation of other operational matrices [14].

Several workers have investigated the properties of integration operational matrices [15,16]. Hence, the characteristic of the basic integration operational matrix **P** deserves deeper investigation.

It was found that in the BPF domain analysis, the use of **P** for integration of any function $f(t)$ and subsequent determination of BPF series coefficients of the function $F(t) = \int f(t)dt$ introduces error. Also, repeated integration in BPF domain causes this error to accumulate, and finally leads to an unreliable result. Chen and Chung have pointed this out [11] and have suggested a new operational matrix for integration, which reduces the error in BPF series analysis compared with that originally derived by Chen et al. [7]. It was also shown by Chen and Chung [11] that integration of a function $f(t)$ in BPF domain by the matrix **P** is equivalent to finding the BPF series coefficients of $F(t)$ by the well-known trapezoidal rule of integration. This fact has led them to use a three-point interpolation polynomial of integration in the Lagrange form for more accurate determination of the BPF series coefficients of $F(t)$ and have derived an improved version of **P** (say, **P1**).

In this section the following facts have been established:

i. When any function $f(t)$ is subjected to BPF series expansion, it becomes a staircase function. Hence, application of the operational matrix **P** upon this function to find the BPF series coefficients of $F(t)$ is as good as exact determination of the coefficients.

ii. The use of a three-point interpolation polynomial in the Lagrange form is equivalent to improving upon the trapezoidal rule by considering only the first term of its remainder [12,13].

iii. Consideration of two terms from the remainder in the trapezoidal rule improves the matrix **P1** further to a newer matrix **P2** [17]. This improvement can be even better if more terms from the remainder are taken into account. But as expected, with inclusion of additional terms from the remainder, the rate of improvement is diminished. We will somewhat generalize this improvement in a sense that the difference between F_i and $\bar{F}_i$ (vide Section 3.1.2) is incurred because of approximation in performing integration.

It will be shown that the improvement can be achieved more and more if we take into account more and more terms of the remainder in the trapezoidal method of numerical integration [12,13].

Thus

$$\bar{F_i} = \frac{1}{h}\left[h\left[\frac{F(ih) + F((i+1)h)}{2} + R\right]\right], \quad i = 0, 1, 2, \ldots, (m-1) \quad (3.22)$$

where R is the remainder term.

Comparing equations (3.16) and (3.22) it is observed that $\frac{1}{h}R$ is assumed to be zero in equation (3.16) and thus error has been introduced in the evaluation of the integral in the LHS of equation (3.22).

Using equation (3.16), we can write:

$$\bar{F_i} = F_i + \frac{1}{h}R \quad (3.23)$$

The remainder term $\frac{1}{h}R$ is given by [12]:

$$\frac{1}{h}R = -\frac{1}{12}\nabla^2 F\{(i+1)h\} - \frac{1}{24}\nabla^2 F\{(i+1)h\} - \ldots$$

$$= \frac{1}{h}[r_1 + r_2 + \ldots] \quad (3.24)$$

where ∇ denotes the backward difference operator.

Let us now evaluate the contribution of the first term $\frac{1}{h}r_1$.

$$\frac{1}{h}r_1 = -\frac{1}{12}\nabla^2 F\{(i+1)h\}$$

$$= -\frac{1}{12}\nabla\left[\nabla F\{(i+1)h\}\right]$$

$$= -\frac{1}{12}\left[\{F\{(i+1)h\} - F(ih)\} - \{F(ih) - F\{(i-1)h\}\}\right] \quad (3.25)$$

$$= -\frac{1}{12}[hf_i - hf_{i-1}]$$

Equation (3.25) is defined for $i \geq 1$, because $F\{(i-1)\,h\}$ is defined for $i \geq 1$. Hence, contribution of $\frac{1}{h}r_1$ will start from $i = 1$. For $i = 0$, $\bar{F}_0 = \frac{1}{2}hf_0$ from equation (3.19) and from $i \geq 1$, it is given by equation (3.25).

Thus, adding only the first term of the remainder in equation (3.19), we have the improved $\bar{F}_i$'s as:

$$\bar{F}_i = h\sum_{j=0}^{i-1} f_j + \frac{1}{2}hf_i - \frac{1}{12}[hf_i - hf_{i-1}]$$

$$= h\sum_{j=0}^{i-2} f_j + hf_{i-1} + \frac{1}{2}hf_i - \frac{1}{12}hf_i + \frac{1}{12}hf_{i-1} \qquad (3.26)$$

$$= h\left[\sum_{j=0}^{i-2} f_j + \frac{13}{12}f_{i-1} + \frac{5}{12}f_i\right]$$

Hence

$$\bar{F}_0 = \frac{5}{12}hf_0$$

$$\bar{F}_1 = \frac{13}{12}hf_0 + \frac{5}{12}hf_1$$

$$\bar{F}_2 = hf_0 + \frac{13}{12}hf_1 + \frac{5}{12}hf_2$$

$$\bar{F}_3 = hf_0 + hf_1 + \frac{13}{12}hf_2 + \frac{5}{12}hf_3$$

and so on.

Hence, the modified integration operational matrix, with the addition of the first term of the remainder, is given by,

$$
\mathbf{P1}_{(m)} = h
\begin{bmatrix}
\dfrac{1}{2} & \dfrac{13}{12} & 1 & 1 & 1 & \cdots & 1 \\[2mm]
 & \dfrac{5}{12} & \dfrac{13}{12} & 1 & 1 & \cdots & 1 \\[2mm]
 & & \dfrac{5}{12} & \dfrac{13}{12} & 1 & \ddots & \vdots \\[2mm]
 & & & \dfrac{5}{12} & \dfrac{13}{12} & \ddots & 1 \\[2mm]
 & & & & \dfrac{5}{12} & \ddots & 1 \\[2mm]
 & \mathbf{0} & & & & \ddots & \dfrac{13}{12} \\[2mm]
 & & & & & & \dfrac{5}{12}
\end{bmatrix}_{(m \times m)}
\tag{3.27}
$$

It is expected that **P1** is somewhat *improved* from **P** and it helps the improved $\bar{F}_i$'s to move closer to the actual values of $\bar{F}_i$. This operational matrix **P1** is identical to that developed by Chen and Chung [11].

Let us now consider the contribution of the first and the second terms of the remainder.

The second term $\dfrac{1}{h} r_2$ is calculated as follows:

$$
\frac{1}{h} r_2 = -\frac{1}{24} \nabla^3 F\{(i+1)h\} = -\frac{1}{24} \nabla^2 \left[\nabla F\{(i+1)h\} \right]
$$

$$
= -\frac{1}{24} \nabla^2 \left[F\{(i+1)h\} - F(ih) \right]
$$

$$
= -\frac{1}{24} \nabla \left[\{F\{(i+1)h\} - F(ih)\} - \{F(ih) - F\{(i-1)h\}\} \right]
$$

$$
= -\frac{1}{24} \left[\{F\{(i+1)h\} - F(ih)\} - \{F(ih) - F\{(i-1)h\}\} \right]
$$

$$
+ \frac{1}{24} \left[\{F(ih) - F\{(i-1)h\}\} - \{F\{(i-1)h\} - F\{(i-2)h\}\} \right]
$$

$$
= -\frac{1}{24} \left[hf_i - hf_{i-1} \right] + \frac{1}{24} \left[hf_{i-1} - hf_{i-2} \right]
$$

$$
= -\frac{1}{24} hf_i + \frac{2}{24} hf_{i-1} - \frac{1}{24} hf_{i-2}
\tag{3.28}
$$

From equation (3.28), it is seen that because of the presence of F $\{(i − 2)\ h\}$, contribution of $\dfrac{1}{h} r_2$ will start from $i \geq 2$ as this equation is defined from $i \geq 2$. Adding both the terms of the remainder, we have:

$$\bar{F}_i = h\sum_{i=0}^{i-1} f_i + \frac{1}{2}hf_i - \frac{1}{12}hf_i + \frac{1}{12}hf_{i-1} - \frac{1}{24}hf_i + \frac{2}{24}hf_{i-1} - \frac{1}{24}hf_{i-2}$$

$$= h\sum_{i=0}^{i-3} f_i + hf_{i-2} + hf_{i-1} + \frac{9}{24}hf_i + \frac{4}{24}hf_{i-1} - \frac{1}{24}hf_{i-2} \qquad (3.29)$$

$$= h\left[\sum_{i=0}^{i-3} f_i + \frac{23}{24}f_{i-2} + \frac{28}{24}f_{i-1} + \frac{9}{24}f_i\right]$$

This equation gives the elements of a *further improved* operational matrix for $i \geq 2$. Once again, for $i = 0$, we use the first element from **P** as $\bar{F}_0 = \dfrac{1}{2}hf_0$. For $i = 1$, we use the element from **P1** as $\bar{F}_1 = \dfrac{13}{12}hf_0 + \dfrac{5}{12}hf_1$ and from then on we use equation (3.29).

Thus

$$\mathbf{P2}_{(m)} = h\begin{bmatrix} \dfrac{1}{2} & \dfrac{13}{12} & \dfrac{23}{24} & 1 & 1 & \cdots & 1 \\[2mm] & \dfrac{5}{12} & \dfrac{28}{24} & \dfrac{23}{24} & 1 & \cdots & 1 \\[2mm] & & \dfrac{9}{24} & \dfrac{28}{24} & \dfrac{23}{24} & \ddots & \vdots \\[2mm] & & & \dfrac{9}{24} & \dfrac{28}{24} & \ddots & 1 \\[2mm] & & & & \dfrac{9}{24} & \ddots & \dfrac{23}{24} \\[2mm] & & \mathbf{0} & & & \ddots & \dfrac{28}{24} \\[2mm] & & & & & & \dfrac{9}{24} \end{bmatrix}_{(m\times m)} \qquad (3.30)$$

The matrix **P2** is the new integral operational matrix which is *further improved* than **P1**.

We arrived at the improved matrix **P1** by taking into account the first term of the remainder R in the trapezoidal rule.

At this stage an important observation regarding the improvements of integral operational matrices (*e.g.* **P**, **P1**, **P2**, etc.) can be made. Consideration of only the first term of the remainder $\frac{1}{h}R$ of equation (3.24) has caused improvement from the *second column* of **P1**. Consideration of first two terms of equation (3.24) has caused improvement from the *third column* of **P2**. Hence, it is apparent that consideration of further terms of the remainder, *i.e.* $\frac{1}{h}r_3$ and so on, will improve the integral operational matrix from the fourth and further respective columns.

That is, if we take into account up to the ith term of $\frac{1}{h}R$ in equation (3.24).

$$\frac{1}{h}R \approx \sum_{j=1}^{i} r_j, \quad i = 1, 2, 3, \ldots$$

then, the analysis will produce an integral operational matrix, say **Pi**, with the improvement effected from its $(i + 1)$th column onward. That is, from column 1 to column i, the elements of the integral operational matrix **Pi** will remain same. However, with higher and higher values of i, further improvement will become less and less significant and at the same time, the analysis will become more involved.

To establish the validity of our method, we make use of the same examples that were used by Chen and Chung [11], and compare the results.

3.3.1 Numerical Examples [11]

Example 3.3

Let $f(t) = \exp(2t)$ for $t \geq 0$.

Then $F(t) = \frac{1}{2}[\exp(2t) - 1]$.

The eight term BPF series expansion of $f(t)$ on $t \in [0, 1]$ is:

$$f(t) \approx \begin{bmatrix} 1.13610167 & 1.45878342 & 1.87311498 & 2.40512725 & 3.08824452 \\ 3.96538445 & 5.09165442 & 6.53781369 \end{bmatrix} \Psi_{(8)}(t)$$

Table 3.3 tabulates the respective BPF coefficients of $F(t)$ using exact integration and integral operational matrices **P**, **P1**, and **P2**.

The improvement in the operational matrices **P1** and **P2** are apparent from the table.

TABLE 3.3

Comparison of Results of Integration of the Function $f(t)$ of Example 3.3 Using the Operational Matrices **P**, **P1**, and **P2**, for $m = 8$ and $T = 1$ s (Vide Appendix B, Program No. 3.3)

t (sec)	Direct BPF Expansion of Exact Integration	Integration by P, Equation (3.10)	Integration by P1, Equation (3.27)	Integration by P2, Equation (3.30)
0.000				
	0.06805083	0.07100635	0.07100635	0.07100635
0.125				
	0.22939171	0.23318667	0.22982540	0.22982540
0.250				
	0.43655749	0.44143032	0.43711437	0.43663703
0.375				
	0.70256362	0.70882046	0.70327867	0.70266575
0.500				
	1.04412226	1.05215620	1.04504039	1.04425339
0.625				
	1.48269223	1.49300801	1.48387113	1.48286060
0.750				
	2.04582721	2.05907294	2.04734096	2.04604340
0.875				
	2.76890685	2.78591469	2.77085053	2.76918444
1.000				

Example 3.4

Consider the differential equation

$$\dot{f}(t) - f(t) = 0, \quad f(0) = 1.$$

The solution is $f(t) = \exp(t)$.
By BPF analysis

$$\mathbf{F} = \begin{bmatrix} 1 & 1 & 1 & \cdots & 1 \end{bmatrix}(\mathbf{I} - \mathbf{P})^{-1} \tag{3.31}$$

where $\mathbf{F} = \begin{bmatrix} f_0 & f_1 & \cdots & f_{(m-1)} \end{bmatrix}$ and $f(t) \approx \mathbf{F}^{\mathrm{T}} \, \mathbf{\Psi}_{(m)}(t)$

Table 3.4 tabulates the BPF series solution F of $f(t)$ for $m = 10$ and $t \in [0, 1)$ using integral operational matrices **P**, **P1**, and **P2** along with its BPF series expansion coefficients obtained from exact integration.

TABLE 3.4

Comparison of Results of Integration of the Function $f(t)$ of Example 3.4 Using the Operational Matrices **P**, **P1**, and **P2**, for $m = 10$ and $T = 1$ s

t (sec)	Direct BPF Expansion of Exact Integration	Integration by P, Equation (3.10)	Integration by P1, Equation (3.27)	Integration by P2, Equation (3.30)
0.0				
	1.05170918	1.05258546	1.05258546	1.05258546
0.1				
	1.16231840	1.16328684	1.16236509	1.16236509
0.2				
	1.28456049	1.28563078	1.28461210	1.28456363
0.3				
	1.41965890	1.42084175	1.41971593	1.41966236
0.4				
	1.56896573	1.57027298	1.56902876	1.56896956
0.5				
	1.73397530	1.73542004	1.73404496	1.73397953
0.6				
	1.91633907	1.91793575	1.91641606	1.91634375
0.7				
	2.11788221	2.11964682	2.11796729	2.11788738
0.8				
	2.34062183	2.34257202	2.34071586	2.34062754
0.9				
	2.58678717	2.58894247	2.58689109	2.58679348
1.0				

3.4 One-Shot Operational Matrices for Repeated Integration

The principle of one-shot operational matrix for repeated integration (**OSOMRI**) was introduced by Rao [18] and Chi-Hsu [19]. The nth order **OSOMRI** of dimension $m \times m$, namely $\mathbf{P}(n)_{(m)}$, is obtained by integrating n times the BPF set of order m and expanding the result in terms of an m-term BPF vector. That is, when $n = 1$, we get the operational matrix $\mathbf{P}(1)$ indicating first-order integration. Thus, $\mathbf{P}(1) \triangleq \mathbf{P}$.

Mathematically, n times repeated integration may be represented as:

$$\underbrace{\int_0^t \int_0^t \int_0^t \cdots \int_0^t}_{n \text{ times}} \mathbf{\Psi}_{(m)}(t)\,d\tau\,d\tau\ldots d\tau \approx \mathbf{P}(n)_{(m)} \mathbf{\Psi}_{(m)}(t), \text{ where, } n = 1, 2, 3\ldots \quad (3.32)$$

In Reference [18], $\mathbf{P}(n)_{(m)}$ is expressed in the following form:

$$\mathbf{P}(n)_{(m)} = \frac{1}{m^n}\left[\frac{\mathbf{I}}{(n+1)!} + \sum_{r=1}^{m-1}\left\{\sum_{q=0}^{n-1}\frac{r^{q+1}-(r-1)^{q+1}}{(q+1)!(n-q)!}\right\}\mathbf{Q}^r\right] \qquad (3.33)$$

where $\mathbf{I}$ is the unit matrix of order m and $\mathbf{Q}$ is the delay matrix of order m defined in Reference [7].

In Reference [18], $\mathbf{P}(n)_{(m)}$ is expressed in the following manner:

$$\mathbf{P}(n)_{(m)} = \frac{h^n}{(n+1)!}\begin{bmatrix} p(n)1 & p(n)2 & p(n)3 & \cdots & p(n)m \\ & p(n)1 & p(n)2 & \cdots & p(n)(m-1) \\ & & \ddots & \ddots & \vdots \\ & \mathbf{0} & & \ddots & \vdots \\ & & & & p(n)1 \end{bmatrix} \qquad (3.34)$$

with

$$p(n)j = \begin{cases} 1, & \text{for } j = 1 \\ j^{n+1} - 2(j-1)^{n+1} + (j-2)^{n+1}, & \text{for } 2 \le j \le m \end{cases} \qquad (3.35)$$

However, an alternative form of $\mathbf{P}(n)_{(m)}$, with respect to equation (3.34), is derived as:

$$p(n)j = \begin{cases} 1, & \text{for } j = 1 \\ (n+1)!\displaystyle\sum_{p=1}^{n}\frac{(j-1)^p - (j-2)^p}{(n-p+1)!}, & \text{for } 2 \le j \le m \end{cases} \qquad (3.36)$$

By using $\mathbf{P}(n)_{(m)}$, any multiple integration operation can be transformed into algebraic operations in one step or one shot. Also, the amount of error introduced by using $\mathbf{P}^n_{(m)}$ for multiple integration is reduced appreciably if we use $\mathbf{P}(n)_{(m)}$ instead [18,20]. Further, when n and m are large, the use of $\mathbf{P}^n_{(m)}$ requires much more storage as well as computational time than that required by $\mathbf{P}(n)_{(m)}$.

3.4.1 Numerical Example

Example 3.5

To establish the validity of one-shot operational matrix for integration, we consider the following function and integrate it twice using both one-shot operational matrix and the first-order operational matrix for integration. The results of such integrations are compared with the exact integration of the function.

$$f(t) = 1 - \exp(-t).$$

That is, we integrate this function twice, using (i) first-order integration operational matrix **P** repeatedly and (ii) one-shot operational matrix for repeated integration **P(n)**.

Figure 3.6 shows the graphical comparison between these two results and compares the same with the direct expansion of exact integration. It is observed that the results obtained using the one-shot operational matrix for integration is better than that obtained using the first-order integration operational matrix. Table 3.5 provides the quantitative comparison between these two methods.

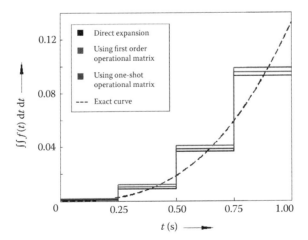

FIGURE 3.6
Double Integration of the function *f(t)* of Example 3.5 using (a) the first-order operational matrix **P** and (b) the one-shot operational matrix **P(n)**, for *m* = 4 and *T* = 1 s (vide Appendix B, Program no. 3.4).

TABLE 3.5

Double Integration of the Function $f(t)$ of Example 3.5 Using (a) First-Order Operational Matrix **P** and (b) One-Shot Operational Matrix **P(2)**, for $m = 4$ and $T = 1$ s (Vide Appendix B, Program No. 3.4)

t (sec)	Direct Expansion of Exact Integration	Integration by P, Equation (3.10)	Integration by P(2), Equation (3.35) or (3.36)
0.00			
	0.00061980	0.00180005	0.00120003
0.25			
	0.00883617	0.01205831	0.01043894
0.50			
	0.03626024	0.04107260	0.03865936
0.75			
	0.09246822	0.09851905	0.09548754
1.00			

3.5 Operational Matrix for Differentiation

From $\mathbf{P}_{(m)}$ of equation (3.12), the operational matrix for differentiation $\mathbf{D}_{(m)}$ may be derived by inverting $\mathbf{P}_{(m)}$ as under:

$$\mathbf{D}_{(m)} = \mathbf{P}_{(m)}^{-1} = \frac{1}{h} \begin{bmatrix} 2 & -4 & 4 & \cdots & \cdots & 4(-1)^{m-1} \\ & 2 & -4 & \ddots & \ddots & 4(-1)^{m-2} \\ & & 2 & \ddots & \ddots & \vdots \\ & & & \ddots & \ddots & \vdots \\ & 0 & & & \ddots & \vdots \\ & & & & & 2 \end{bmatrix}_{(m \times m),\, m \geq 2} \tag{3.37}$$

$\mathbf{D}_{(m)}$ operates on a BPF vector in a manner similar to $\mathbf{P}_{(m)}$.

From equation (3.12), the operational matrix for differentiation, $\mathbf{D}_{(m)}$ may be obtained by inverting the expression on the RHS. It should be noted that the product on the RHS of equation (3.12) is commutative [10].

Thus

$$\mathbf{D}_{(m)} = \frac{2}{h} \left[\mathbf{I}_{(m)} - \mathbf{Q}_{(m)} \right] \left[\mathbf{I}_{(m)} + \mathbf{Q}_{(m)} \right]^{-1} \tag{3.38}$$

3.5.1 Numerical Example

Example 3.6

Let us consider the function $f(t) = \dfrac{t^2}{2}$. We differentiate this function once via the operational matrix $\mathbf{D}$, for $m = 10$ and $T = 1$ s. Results obtained are compared with the direct BPF expansion of the differentiation of $f(t)$. Table 3.6 shows the quantitative comparison of the results along with percentage errors for each coefficient. In Figure 3.7, the results obtained for $m = 10$ are compared graphically with direct BPF expansion of the differentiation of $f(t)$.

Figure 3.7 and the percentage error column of Table 3.6 reveal that use of the operational matrix $\mathbf{D}$ shows a slight tendency of *oscillation* in the output.

TABLE 3.6

Comparison of Results of Differentiation of the Function $f(t)$ of Example 3.6 Obtained via the Operational Matrix $\mathbf{D}_{(m)}$ and Direct Block Pulse Function (BPF) Expansion of the Differentiated Function along with Percentage Errors, for $m = 10$ and $T = 1$ s (Vide Appendix B, Program No. 3.5)

t (sec)	BPF Coefficients Obtained Using Operational Matrix D	BPF Coefficients Obtained from Direct BPF Expansion of the Differentiated Function	Percentage Error
0.0			
	0.03333333	0.05000000	33.33333333
0.1			
	0.16666667	0.15000000	– 11.11111111
0.2			
	0.23333333	0.25000000	6.66666667
0.3			
	0.36666667	0.35000000	– 4.76190476
0.4			
	0.43333333	0.45000000	3.70370370
0.5			
	0.56666667	0.55000000	– 3.03030303
0.6			
	0.63333333	0.65000000	2.56410256
0.7			
	0.76666667	0.75000000	– 2.22222222
0.8			
	0.83333333	0.85000000	1.96078431
0.9			
	0.96666667	0.95000000	– 1.75438596
1.0			

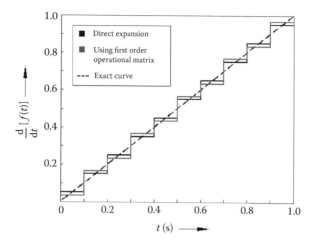

FIGURE 3.7

Differentiation of the function $f(t)$ of Example 3.6 using the first-order operational matrix $\mathbf{D}_{g(m)}$ in block pulse function (BPF) domain, for $m = 10$ and $T = 1$ s (vide Appendix B, Program no. 3.5).

3.6 Operational Matrices for Differentiation in Generalized Block Pulse Function Domain

From $\mathbf{P}_{g(m)}$ of equation (3.21), the general form of the operational matrix for differentiation $\mathbf{D}_{g(m)}$ may be derived by inverting $\mathbf{P}_{g(m)}$ as under:

$$\mathbf{D}_{g(m)} = \mathbf{P}_{g(m)}^{-1} = \begin{bmatrix} \dfrac{2}{h_0} & -\dfrac{4}{h_1} & \dfrac{4}{h_2} & \cdots & \cdots & \dfrac{4}{h_{(m-1)}}(-1)^{m-1} \\ & \dfrac{2}{h_1} & -\dfrac{4}{h_2} & \ddots & \ddots & \dfrac{4}{h_{(m-1)}}(-1)^{m-2} \\ & & \dfrac{2}{h_2} & \ddots & \ddots & \vdots \\ & & & \ddots & \ddots & \vdots \\ & 0 & & & \ddots & \dfrac{4}{h_{(m-1)}}(-1) \\ & & & & & \dfrac{2}{h_{(m-1)}} \end{bmatrix}_{(m \times m)}$$

(3.39)

$\mathbf{D}_{g(m)}$ operates on a GBPF vector in a manner similar to $\mathbf{P}_{g(m)}$.

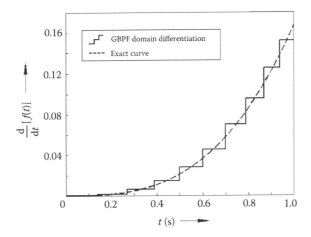

FIGURE 3.8
Differentiation of the function $f(t)$ of Example 3.7 using the first-order operational matrix $\mathbf{D}_{g(m)}$ in generalized block pulse function (GBPF) domain, for $m = 10$ and $T = 1$ s (vide Appendix B, Program no. 3.6).

3.6.1 Numerical Example

Example 3.7

Let us consider the function $f(t) = \dfrac{t^4}{24}$. We differentiate this function via the operational matrix $\mathbf{D}_g$ in GBPF domain, for $m = 10$ and $T = 1$ s. The subintervals chosen are:

$$h_0 = 0.14 \text{ s, } h_1 = 0.13 \text{ s, } h_2 = 0.12 \text{ s, } h_3 = 0.11 \text{ s, } h_4 = 0.1 \text{ s, } h_5 = 0.1 \text{ s,}$$
$$h_6 = 0.09 \text{ s, } h_7 = 0.08 \text{ s, } h_8 = 0.07 \text{ s, } h_9 = 0.06 \text{ s.}$$

Figure 3.8 shows the function differentiated in GBPF domain, along with the exact curve of the differentiation of $f(t)$. The MISE computed in GBPF domain is only 0.00026492. Table 3.7 shows the quantitative comparison of the results along with percentage errors.

3.7 One-Shot Operational Matrices for Repeated Differentiation

From $\mathbf{P}(n)_{(m)}$ of equation (3.35) or (3.36), the one-shot operational matrix for differentiation $\mathbf{D}(n)_{(m)}$ for m number of block pulse coefficients can be directly obtained by inverting $\mathbf{P}(n)_{(m)}$ as:

$$\mathbf{D}(n)_{(m)} = \mathbf{P}(n)_{(m)}^{-1} \tag{3.40}$$

TABLE 3.7

Comparison of Results of Differentiation of the Function $f(t)$ of Example 3.7 Obtained via the Operational Matrix $\mathbf{D}_{g(m)}$ and Direct Generalized Block Pulse Function (GBPF) Domain Expansion of the Differentiated Function along with Percentage Errors, for $m = 10$ and $T = 1$ s (Vide Appendix B, Program No. 3.6)

t (sec)	GBPF Coefficients Using Operational Matrix $\mathbf{D}_{g(m)}$	GBPF Coefficients Using Direct Expansion of Differentiated Function	Percentage Error
0.00			
	0.00004573	0.00011433	−9.76000000
0.14			
	0.00126354	0.00158021	−102.16594872
0.27			
	0.00593749	0.00618750	−119.23030769
0.39			
	0.01455903	0.01491121	−99.66668052
0.50			
	0.02782358	0.02795833	−80.96639566
0.60			
	0.04517642	0.04604167	−61.58479881
0.70			
	0.06987937	0.06916704	−51.77420211
0.79			
	0.09480944	0.09551917	−34.24344838
0.87			
	0.12500429	0.12372104	−21.70802898
0.94			
	0.15163073	0.15225767	−5.81964525
1.00			

3.7.1 Numerical Example

Example 3.8

Now we consider three different functions to check the performance efficiency of one-shot operational matrix for double differentiation.

    i. $f_1(t) = \dfrac{t^4}{24}$,

    ii. $f_2(t) = \dfrac{t^3}{6}$ and

    iii. $f_3(t) = \dfrac{t^2}{2}$

We perform BPF domain double derivative operation on all the three functions using (i) the first-order operational matrix for differentiation $\mathbf{D}$ repeatedly, and (ii) the one-shot operational matrix for differentiation

$D(n)$ with $n = 2$, as obtained from equation (3.40) above. Finally, we compare the results with the results obtained using the first-order operational matrix for differentiation in Figures 3.9 through 3.11.

For the first two functions $f_1(t)$ and $f_2(t)$, we note from Figures 3.9 and 3.10 that undesired oscillations occur in the results of differentiation. This may be due to the structure of the matrix **D**.

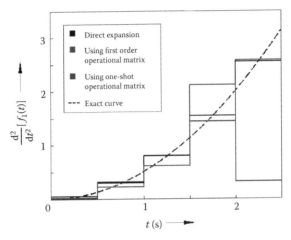

FIGURE 3.9

Double differentiation of the function $f_1(t) = \dfrac{t^4}{24}$ using (a) the first-order operational matrix **D** and (b) the one-shot operational matrix **D(2)**, for $m = 5$ and $T = 2.5$ s (vide Appendix B, Program no. 3.7).

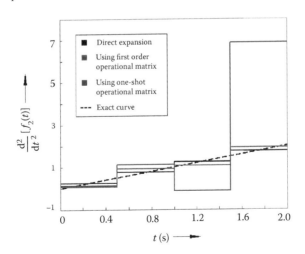

FIGURE 3.10

Double differentiation of the function $f_2(t) = \dfrac{t^3}{6}$ of Example 3.8 using (a) the first-order operational matrix **D** and (b) the one-shot operational matrix **D(2)**, for $m = 4$ and $T = 2$ s (vide Appendix B, Program no. 3.7).

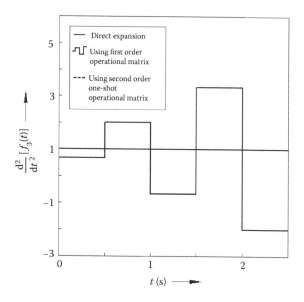

FIGURE 3.11

Double differentiation of the function $f_3(t) = \dfrac{t^2}{2}$ using (a) the first-order operational matrix **D** and (b) the one-shot operational matrix **D(2)**, for $m = 5$ and $T = 2.5$ s. The results are compared with the direct expansion. It is observed that the direct expansion curve merges with the curve obtained via **D(2)** (vide Appendix B, Program no. 3.7).

We also observe that for a fixed time period, with increasing number of segments, the amplitude of oscillation increases, possibly because of the term $\dfrac{1}{h}$ in the operational matrix **D**. But for the function $f_3(t) = \dfrac{t^2}{2}$, no oscillation is noted for its double differentiation using the one-shot matrix **D(2)**. However, oscillation is predominant when **D²** is used instead of **D(2)**. This is seen from Figure 3.11.

3.8 Conclusion

In this Chapter, the operational matrices for integration in BPF as well as in GBPF domain have been reviewed. Integration in both BPF and GBPF domains are illustrated via examples. Qualitative comparisons with direct expansion of the functions after exact integration are also provided. Based on the nature of the function, we can decide whether GBPF domain operations will produce results with less MISE. This has been established quantitatively by comparing the MISE for both the cases via examples.

This Chapter also reviews the operational matrix for integration **P2** [17] which is an improvement on the conventional integration operational matrix **P** and the improved operational matrix **P1** [11].

In deriving the integral operational matrix **P1** Chen and Chung [11] banked upon three-point interpolation polynomial in the Lagrange form. But it has been shown that it is equivalent to an extension of the trapezoidal rule where only the first term of the remainder **R** was considered for the evaluation of the integration. Consideration of both the first and second terms of the remainder has produced a further improved integral operational matrix **P2**, which has been employed to treat Examples 3.3 and 3.4, considered by Chen and Chung [11]. Improvement in the results is apparent from the values of coefficients given in Tables 3.3 and 3.4. Inclusion of still further terms of the remainder will improve upon **P2** further, but the rate of improvement will diminish gradually.

Also in this Chapter, the one-shot operational matrix for integration in BPF domain is discussed. One illustration of double integration is presented through Example 3.5. This operation is performed using (i) repeated use of the first-order operational matrix for integration **P** and (ii) the one-shot operational matrix for integration. Figure 3.6 and Table 3.5 compared these two results and it is observed that the results obtained using the one-shot operational matrix are better than that obtained using first-order integration operational matrix repeatedly.

The first-order operational matrix for differentiation in BPF domain is also reviewed. A basic function has been considered in Example 3.6, for this operation and the results are compared with the direct expansion of the differentiated function in BPF domain. However, some oscillations are noted in the results vide Figure 3.7 and Table 3.6.

Next the operational matrix for differentiation in GBPF domain is presented. One typical numerical example has been provided (Example 3.7). The MISE has been estimated and is found to be only 0.00026492, which is reasonably small. Figure 3.8 shows the differentiated function in GBPF domain, for $m = 10$ and $T = 1$ s, with its exact curve and it is presented in a tabular form along with its percentage errors in Table 3.7.

Finally, this Chapter presents the one-shot operational matrix for differentiation **D(n)** in BPF domain, obtained by inversion of the one-shot operational matrix for integration **P(n)**. Like repeated integration, these matrices are useful for repeated differentiation. In Example 3.8, three different functions have been considered for double differentiation with one-shot operational matrix and the results are studied separately. These results are then compared with the results obtained using first-order operational matrix for differentiation in BPF domain and are presented in Figures 3.9 through 3.11. Except for the function $f_3(t) = \dfrac{t^2}{2}$, in Figure 3.11, we find undesired oscillations in the result. This is possibly because of the structure of the operational matrix for differentiation as well as in GBPF domain. We observe that for a fixed time period, with increasing m, the amplitude of oscillation increases rapidly, because of the term $\dfrac{1}{h}$ in the operational matrix $\mathbf{D}$ and the terms $\dfrac{1}{h_0}, \dfrac{1}{h_1}, \cdots, \dfrac{1}{h_i}$ in the operational matrix $\mathbf{D_g}$ in GBPF domain.

References

1. T. H. Turney, *Heaviside's operational calculus made easy*, Chapman & Hall Ltd., London, 1946.
2. J. Mikusinski, *Operational calculus*, Pergamon Press, London, 1959.
3. C. F. Chen and C. H. Hsiao, A state space approach to Walsh series solution of linear systems, Int. J. Syst. Sci., vol. **6**, no. 9, pp. 833–858, 1975.
4. C. F. Chen and C. H. Hsiao, Design of piecewise constant gains for optimal control via Walsh functions, IEEE Trans. Automatic Control, vol. **AC-20**, no. 5, pp. 596–603, 1975.
5. T. Le Van, L. D. C. Tam, and N. Van Houtte, On direct algebraic solutions of linear differential equations using Walsh transforms, IEEE Trans. Circuits and Syst., vol. **CAS-22**, no. 5, pp. 419–422, 1975.
6. G. P. Rao and L. Sivakumar, System identification via Walsh functions, Proc. IEE, vol. **122**, no. 10, pp. 1160–1161, 1975.
7. C. F. Chen, Y. T. Tsay, and T. T. Wu, Walsh operational matrices for fractional calculus and their application to distributed systems, J. Franklin Inst., vol. **303**, no. 3, pp. 267–284, 1977.
8. S. G. Mouroutsos and P. D. Sparis, Taylor series approach to system identification, analysis and optimal control, J. Franklin Inst., vol. **319**, no. 3, pp. 359–371, 1985.
9. P. N. Paraskevopoulos, P. D. Sparis, and S. G. Mouroutsos, The Fourier series operational matrix of integration, Int. J. Syst. Sci., vol. **16**, no. 2, pp. 171–176, 1985.
10. Anish Deb, Gautam Sarkar, and Sunit K. Sen, Block pulse functions, the most fundamental of all piecewise constant basis functions, Int. J. Syst. Sci., vol. **25**, no. 2, pp. 351–363, 1994.
11. W. L. Chen and C. Y. Chung, New integral operational matrix in block pulse series analysis, Int. J. Syst. Sci., vol. **18**, no. 3, pp. 403–408, 1987.
12. A. Constantinides, *Applied numerical methods with personal computers*, McGraw-Hill Inc., New York, 1987.
13. S. C. Chapra and R. P. Canale, *Numerical methods for engineers* (6th Ed.), McGraw-Hill Education (India) Pvt. Ltd., New Delhi, 2012.
14. Z. H. Jiang and W. Schaufelberger, *Block pulse functions and their applications in control systems*, LNCIS, vol. **179**, Springer-Verlag, Berlin, 1992.
15. W. Marszalek, On the nature of block-pulse operational matrices, Int. J. Syst. Sci., vol. **15**, no. 9, pp. 983–989, 1984.
16. W. Marszalek, On the nature of block-pulse operational matrices: Some further results, Int. J. Syst. Sci., vol. **16**, no. 6, pp. 727–743, 1985.

17. Anish Deb, G. Sarkar, S. K. Sen, and M. Bhattacharjee, On improvement of the integral operational matrix in block pulse function analysis, J. Franklin Inst., vol. **332B**, no. 4, pp. 469–478, 1995.
18. G. P. Rao, *Piecewise constant orthogonal functions and their application to systems and control*, LNC1S, vol. **55**, Springer-Verlag, Berlin, 1983.
19. W. Chi-Hsu, On the generalization of block pulse operational matrices for fractional and operational calculus, J. Franklin Inst., vol. **315**, no. 2, pp. 91–102, 1983.
20. Anish Deb, G. Sarkar, M. Bhattacharjee, and S. K. Sen, All-integrator approach to linear SISO control system analysis using block pulse function (BPF), J. Franklin Inst., vol. **334B**, no. 2, pp. 319–335, 1997.

Study Problems

3.1   In BPF domain, using the operational matrix **P**, we can integrate any time function expanded via block pulse functions. Why this matrix **P** is called an 'Operational Matrix'?

To integrate a function $f(t)$ in BPF domain, taking 10 subintervals within a span of 1 s, what should be the dimension of the operational matrix for integration, **P**?

3.2   Decompose the $(i + 1)$th block pulse function component using two delayed unit step functions, and graphically prove that the integration of these component functions finally result in the figure shown in Figure 3.1(b).

3.3   Explain why the operational matrix for integration **P**, is of an upper triangular form. What are the mathematical advantages of an upper triangular matrix?

3.4   Integrate a function $f(t) = \sin(\pi t)$ in BPF domain, considering $m = 10$ and $T = 1$ s, and estimate the MISE.

3.5   Integrate a function $f(t) = \exp(-t)$ in GBPF domain, considering $h_0 = 0.55$ s, $\delta = 0.1$ s, $m = 10$ and $T = 10$ s. Then estimate the MISE.

3.6   Integrate a function $f(t) = t$ in BPF domain, using the operational matrix for integration and its improved versions **P1** and **P2**, considering $m = 10$ and $T = 1$ s. Estimate the MISE in each case.

3.7   Integrate the function $f(t) = t$ twice using the operational matrix **P** and the one-shot operational matrix **P(2)** in BPF domain, considering $m = 10$ and $T = 1$ s, and compare the results after computing respective MISEs.

3.8   Differentiate the function $f(t) = t^3$ in BPF domain using the **D** matrix considering $m = 10$ and $T = 1$ s. Then find out the percentage error of the BPF coefficients obtained using the **D** matrix with respect to the direct BPF expansion of the differentiated function.

3.9 Differentiate the function $f(t) = t^4$ twice in BPF domain using the first-order operational matrix **D** twice and the one-shot operational matrix **D(2)** considering $m = 10$ and $T = 1$ s. Find out the percentage error of the BPF coefficients obtained using the above two methods with respect to the direct BPF expansion of the differentiated function.

4

Operational Transfer Functions for System Analysis

Keeping in mind the flexibility of block pulse functions (BPFs), the present chapter makes use of the block pulse domain operational matrix for differentiation $\mathbf{D}_{(m)}$ [1] to find an operational transfer function of any linear control system in block pulse domain [2,3]. This is done by sensing the equivalence of the Laplace operator s with the matrix $\mathbf{D}_{(m)}$. Analysis of simple control systems using this Block Pulse Operational Transfer Function (BPOTF) [2] shows that the results are not so accurate when compared with the direct expansion of the solution in the BPF domain. This is due to the fact that the evaluation of the **BPOTF** contains second and higher powers of the BPF domain integration operational matrix $\mathbf{P}_{(m)}$. These higher powers of $\mathbf{P}_{(m)}$ introduce errors because the inherent error in $\mathbf{P}_{(m)}$ is compounded when the power of $\mathbf{P}_{(m)}$ is raised.

To remove this defect, one-shot operational matrices for repeated integration (OSOMRI) were derived by integrating each member of the BPF set repeatedly, and expanding the result in terms of an m-set BPFs. These matrices are now used to develop a modified Block Pulse Operational Transfer Function (**MBPOTF**) for linear single-input, single output (SISO) control system analysis in the block pulse domain.

A few linear SISO control systems, open loop as well as closed loop having different plant transfer functions, are analyzed using the developed **MBPOTF**s as illustrative examples. The results are found to match exactly with the direct BPF expansions of the exact solutions.

4.1 Walsh Operational Transfer Function (WOTF) [4,5]

The Walsh Operational Transfer Function (**WOTF**) is a special kind of transfer function which works in Walsh function domain in an operational manner. The **WOTF** has been derived from the conventional input-output relation in Laplace domain.

The input-output relationship of a linear time-invariant SISO system in Laplace domain is well known, and is given by

$$C(s) = G(s)R(s) \qquad (4.1)$$

where $G(s)$ is the transfer function of the system and $C(s)$ and $R(s)$ are the Laplace transformed output and input, respectively.

Since the operational matrix $\mathbf{D}_{w(m)}$ performs in the Walsh domain like the Laplace operator s in the Laplace domain, we can convert $G(s)$ in equation (4.1) to the $\mathbf{WOTF}_{(m)}$ by simply replacing s by the operational matrix $\mathbf{D}_{w(m)}$. That is

$$G(s)\Big|_{s=\mathbf{D}_{W(m)}} = G(\mathbf{D}_{W(m)}) = \mathbf{WOTF}_{(m)} \qquad (4.2)$$

By the transformation of equation (4.2), the system transfer function is converted to an operational transfer function defined in the Walsh domain.

Since Walsh functions are defined in time domain, representation of any time function $f(t)$ by a series combination of Walsh functions does not obscure the time variation of $f(t)$ and we do not need any inverse transformation like we do in case of Laplace domain.

Hence, our next step would be to convert $R(s)$ or the corresponding time function $r(t)$, to the Walsh domain and represent it by an input Walsh vector $\mathbf{r}(\phi)$. At this point we seek an equation having a form similar to that of equation (4.1) but which will relate the input-output of a linear system in the Walsh domain instead of the Laplace domain.

To achieve this, we simply write

$$\mathbf{c}(\phi) = \mathbf{WOTF}_{(m)}\mathbf{r}(\phi) \qquad (4.3)$$

Thus $\mathbf{c}(\phi)$ in equation (4.3) will actually give the output time response in the form of a series combination of Walsh functions.

The points to be noted about equation (4.3) are as follows:

a. $\mathbf{WOTF}_{(m)}$ is always a square matrix having the same order as that of $\mathbf{D}_{W(m)}$ and it operates upon Walsh functions only.

b. $\mathbf{r}(\phi)$ and $\mathbf{c}(\phi)$ are row matrices associated with a column Walsh vector.

c. Due to the above properties, the matrix multiplication on the right-hand side of equation (4.3) is rather simple.

d. To find the output response, no inverse transformation is necessary.

It is evident that the output response obtained in Walsh domain gives the time response in a piecewise constant manner, i.e., in the form of a staircase waveform.

Superiority of equation (4.3) over equation (4.1) may be judged from the following facts:

a. Determination of inverse transform of $G(s)R(s)$ to obtain the output time response for a higher order system with complex input waveform is very difficult and tedious.

b. Any input $r(t)$ may be expressed by a Walsh series $\mathbf{r}(\phi)$. In particular, pulse-trains, square wave functions, and inputs of a similar nature could be expressed in the Walsh domain with more ease. The same task is somewhat difficult in the Laplace domain, where the power electronic and other similar complex waveforms cannot be easily expressed.

c. Only the computation of $\mathbf{WOTF}_{(m)}$ needs the help of a computer and a system with a particular $\mathbf{WOTF}_{(m)}$ can be tested for n number of inputs by making $\mathbf{r}$ a rectangular matrix of dimension $(n \times m)$ and the simple multiplication operation on the RHS of equation (4.3) will yield $\mathbf{c}_{(nxm)}$ giving n number of outputs. That is, input-output correspondence exists between each row of $\mathbf{r}_{(nxm)}$ and each row of $\mathbf{c}_{(nxm)}$ through a single matrix multiplication.

But in Laplace domain analysis, for each type of input waveform, the inverse transform of the product $[G(s)R(s)]$ has to be computed separately.

4.2 Block Pulse Operational Transfer Function (BPOTF) for System Analysis

Analyses of control systems have already been carried out by Chen et al. [6], Rao [7] and others using Walsh and BPFs. However, the operational transfer function technique was introduced by Deb et al. [8,9], where Walsh functions were used. Since BPFs are connected with Walsh functions by similarity transformation, it is easy to conclude that the operational transfer function technique will be effective with BPFs as well. For the sake of clarity, we illustrate briefly the BPOTF technique following Deb and Dutta [5].

For a linear SISO system, we have the well-known relation (4.1).

The operational matrix $\mathbf{D}$ performs differentiation in the BPF domain as the Laplace operator s in the Laplace domain. To take advantage of the operational property of $\mathbf{D}$, we replace s in $G(s)$ by the operational matrix $\mathbf{D}$.

Thus,

$$G(s)\big|_{s=D} = G[\mathbf{D}] = \mathbf{BPOTF} \tag{4.4}$$

$\mathbf{D}$ is a square matrix of order m, then $\mathbf{BPOTF}$ is also a square matrix of order m, known as the BPOTF in the BPF domain. $\mathbf{BPOTF}$ is obviously equivalent to $G(s)$ in the Laplace domain and it is operative upon BPFs only.

Following equation (1.13), we can express the input $r(t)$ and the output $c(t)$ in terms of an m-set BPF as

$$\left. \begin{aligned} r(t) &\approx \begin{bmatrix} r_0 & r_1 & r_2 & \cdots & r_{(m-1)} \end{bmatrix} \mathbf{\Psi}_{(m)}(t) \triangleq \mathbf{R}^{\mathrm{T}} \mathbf{\Psi}_{(m)}(t) \\ c(t) &\approx \begin{bmatrix} c_0 & c_1 & c_2 & \cdots & c_{(m-1)} \end{bmatrix} \mathbf{\Psi}_{(m)}(t) \triangleq \mathbf{C}^{\mathrm{T}} \mathbf{\Psi}_{(m)}(t) \end{aligned} \right\} \tag{4.5}$$

Following equation (4.1), we use equations (4.4) and (4.5) to obtain

$$\mathbf{C}^{\mathrm{T}} = \mathbf{R}^{\mathrm{T}} \; \mathbf{BPOTF}_{(m)} \tag{4.6}$$

Equation (4.6) is similar to equation (4.3) and has similar advantages already mentioned.

The only disadvantage of equation (4.6) is that like of equation (4.3), it offers a piecewise solution for $c(t)$ instead of a continuous one. However, by making m large, the accuracy of computation may be increased.

4.2.1 Numerical Examples

Example 4.1

Consider an open loop system having a transfer function

$$G1(s) = (s+1)^{-1}$$

For a step input, the output is given by

$$c1(t) = \begin{bmatrix} 1 - \exp(-t) \end{bmatrix}, \quad t \geq 0 \tag{4.7}$$

The BPOTF for this system is

$$\mathbf{BPOTF1} = [\mathbf{D} + \mathbf{I}]^{-1} \tag{4.8}$$

where $\mathbf{D}$ is given by equation (3.38) and $\mathbf{I}$ is the unit matrix of order m.

While equation (4.7) is the exact solution for the output $c1(t)$, the BPF domain analysis, as indicated by equation (4.6), yields the equivalent block pulse domain output **C1** as

$$\mathbf{C1}^{T} = \mathbf{R1}^{T}[\mathbf{D}+\mathbf{I}]^{-1} \tag{4.9}$$

where $\mathbf{R1}^{T} = [\,1 \quad 1 \quad \cdots \quad 1\,]$ for a step input.
For $m = 10$, $T = 10$ s, $h = 1$ s, the BPF vector **C1** is given by

$$
\begin{aligned}
C1 = [0.33333333 \quad & 0.77777778 \quad 0.92592593 \quad 0.97530864 \quad 0.99176955 \\
0.99725652 \quad & 0.99908551 \quad 0.99969517 \quad 0.99989838 \quad 0.99996613]^{T}
\end{aligned}
$$

$$\tag{4.10}$$

In Table 4.1, the direct BPF expansion of the exact solution $c1(t)$ is compared with the BPF domain solution **C1** by the existing operational method and the percentage error of respective coefficients are also tabulated. Also qualitative comparison of the direct BPF expansion with the exact solution is presented in Figure 4.1.

Example 4.2

Consider the feedback control system with input $r_2(t)$ and forward path transfer function $G1(s)$, shown in Figure 4.2.
The equivalent open loop transfer function of the unity feedback system is

$$G2(s) = G1(s)\left[\,1+G1(s)\,\right]^{-1}$$

Hence, the related BPOTF is

$$\mathbf{BPOTF2} = \mathbf{BPOTF1}[\mathbf{I}+\mathbf{BPOTF1}]^{-1} \tag{4.11}$$

Using equation (4.8), **BPOTF2** is given by

$$\mathbf{BPOTF2} = [\mathbf{D}+2\mathbf{I}]^{-1}$$

For a step input, the BPF representation of $r_2(t)$ is

$$\mathbf{R2}^{T} = [\,1 \quad 1 \quad \cdots \quad 1\,]$$

TABLE 4.1

Comparison of the Exact Solution, Solution by **BPOTF** Approach and Direct BPF
Expansion of the Exact Solution of the First Order Plant of Example 4.1, Having the
Transfer Function $G1(s) = (s + 1)^{-1}$ with Step Input for $m = 10$ and $T = 10$ s (Vide
Appendix B, Program No. 4.1)

t (sec)	$c1(t)$	C1	Direct Expansion of $c1(t)$ in BPF	Percentage Error
0	0.00000000			
		0.33333333	0.36787944	9.39060572
1	0.63212056			
		0.77777778	0.76745584	− 1.34495500
2	0.86466472			
		0.92592593	0.91445179	− 1.25475624
3	0.95021293			
		0.97530864	0.96852857	− 0.70003835
4	0.98168436			
		0.99176955	0.98842231	− 0.33864464
5	0.99326205			
		0.99725652	0.99574081	− 0.15221939
6	0.99752125			
		0.99908551	0.99843313	− 0.06533993
7	0.99908812			
		0.99969517	0.99942358	− 0.02717444
8	0.99966454			
		0.99989839	0.99978795	− 0.01104657
9	0.99987659			
		0.99996613	0.99992199	− 0.00441431
10	0.99995460			

and the output response in BPF domain, **C2**, is

$$\mathbf{C2}^{\mathsf{T}} = \mathbf{R2}^{\mathsf{T}}[\mathbf{D} + 2\mathbf{I}]^{-1} \qquad (4.12)$$

The exact solution is

$$c2(t) = \frac{1}{2}\big[1 - \exp(-2t)\big], \quad t \geq 0 \qquad (4.13)$$

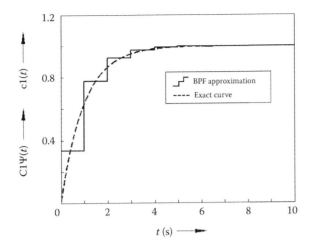

FIGURE 4.1
Comparison of the exact solution with the BPF domain solution for a first-order plant of Example 4.1 (vide Appendix B, Program no. 4.1).

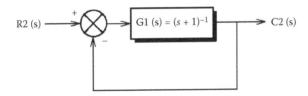

FIGURE 4.2
A simple feedback control system.

For $m = 10$, $T = 10$ s and $h = 1$ s, **C2** is given by

$$C2 = [0.25000000 \quad 0.50000000 \quad 0.5000000 \quad 0.50000000 \quad 0.50000000$$
$$0.50000000 \quad 0.50000000 \quad 0.50000000 \quad 0.50000000 \quad 0.50000000]^T$$

(4.14)

Table 4.2 compares the results obtained from BPF domain operational analysis along with the direct BPF expansion of the exact solution $c2(t)$, and the percentage errors for different coefficients are computed. Figure 4.3 shows the graphical comparison of the direct BPF expansion of output with the exact solution $c2(t)$.

TABLE 4.2

Comparison of the Exact Solution, Solution by **BPOTF** Approach and Direct BPF Expansion of the Exact Solution for the Feedback Control System of Example 4.2 (Shown in Figure 4.1), with Step Input for $m = 10$ and $T = 10$ s (Vide Appendix B, Program No. 4.2)

t (sec)	*c2(t)*	C2	Direct Expansion of *c1(t)* in BPF	Percentage Error
0	0.00000000			
		0.25000000	0.28383382	11.92029220
1	0.43233236			
		0.50000000	0.47074509	– 6.21459719
2	0.49084218			
		0.50000000	0.49604078	– 0.79816456
3	0.49876062			
		0.50000000	0.49946418	– 0.10727944
4	0.49983227			
		0.50000000	0.49992748	– 0.01450524
5	0.49997730			
		0.50000000	0.49999019	– 0.00196282
6	0.49999693			
		0.50000000	0.49999867	– 0.00026563
7	0.49999958			
		0.50000000	0.49999982	– 0.00003595
8	0.49999994			
		0.50000000	0.49999998	– 0.00000487
9	0.49999999			
		0.50000000	0.50000000	– 0.00000066
10	0.50000000			

It transpires from Tables 4.1 and 4.2 that the operational technique introduces some error in the analysis because the integration is not *exact*. Also, evaluation of the **BPOTF** involves the use of higher order integration operational matrices such as $\mathbf{P}_{(m)}^2$, $\mathbf{P}_{(m)}^3$, etc. and consequently the error is compounded.

In Section 4.3, the nature of **BPOTF** for a first-order plant is explored and it is shown that the inverse operation of equation (4.8) is effectively the sum of an infinite matrix power series involving the matrix $\mathbf{P}_{(m)}$.

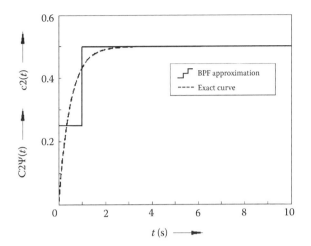

FIGURE 4.3
Comparison of the exact solution with the BPF domain solution for the feedback control system of Example 4.2 (vide Appendix B, Program no. 4.2).

4.3 Oscillatory Phenomenon in Block Pulse Domain Analysis of First-Order Systems [4]

Relating to approximation of a particular function by partial sums of a Fourier series, the Gibbs phenomenon [10] is a well-known feature. If the function $f(t)$ to be approximated has discontinuities, the partial sum approximating the function at such discontinuities shows a noticeable oscillation with a maximum overshoot of 9%. If the number of terms in the Fourier series used for approximating $f(t)$ is increased to infinity, even then there is no reduction to the figure of 9% mentioned above.

A similar phenomenon has been observed by J. L. Walsh [11] when he used a new closed set of orthonormal functions, later named Walsh functions, for approximating a function $f(t)$ at discontinuities. Supposing that $f(t)$ has a discontinuity at $t = t_1$, if the point t_1 is dyadically irrational, $f(t)$ cannot be expanded in terms of Walsh functions, and occurs a phenomenon quite analogous to Gibbs phenomenon for Fourier series. Walsh, in his derivation, used a function

$$f(t_1) = \begin{cases} 1, 0 \le t < t_1 \\ 0, t_1 \le t < 1 \end{cases}$$

to analyze its nonconvergence at $t = t_1$, when t_1 is dyadically irrational.

In Reference [11], it was claimed that Walsh (or block pulse) domain analysis handles functions with discontinuities more elegantly than Fourier analysis due to the fact that there is no oscillation like Gibbs phenomenon. However, our results are contrary to this claim, and this section establishes a new condition that guides the occurrence of such oscillations for Walsh/block pulse domain solution of first-order systems with unit step input. A typical representative curve is presented to show the nature of such oscillations with some observations.

Consider a first-order linear time-invariant SISO system having a transfer function given by

$$G(s) = \frac{1}{(s+a)} \tag{4.15}$$

where $\frac{1}{a}$ is the time constant of the system.

The **BPOTF** for the same system may be written as

$$\mathbf{BPOTF3} = [\mathbf{D} + a\mathbf{I}] \tag{4.16}$$

where **I** is a unit matrix of the same order as that of **D**.

We consider m BPFs of the series, and if we are interested in the solution over a region of $0 \le t < \lambda$ seconds, then **BPOTF** becomes

$$\mathbf{BPOTF} = \left[\frac{\mathbf{D}}{\lambda} + a\mathbf{I}\right] \tag{4.17}$$

The response for such a system due to a unit step function input is given by

$$\mathbf{C}^\mathbf{T}\mathbf{\Psi}_{(m)} = [\,\underbrace{1 \quad 0 \quad 0 \quad \cdots \quad 0}_{(m-1)\,\text{zeros}}\,]\left[\frac{\mathbf{D}}{\lambda} + a\mathbf{I}\right]^{-1}\mathbf{\Psi}_{(m)} \tag{4.18}$$

4.3.1 Numerical Example

Example 4.3

Consider a first-order linear time-invariant SISO system having a transfer function given by

$$G(s) = \frac{1}{(s+5)} \tag{4.19}$$

For the determination of output, the RHS of equation (4.18) has been evaluated for $\lambda = 4$ and for different values of m and a. Three values of m— namely, 4, 8 and 16—have been considered and by varying the value of a, the output responses for different combinations of λ, m, and a have been obtained. It was found from the computed results that the output response showed distinct oscillations like the Gibbs phenomenon for different combinations of λ, m, and a if

$$a\lambda > 2m \tag{4.20}$$

and no oscillations were observed for

$$a\lambda \le 2m \tag{4.21}$$

This is illustrated by the time-domain solutions obtained from BPF analysis and the actual solution, shown in Figure 4.4, for $m = 8$, $\lambda = 4$, and $a = 5$. It may be noted that the actual solution for a first-order system with transfer function given by equation (4.19) and a unit step input is

$$c(t) = \frac{1}{5}\left[1 - \exp(-5t)\right], t \ge 0 \tag{4.22}$$

The reason for such oscillations has been investigated in detail in Reference [4].

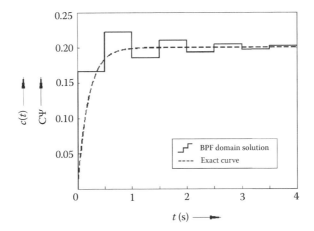

FIGURE 4.4

Comparison between the exact solution $c(t)$ and the BPF domain solution $\mathbf{C\Psi}(t)$ for the first-order system for $m = 8$, $\lambda = 4$, and $a = 5$, respectively (vide Appendix B, Program no. 4.3).

4.4 Nature of Expansion of the BPOTF of a First-Order Plant

Let us consider the **BPOTF1** given by equation (4.8). It is evident that the evaluation of the inverse (provided the inverse exists) on the RHS of equation (4.8) gives the requisite **BPOTF** and hence, the error found in the BPF domain output **C1** is due to the error in **BPOTF1**.

Let

$$f(\mathbf{D}) = [\mathbf{D} + a\mathbf{I}]^{-1} \tag{4.23}$$

where $f(\mathbf{D})$ is a function of the matrix $\mathbf{D}$ of order m, and $a \geq 0$ $[a \in \mathbb{R}]$.

It is well-known [12] that any function of a matrix $f(\mathbf{A})$ is defined as being equal to a matrix polynomial $h(\mathbf{A})$ when the associated scalar function $f(\mu)$ and the scalar polynomial $h(\mu)$ and all their derivatives up to order $(n_i - 1)$ have the same values at the eigenvalues μ_i of $\mathbf{A}$, $i = 1,2,3,\ldots,k$, where μ_i is the multiplicity of μ_i in the minimum polynomial of $\mathbf{A}$.

Expanding equation (4.23), we can write

$$f(\mathbf{D}) \triangleq g(\mathbf{P}) = \mathbf{P} - a\mathbf{P}^2 + a^2\mathbf{P}^3 - a^3\mathbf{P}^4 + \cdots + (-1)^{r-1}a^{r-1}\mathbf{P}^r + \cdots \tag{4.24}$$

where $\mathbf{P} = \mathbf{D}^{-1}$.

A simple scrutiny of equation (4.24) reveals that the RHS of the matrix polynomial involves evaluation of second and higher powers of the operational matrix $\mathbf{P}$. Since the integer integration by the operational matrix $\mathbf{P}$ in the BPF domain is fairly approximate (in fact, it is equivalent to the well-known trapezoidal rule, as mentioned in Chapter 3), evaluation of higher powers of $\mathbf{P}$ magnifies the error in a nonlinear manner. Hence, much error is introduced in the evaluation of $g(\mathbf{P})$.

Since $\mathbf{D}$ is of upper triangular form, evaluation of the RHS of equation (4.23) gives an upper triangular matrix of the form

$$f(\mathbf{D}) = \begin{bmatrix} p_{11} & p_{12} & \cdots & p_{1r} & \cdots & p_{1m} \end{bmatrix}_{m \times m} \tag{4.25}$$

where

$$\begin{bmatrix} a & b & c \end{bmatrix} = \begin{bmatrix} a & b & c \\ 0 & a & b \\ 0 & 0 & a \end{bmatrix}$$

The rth term p_{1r} of equation (4.25) is calculated as

$$p_{1r} = h\left[\sum_{i=1}^{N}\sum_{j=1}^{i}\left(-\frac{ah}{2}\right)^{i-1}\binom{r-2}{j-1}\binom{r+i-(j+1)}{r-1}\right], \quad 2 \le r \le m \qquad (4.26)$$
$$N \to \infty$$

Similarly, evaluation of the RHS of equation (4.24) gives an upper triangular matrix given by

$$g(\mathbf{P}) = \begin{bmatrix} q_{11} & q_{12} & \cdots & q_{1r} & \cdots & q_{1m} \end{bmatrix}_{m \times m} \qquad (4.27)$$

The rth term q_{1r} of equation (4.27) is given by

$$q_{1r} = h\left[\sum_{i=1}^{N}\sum_{j=1}^{i+1}\left(-\frac{ah}{2}\right)^{i-1}\frac{1}{2}\binom{i}{j-1}\binom{r+i-(j+1)}{i-1}\right], 2 \le r \le m \qquad (4.28)$$
$$N \to \infty$$

Comparing equations (4.26) and (4.28), it is observed that there is a term to term correspondence between p_{1r} and q_{1r}.

The matrix power series of equation (4.24) is convergent [12] if $|ah/2|<1$, where $ah/2$ is the eigenvalue of $\mathbf{P}_{(m)}$, repeated m times.

Now, for convenience, we write $\mathbf{P}$ as $\mathbf{P(1)}$ and in the power series of equation (4.24), we replace $\mathbf{P}_{(m)}^{n}$ by $\mathbf{P(n)}_{(m)}$, where $n = 2, 3, 4, \ldots$ Then the series sum differs from the result of equation (4.23) and we call this sum the Modified Block Pulse Operational Transfer Function (**MBPOTF**). If we put $a = 1$ in the RHS of equation (4.24), the sum of the series with $\mathbf{P(n)}_{(m)}$, e.g., $\mathbf{P(1)}_{(m)}$, $\mathbf{P(2)}_{(m)}$, $\mathbf{P(3)}_{(m)}$ etc. gives the **MBPOTF1** while evaluation of the RHS of equation (4.23) gives **BPOTF1**. Obviously, **MBPOTF** is more accurate than **BPOTF**, because of the use of one-shot integrator operational matrices.

4.5 Modified BPOTF (MBPOTF) Using All-Integrator Approach for System Analysis [2]

In this section, we develop modified BPF operational transfer functions (**MBPOTFs**) for several standard plants. These **MBPOTFs** are BPF domain equivalents of the Laplace domain transfer functions. Since the set of BPF is described in time domain, no inverse transformations are necessary to determine the outputs. It will be evident from the following that any system having any type of complex transfer function can be analyzed using the results obtained for simple plants with the help of the partial fraction technique.

4.5.1 First-Order Plant

Consider a first-order plant having a transfer function given by

$$G3(s) = (s+a)^{-1}$$

where a is real and greater than zero.

$G3(s)$ can be expressed as a power series involving negative powers of the Laplace operator s. That is

$$G3(s) = (s+a)^{-1} = a^{-1}[as^{-1} - a^2 s^{-2} + a^3 s^{-3} - a^4 s^{-4} + \cdots \text{to } \infty]$$

Since the terms within the square bracket involve integrators of different orders, it is logical that these can be replaced by one-shot operational matrices of respective orders to obtain an equivalent operational transfer function in the BPF domain. Calling this operational transfer function **MBPOTF3** and using first-order integration operational matrix from equation (3.10) and any one of equations (3.33) and (3.34) for one-shot operational matrices for integration, we can write

$$\textbf{MBPOTF3}_{(m)} = a^{-1}\left[a\textbf{P}(1)_{(m)} - a^2\textbf{P}(2)_{(m)} + a^3\textbf{P}(3)_{(m)} - a^4\textbf{P}(4)_{(m)} + \cdots \text{to } \infty \right]$$

where m basis functions have been considered.

Upon simplification, we have

$$\textbf{MBPOTF3}_{(m)} =$$

$$\frac{\left\{1-\exp(-ah)\right\}^2}{a^2 h}
\begin{bmatrix}
p11 & p12 & p13 & p14 & \cdots & p1j & \cdots & p1m \\
 & p11 & p12 & p13 & \cdots & \cdots & \cdots & p1(m-1) \\
 & & p11 & p12 & \cdots & \cdots & \cdots & p1(m-2) \\
 & & & \ddots & \ddots & & & \vdots \\
 & & & & \ddots & \ddots & & \vdots \\
 & \mathbf{0} & & & & \ddots & \ddots & \vdots \\
 & & & & & & \ddots & p12 \\
 & & & & & & & p11
\end{bmatrix}_{m \times m}$$

$$(4.29)$$

where

$$p1j = \begin{cases} \left\{\exp\,(-ah)+ah-1\right\}/\left\{1-\exp\,(-ah)\right\}^2, & \text{for } j=1 \\ \exp\left\{-(j-2)ah\right\}, & \text{for } 2 \le j \le m \end{cases}$$

If we consider $a = 1$, then from equation (4.29), the **MBPOTF** for G1(s) = $(s + 1)^{-1}$ is obtained as

$$\textbf{MBPOTF1}_{(m)} =$$

$$\frac{\left\{1-\exp(-h)\right\}^2}{h}
\begin{bmatrix}
q11 & q12 & q13 & q14 & \cdots & q1j & \cdots & q1m \\
 & q11 & q12 & q13 & \cdots & \cdots & \cdots & q1(m-1) \\
 & & q11 & q12 & \cdots & \cdots & \cdots & q1(m-2) \\
 & & & \ddots & \ddots & & & \vdots \\
 & & & & \ddots & \ddots & & \vdots \\
 & \mathbf{0} & & & & \ddots & \ddots & \vdots \\
 & & & & & & \ddots & q12 \\
 & & & & & & & q11
\end{bmatrix}_{m \times m}$$

$$(4.30)$$

where

$$q1j = \begin{cases} \{\exp(-h) + h - 1\} / \{1 - \exp(-h)\}^2, & for\ j = 1 \\ \exp\{-(j-2)h\}, & for\ 2 \leq j \leq m \end{cases}$$

Now we analyze systems having transfer functions $G1(s)$ and $G2(s)$, described in Section 4.2.1 for unit step inputs, considering $m = 10$, $T = 10$ s, and $h = 1$ s.

The unit step input expressed in BPF domain is given by

$$u(t) = \mathbf{R1}^T \mathbf{\Psi}_{(10)}(t) = [\ 1\ \ 1\ \ 1\ \ 1\ \ 1\ \ 1\ \ 1\ \ 1\ \ 1\ \ 1\ ]\mathbf{\Psi}_{(10)}(t)$$

Using **MBPOTF1** of equation (4.30), the output **C1′** is given by

$$\mathbf{C1'}^T = \mathbf{R1}^T \mathbf{MBPOTF1} \tag{4.31}$$

Table 4.3 compares the results obtained above with the direct BPF expansion of the exact solution $c1(t)$ given by equation (4.7).

It is observed that the percentage error is almost zero (up to eight decimal places the digits are all zeros) proving the accuracy of the one-shot all-integrator operational approach.

For the closed loop system shown in Figure 4.1, the equivalent open loop transfer function is

$$G2(s) = G1(s)\big[1 + G1(s)\big]^{-1} = (s+2)^{-1}$$

and its modified block pulse operational transfer function **MBPOTF2** may be derived from equation (4.29) by substituting $a = 2$.

$$\mathbf{MBPOTF2}_{(m)} =$$

$$\frac{\{1 - \exp(-2h)\}^2}{4h} \begin{bmatrix} r11 & r12 & r13 & r14 & \cdots & r1j & \cdots & r1m \\ & r11 & r12 & r13 & \cdots & \cdots & \cdots & r1(m-1) \\ & & r11 & r12 & \cdots & \cdots & \cdots & r1(m-2) \\ & & & \ddots & \ddots & & & \vdots \\ & & & & \ddots & \ddots & & \vdots \\ & \mathbf{0} & & & & \ddots & \ddots & \vdots \\ & & & & & & \ddots & r12 \\ & & & & & & & r11 \end{bmatrix}_{(m \times m)} \tag{4.32}$$

TABLE 4.3

Comparison of the Results Obtained via **MBPOTF** Approach with the Direct BPF Expansion of the Exact Output for the First-Order Plant of Example 4.1, Having the Transfer Function G1(s) = (s + 1)$^{-1}$ with Step Input for m = 10 and T = 10 s (Vide Appendix B, Program No. 4.4)

t (sec)	C1′	Direct Expansion of c1(t) in BPF	Percentage Error
0			
	0.36787944	0.36787944	0.00000000
1			
	0.76745584	0.76745584	0.00000000
2			
	0.91445179	0.91445179	0.00000000
3			
	0.96852857	0.96852857	0.00000000
4			
	0.98842231	0.98842231	0.00000000
5			
	0.99574081	0.99574081	0.00000000
6			
	0.99843313	0.99843313	0.00000000
7			
	0.99942358	0.99942358	0.00000000
8			
	0.99978795	0.99978795	0.00000000
9			
	0.99992199	0.99992199	0.00000000
10			

where

$$r1j = \begin{cases} \{\exp(-2h)+2h-1\}/\{1-\exp(-2h)\}^2, & \text{for } j=1 \\ \exp\{-(j-2)2h\}, & \text{for } 2 \leq j \leq m \end{cases}$$

Then, following equation (4.6), the output **C2′** in the BPF domain is

$$\mathbf{C2'}^{\mathrm{T}} = \mathbf{R1}^{\mathrm{T}}\mathbf{MBPOTF2} \tag{4.33}$$

Table 4.4 compares the results obtained above with the direct BPF expansion of the exact solution c2(t).

TABLE 4.4

Comparison of the Results Obtained via **MBPOTF** Approach with the Direct
BPF Expansion of the Output for the Feedback Control System of Example 4.2,
with Step Input for m = 10 and T = 10 s (Vide Appendix B, Program No. 4.5)

t (sec)	C2′	Direct Expansion of $c2(t)$ in BPF	Percentage Error
0			
	0.28383382	0.28383382	0.00000000
1			
	0.47074509	0.47074509	0.00000000
2			
	0.49604078	0.49604078	0.00000000
3			
	0.49946418	0.49946418	0.00000000
4			
	0.49992748	0.49992748	0.00000000
5			
	0.49999019	0.49999019	0.00000000
6			
	0.49999867	0.49999867	0.00000000
7			
	0.49999982	0.49999982	0.00000000
8			
	0.49999998	0.49999998	0.00000000
9			
	0.50000000	0.50000000	0.00000000
10			

It is observed that up to 8th decimal place, the percentage error is zero.

4.5.2 Second-Order Plant with Imaginary Roots

Consider a system having the transfer function $G4(s) = (s^2 + a^2)^{-1}$.
The transfer function can be decomposed as

$$G4(s) = \frac{1}{2ja}\left[\frac{1}{s - ja} - \frac{1}{s + ja}\right], \text{where } j = \sqrt{-1}$$

Using equation (4.29), the **MBPOTF** is given by

$$\mathbf{MBPOTF4}_{(m)} = \begin{bmatrix} v11 & v12 & v13 & v14 & \cdots & v1j & \cdots & v1m \\ & v11 & v12 & v13 & \cdots & \cdots & \cdots & v1(m-1) \\ & & v11 & v12 & \cdots & \cdots & \cdots & v1(m-2) \\ & & & \ddots & \ddots & & & \vdots \\ & & 0 & & \ddots & \ddots & & \vdots \\ & & & & & \ddots & \ddots & \vdots \\ & & & & & & \ddots & v12 \\ & & & & & & & v11 \end{bmatrix}_{m \times m}$$

(4.34)

where

$$v1j = \begin{cases} \{ah - \sin(ah)\}/a^3h, & \text{for } j = 1 \\ (4/a^3h)\left[\sin^2(ah/2)\sin\{(j-1)ah\}\right], & \text{for } 2 \le j \le m \end{cases}$$

For a step input, the exact output response of this system, $c4(t)$, is determined for $a = \sqrt{2}$ by the conventional analysis. This result is expanded into an m-term BPF series and compared in Table 4.5 with the BPF domain solution **C4′**, obtained by using **MBPOTF4**$_{(m)}$ and equation (4.6) for $m = 10$, $T = 10$ s, and $h = 1$ s. It is observed that the percentage error is *almost* zero.

4.5.3 Second-Order Plant with Complex Roots

Let us consider a system with complex conjugate poles having the transfer function

$$G5(s) = (s^2 + as + b)^{-1}, a, b > 0.$$

$G5(s)$ can be decomposed as

$$G5(s) = \frac{1}{2j\beta}\left[\frac{1}{s+\alpha-j\beta} - \frac{1}{s+\alpha+j\beta}\right]$$

where

$$\alpha = \frac{1}{2}a \text{ and } \beta = \frac{1}{2}\sqrt{(4b - a^2)}$$

TABLE 4.5

Comparison of the Results Obtained via **MBPOTF** Approach with the Direct BPF Expansion of the Exact Output for the Open Loop System Having the Transfer Function $(s^2 + 2)^{-1}$, with Step Input for $m = 10$ and $T = 10$ s (Vide Appendix B, Program No. 4.6)

t (sec)	C4′	Direct Expansion of $c4(t)$ in BPF	Percentage Error
0			
	0.15077200	0.15077200	0.00000000
1			
	0.74030819	0.74030819	0.00000000
2			
	0.92417709	0.92417709	0.00000000
3			
	0.39198730	0.39198730	0.00000000
4			
	0.04213511	0.04213511	0.00000000
5			
	0.46521042	0.46521042	0.00000000
6			
	0.94701446	0.94701446	0.00000000
7			
	0.67420776	0.67420776	0.00000000
8			
	0.10731874	0.10731874	0.00000000
9			
	0.20331991	0.20331991	0.00000000
10			

Again, using equation (4.29), the **MBPOTF5**$_{(m)}$ is given by

$$\mathbf{MBPOTF5}_{(m)} =$$

$$= \frac{1}{\beta h(\alpha^2 + \beta^2)^2}
\begin{bmatrix}
w11 & w12 & w13 & w14 & \cdots & w1j & \cdots & w1m \\
 & w11 & w12 & w13 & \cdots & \cdots & \cdots & w1(m-1) \\
 & & w11 & w12 & \cdots & \cdots & \cdots & w1(m-2) \\
 & & & \ddots & \ddots & & & \vdots \\
 & & & & \ddots & \ddots & & \vdots \\
 & \mathbf{0} & & & & \ddots & \ddots & \vdots \\
 & & & & & & \ddots & w12 \\
 & & & & & & & w11
\end{bmatrix}_{m \times m}$$

$$(4.35)$$

where

$$
w1j = \begin{cases}
\exp(-\alpha h)\left\{(\alpha^2 - \beta^2)\sin(\beta h) + 2\alpha\beta\cos(\beta h)\right\} \\
\quad + \beta h(\alpha^2 + \beta^2) - 2\alpha\beta, \qquad \text{for } j = 1 \\
\exp(-j\alpha h)\left[\exp(2\alpha h)\left[(\alpha^2 - \beta^2)\sin\left\{(j-2)\beta h\right\}\right.\right. \\
\quad + 2\alpha\beta\cos\left\{(j-2)\beta h\right\} - 2\exp(\alpha h)\left[(\alpha^2 - \beta^2)\sin\left\{(j-1)\beta h\right\}\right. \\
\quad + 2\alpha\beta\cos\left\{(j-1)\beta h\right\}\right] + \left\{(\alpha^2 - \beta^2)\sin\left(j\beta h\right)\right. \\
\quad + 2\alpha\beta\cos\left(j\beta h\right)\right], \qquad \text{for } 2 \le j \le m
\end{cases}
$$

Let us consider an open loop control system having the plant transfer function $G5(s)$. For a step input, the output $c5(t)$ of this system is obtained for $a = b = 1$ using the Laplace transform method and the output subsequently expanded in an m-term BPF series. The results obtained from direct BPF expansion are compared in Table 4.6 with the results obtained using **MBPOTF5** for $m = 10$, $T = 10$ s, and $h = 1$ s. As before, the results are found to be in good agreement with each other.

4.6 Error Due to MBPOTF Approach

In Chapter 3, we investigated [13] in depth the imperfections of the block pulse domain integrator **P** and found ways to improve this integrator in a continuous general manner.

It should be noted that the effectiveness of the integrators of different orders, e.g., **P(2)**, **P(3)**, etc. used in the presented **MBPOTF** method remains to be estimated.

To accomplish this task, we note a special feature of the **MBPOTF** approach and with the help of this feature estimate the upper bound of MISE for **MBPOTF** analysis in an indirect manner.

We note from different Tables that for the **MBPOTF** analysis, the computed results match *almost* exactly with the direct BPF expansion of the output time functions. Thus, it may be concluded that the upper bound of error of the **MBPOTF** approach is *equal* to the upper bound of representational error, discussed in Chapter 2.

TABLE 4.6

Comparison of the Results Obtained via **MBPOTF** Approach with the Direct BPF Expansion of the Exact Output for the Open Loop System Having the Transfer Function $(s^2 + s + 1)^{-1}$, with Step Input for $m = 10$ and $T = 10$ s (Vide Appendix B, Program No. 4.7)

t (sec)	C5′	Direct Expansion of $c5(t)$ in BPF	Percentage Error
0			
	0.12619296	0.12619296	0.00000000
1			
	0.60510178	0.60510178	0.00000000
2			
	1.01110785	1.01110785	0.00000000
3			
	1.15400452	1.15400452	0.00000000
4			
	1.11694474	1.11694474	0.00000000
5			
	1.03525097	1.03525097	0.00000000
6			
	0.98468193	0.98468193	0.00000000
7			
	0.97499353	0.97499353	0.00000000
8			
	0.98598279	0.98598279	0.00000000
9			
	0.99818334	0.99818334	0.00000000
10			

4.7 Conclusion

For the analysis of different linear SISO control systems, the block pulse domain operational matrix for differentiation $\mathbf{D}_{(m)}$ is used to find an operational transfer function. It is observed that the analysis based upon this **BPOTF** gives results which are not so accurate when compared with the direct expansion of the exact solution in the BPF domain. To remove this defect, **OSOMRI** are used to develop a **MBPOTF** for linear SISO control system analysis in the BPF domain.

A few linear SISO control systems, open loop as well as closed loop, having different plant transfer functions, e.g., $(s + a)^{-1}$, $(s^2 + a^2)^{-1}$, $(s^2 + \alpha s + \beta)^{-1}$ are analyzed for step input using the developed **MBPOTF**s as illustrative examples. The results are compared in various Tables and are found to *match* *exactly* (the maximum percentage error being $- 0.47779145 \times 10^{-12}$ only) with the direct BPF expansions of the exact solutions.

Also, by the presented technique it is easy to analyze any plant having a specific **MBPOTF** for any number of inputs in one mathematical operation using the spirit of equations (4.6) or (4.31).

Since the BPF is defined in the time domain, no inverse transformations were necessary for obtaining the solution. Simple matrix manipulations were effective to yield the desired results.

References

1. Z. H. Jiang and W. Schaufelberger, *Block pulse functions and their applications in control systems*, LNCIS, vol. **179**, Springer-Verlag, Berlin, 1992.
2. Anish Deb, G. Sarkar, M. Bhattacharjee, and S. K. Sen, All-integrator approach to linear SISO control system analysis using block pulse function (BPF), J. Franklin Inst., vol. **334B**, no. 2, pp. 319–335, 1997.
3. Anish Deb, Gautam Sarkar, and Sunit K. Sen, Block pulse functions, the most fundamental of all piecewise constant basis functions, Int. J. Syst. Sci., vol. **25**, no. 2, pp. 351–363, 1994.
4. Anish Deb and Suchismita Ghosh, *Power electronic systems: Walsh analysis with MATLAB*, CRC Press, Boca Raton, 2014.
5. Anish Deb and Asit K. Dutta, Analysis of a continuously variable pulse-width modulated system via Walsh functions, Int. J. Syst. Sci., vol. **23**, no. 2, pp. 151–166, 1992.
6. C. F. Chen, Y. T. Tsay, and T. T. Wu, Walsh operational matrices for fractional calculus and their application to distributed systems, J. Franklin Inst., vol. **303**, pp. 267–284, 1977.
7. G. P. Rao, *Piecewise constant orthogonal functions and their application to systems and control*, LNC1S, vol. **55**, Springer-Verlag, Berlin, 1983.
8. Anish Deb and Asit K. Datta, Analysis of pulse-fed power electronic circuits using Walsh Function, Int. J. Electronic, vol. **62**, pp. 449–459, 1987.
9. Anish Deb and D. W. Fountain, A note on oscillations in Walsh domain analysis of first order systems, IEEE Trans. Circuits and Syst., vol. **CAS-38**, no. 8, pp. 945–948, 1991.
10. G. P. Tolstov, *Fourier series (In Russian. Translated in English by R. A. Silverman)*, Dover Publications, New York, 1976.
11. J. L. Walsh, A closed set of normal orthogonal functions, Am. J. Math., vol. **45**, pp. 5–24, 1923.
12. K. B. Dutta, *Matrix and Linear Algebra*, Prentice-Hall of India Pvt. Ltd., New Delhi, 1991.
13. Anish Deb, Gautam Sarkar, Sunit K. Sen, and Manabrata Bhattacharjee, On improvement of the integral operational matrix in block pulse function analysis, J. Franklin Inst., vol. **332B**, pp. 469–478, 1995.

Study Problems

4.1   What are the advantages of using block pulse operational transfer function (BPOTF) in control system analysis?

4.2   Consider an open loop system having a transfer function $G(s) = (s + 4)^{-1}$. Find its output $c(t)$ in BPF domain for a step input $u(t)$ using block pulse operational transfer function (**BPOTF**). Consider $m = 4$ and $T = 1$ s.

4.3   Compare the results of **Problem 4.2** with direct BPF expansion of the exact time domain output by estimating the mean integral square error (MISE). Plot the exact solution along with the two BPF solutions obtained from **Problems 4.2** and **4.3**.

4.4   Solve **Problem 4.2** using Walsh Operational Transfer Function (**WOTF**) and prove that the results are identical with that obtained via **BPOTF**.

4.5   Consider an open loop system having a transfer function $G(s) = (s^2 + 4)^{-1}$. Find its output $c(t)$ in BPF domain for a step input $u(t)$ using block pulse operational transfer function (**BPOTF**). Consider $m = 5$ and $T = 1$ s.

4.6   Compare the results of **Problem 4.5** with the direct BPF expansion of the exact time domain output by estimating the mean integral square error (MISE) and comment on the effectiveness of the **BPOTF** approach.

4.7   Consider a closed loop system having a loop transfer function $G(s)H(s) = (s^2 + s + 4)^{-1}$. Find its output $c(t)$ in BPF domain for a step input $u(t)$, using block pulse operational transfer function (**BPOTF**). Consider $m = 6$ and $T = 1$ s.

5

System Analysis and Identification Using Convolution and "Deconvolution" in BPF Domain

System identification is a common issue encountered by control engineers in designing control systems. The unknown components, usually the plant under control, are described satisfactorily by their respective models. Thus, the problem of identification is the characterization of the assumed model based upon some observations or measurements. It is well known that one may set up more than one model for a specific dynamic system. For instance, in linear control problems, a parametric model, such as a state model, is usually adopted in "modern" design while in "classical" design a non-parametric model such as an impulse-response function is more appropriate [1].

In this chapter, we describe a method of system analysis and identification [2] based upon the process of convolution. Here, the unknown plant is modeled by impulse response functions.

The system block diagram is shown in Figure 5.1. A continuous time feedback control system is formed with $g(t)$ as the feedback path impulse-response function and $h(t)$ as the forward impulse response function and $h(t)$ as the feedback impulse response function. Given $h(t)$, the input $r(t)$, and the output $c(t)$, it is desired to find $g(t)$ without breaking the feedback loop.

In this method, all the continuous-time functions are approximated via block pulse functions (BPFs). The approximation enables the use of digital computation in the identification process using special operational matrices, which is the main advantage of this approach.

5.1 The Convolution Process in BPF Domain

In control system analysis, the well-known relation [3] involving the input and the output of a linear time-invariant system is given by

$$C(s) = G(s)R(s) \tag{5.1}$$

where $C(s)$ is the Laplace transform of the output,
$G(s)$ is the transfer function of the plant,

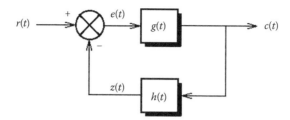

FIGURE 5.1
A linear feedback control system with a plant modeled by impulse response functions.

$R(s)$ is the Laplace transform of the input
and s is the Laplace operator.

In time domain, equation (5.1) takes the form

$$c(t) = \int_0^\infty g(\tau)r(t-\tau)d\tau \qquad (5.2)$$

That is, the output $c(t)$ is determined by evaluating the convolution integral
[4] on the RHS of equation (5.2), where it has been assumed that the integral
exists.

Convolution [4] of two functions is a significant operation pivotal in many
diverse scientific fields. However, as in the case of many important mathematical relationships, the convolution integral does not readily unveil itself
as to its true implications. The convolution integral of equation (5.2) is sym-
bolically represented as

$$c(t) = \int_{-\infty}^\infty g(\tau)r(t-\tau)d\tau = g(t) * r(t) \qquad (5.3)$$

where * indicates the convolution operation.

For convolution of two functions, one function may be considered to be
the *static function* or, say, STF (in our case, $g(t)$), and the other can be termed
as the *scanning function* or, say, SCF (in our case, $r(t)$).

If $g(t)$ and $r(t)$ are to be convolved in BPF domain, the first task is to expand
the functions $g(t)$ and $r(t)$ in terms of BPFs. Then, we need the partial results
of convolution of component BPFs and use those results to determine the
final result, namely $c(t)$, expressed in block pulse domain.

First of all, we state clearly our basic assumptions for the system of Figure 5.1:

i. The single-input single-output plant and measurement can be modeled by their respective impulse response functions. This implies
that the system is linear and causal.

ii. The input is applied at $t = 0$.

iii. The system is continuous and time-invariant.

iv. The time interval considered is $t \in [0, T)$ where T is finite. All the time functions including $g(t)$ and $h(t)$, are absolutely integrable over $t \in [0, T)$. Thus, it implies that each of the time functions can be approximated by a block pulse series given in Chapter 2.

Referring to Figure 5.1, considering only the open loop part of the system, the output $c(t)$ is given by the convolution integral [3,4].

$$c(t) = \int_0^\infty e(\tau)g(t - \tau)d\tau \qquad (5.4)$$

We approximate $e(\tau)$ and $g(\tau)$ by the block pulse series

$$e(\tau) \approx \sum_{i=0}^{m-1} e_i \psi_i(\tau) \qquad (5.5)$$

$$g(\tau) \approx \sum_{i=0}^{m-1} g_i \psi_i(\tau) \qquad (5.6)$$

The idea of developing the solution is as follows:

We first consider the result of convolving two single blocks. Secondly, we derive the block pulse coefficients for the block pulse approximation of the result of convolving a single block with m blocks. Finally, the result of convolving two m-block functions is obtained by the principle of superposition and considering different delays of different BPF components.

The following are the steps of the development:

Step 1: Figure 5.2(a)–5.2(c) depict the results of convolution between the BPF component of height c_0 (scanning function) and three BPF components having heights c_1, c_2, and c_3 (static functions). The convolution results obtained are exact and in each case the result is a triangular waveform.

Step 2: By the use of the superposition theorem and the results of *Step 1*, the convolution of a single BPF (scanning function) with three consecutive BPFs (nos. 1, 2, and 3) is shown in Figure 5.2(d).

The block pulse equivalent of the resulting piecewise linear waveform of Figure 5.2(d) can easily be derived and has been shown in Figure 5.3.

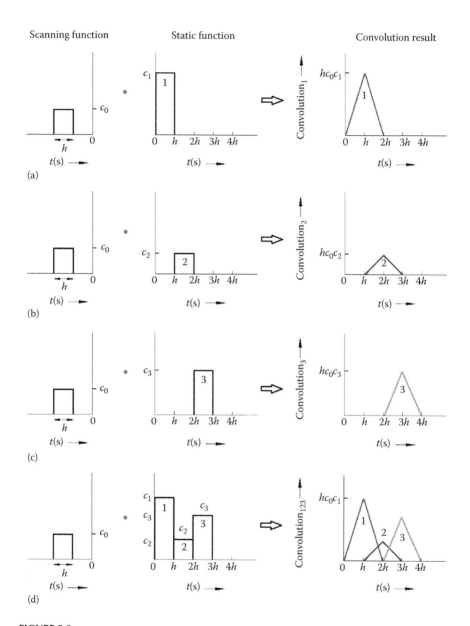

FIGURE 5.2
Convolution of block pulse functions, where * denotes the convolution operation of two time functions: (a) convolution between the scanning function and static function BPF1, (b) convolution between the scanning function and static function BPF2, (c) convolution between the scanning function and static function BPF3, and (d) sum of all three partial results by applying the principle of superposition.

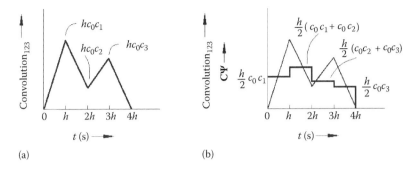

FIGURE 5.3
(a) Piecewise linear waveform of Figure 5.2(d) by adding the three partial results and (b) its block pulse approximation.

Step 3: Consider the convolution of the two BPF trains containing m BPF components each, as given in equations (5.5) and (5.6). Let us define

$$E \triangleq \begin{bmatrix} e_0 & e_1 & \cdots & e_{(m-1)} \end{bmatrix}^\mathsf{T} \tag{5.7}$$

$$G \triangleq \begin{bmatrix} g_0 & g_1 & \cdots & g_{(m-1)} \end{bmatrix}^\mathsf{T} \tag{5.8}$$

We also define the BPF coefficients for the piecewise linear waveforms $\hat{c}(t)$, obtained by convolving the functions (5.7) and (5.8), as

$$\hat{c}(t) \triangleq \sum_{i=0}^{m-1} \hat{c}_i \psi_i(t) \triangleq \hat{\mathbf{C}}^\mathsf{T} \mathbf{\Psi}_{(m)} \tag{5.9}$$

where

$$\hat{\mathbf{C}}^\mathsf{T} \triangleq \begin{bmatrix} \hat{c}_0 & \hat{c}_1 & \cdots & \hat{c}_{(m-1)} \end{bmatrix} \tag{5.10}$$

If we expand the output function $c(t)$ in a block pulse series directly, then we can write

$$c(t) \triangleq \sum_{i=0}^{m-1} c_i \psi_i(t) \tag{5.11}$$

It should be noted that $c(t)$ and $\hat{c}(t)$ are different but for large m, $\hat{c}(t)$ is a close approximation of $c(t)$.

Following the principle explained in *Steps 1* and *2* above, and considering the different delays of different block pulse components into account, we have, by the superposition theorem the triangles obtained are of different altitudes but their base is the same ($2h$).

The components of equation (5.10) may be expressed as

$$
\left.
\begin{aligned}
\hat{c}_0 &= \frac{h}{2} g_0 e_0 \\
\hat{c}_1 &= \frac{h}{2}\left[g_0(e_0 + e_1) + g_1 e_0\right] \\
&\vdots \\
\hat{c}_{(m-1)} &= \frac{h}{2}\left[g_0\left(e_{(m-2)} + e_{(m-1)}\right) + \ \cdots \ + g_{(m-2)}\left(e_0 + e_1\right) + g_{(m-1)}e_0\right]
\end{aligned}
\right\}
\tag{5.12}
$$

Let us define

$$
\mathbf{C}_c \triangleq
\begin{bmatrix}
e_0 & e_0 + e_1 & \cdots & \cdots & e_{(m-2)} + e_{(m-1)} \\
 & e_0 & \cdots & \cdots & e_{(m-3)} + e_{(m-2)} \\
 & & \ddots & & \vdots \\
 & 0 & & \ddots & \vdots \\
 & & & & e_0
\end{bmatrix}_{(m \times m)}
\tag{5.13}
$$

as the *convolution matrix*. Then we have

$$
\hat{\mathbf{C}}^{\mathrm{T}} = \frac{h}{2}\mathbf{G}^{\mathrm{T}}\mathbf{C}_c
\tag{5.14}
$$

which is the solution of the block pulse coefficients for $\hat{c}(t)$.

Referring to Figure 5.2(d), we note that the spread of the convolution spectrum is $4h$, which is the sum of the spans of the convolving functions (h and $3h$), h being the width of each block pulse.

So, it is apparent that when two functions $f_1(t)$ and $f_2(t)$, each expanded in m number of BPFs, convolve with each other, the resulting convolution spectrum will have a span of $2mh$, meaning, it will be comprised of $2m$ number of component BPFs. But, while using the convolution matrix $\mathbf{C}_c$ of equation (5.13), we note that it has a dimension of $m \times m$, implying the spread of the resulting convolution spectrum to be limited to mh only. That is, since we have chosen a specific m for determining the convolution spectrum, m is constant throughout the analysis. This constraint leads to *truncation* of half ($= mh$) of the convolution spectrum which actually spans a total period of $2mh$ as mentioned above. This truncation seems to be the only limitation of the method.

However, to overcome this limitation, we can expand each of the functions $f_1(t)$ and $f_2(t)$ via m number of BPFs and *add* m number of zeros at the end of each of the BPF coefficient vectors of the convolving functions $f_1(t)$ and $f_2(t)$. Then, the convolution matrix $\mathbf{C_c}$ of (5.13) will be of dimension $2m \times 2m$, and equation (5.14) will yield the *complete* convolution spectrum spanning a period of $2mh$.

Equation (5.12) may be manipulated to give a simple recursive formula which performs the convolution operation of equation (5.14), shown below

$$\hat{c}_k = \sum_{i=0}^{k-1}(-1)^{k+i+1}\hat{c}_i + \frac{h}{2}\sum_{i=0}^{k}g_i e_{k-i} \tag{5.15}$$

5.1.1 Numerical Examples

Example 5.1 [5]

Consider a simple open loop system with unit step input and the impulse response $g(t) = \exp(-4t)$.

The output of the system is $c(t) = \frac{1}{4}[1 - \exp(-4t)]$.

For $m = 8$ and $T = 1$ s, the convolution result in BPF domain using equation (5.14) is quantitatively compared with the direct expansion of the exact output. Also, we determine the convolution result using the recursive formula of (5.15). It is noted that equations (5.14) and (5.15) yield identical result. The comparison of the convolution result is presented in Table 5.1. It is noted that the percentage error is 7.66288753 at the beginning but it reduces to 0.05050232 in the last sub-interval.

Figure 5.4 shows graphical comparison of the convolution result in BPF domain with the direct BPF expansion of the exact output.

Example 5.2

Consider the example treated by Kwong and Chen [2].

Referring to Figure 5.1, $r(t) = u(t)$, a unit step function, the impulse response $g(t) = 2\exp(-4t)$ and $h(t) = 4u(t)$.

It is apparent that the impulse response of the equivalent open loop system is

$$g_1(t) = 2\exp(-2t)\left[\cos(2t) - \sin(2t)\right].$$

TABLE 5.1

Output of the Open Loop System of Example 5.1 for $m = 8$ and $T = 1$ s in Block Pulse Function (BPF) Domain Using the Convolution Matrix and the Recursive Formula and It Is Compared with the Direct BPF Expansion of the Output (Vide Appendix B, Program No. 5.1)

t (sec)	Direct BPF Expansion of the Output	Output After Convolution in BPF Domain Using Equation (5.14) or (5.15)	Percentage Error
0.000			
	0.05326533	0.04918367	7.66288753
0.125			
	0.13067439	0.12819874	1.89452067
0.250			
	0.17762536	0.17612380	0.84535207
0.375			
	0.20610256	0.20519182	0.44188775
0.500			
	0.22337486	0.22282246	0.24729414
0.625			
	0.23385103	0.23351599	0.14327208
0.750			
	0.24020516	0.24000194	0.08460018
0.875			
	0.24405913	0.24393587	0.05050232
1.000			

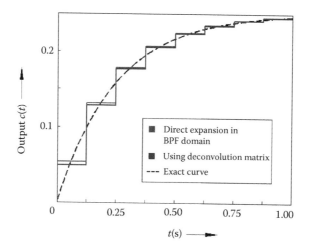

FIGURE 5.4

Output of the open loop system of Example 5.1 using the convolution process, and its comparison with the direct BPF expansion for $m = 8$ and $T = 1$ s (vide Appendix B, Program no. 5.1).

Analysis of the system yields the output as

$$c(t) = \exp(-2t)\sin(2t)$$

For $m = 16$ and $T = 1$ s, the BPF expansion coefficients of the output of the equivalent open loop system are

$$c(t) \approx \begin{bmatrix} 0.05745426 & 0.15347314 & 0.22402022 & 0.27278431 \\ 0.30323895 & 0.31859432 & 0.32176688 & 0.31536397 \\ 0.30168026 & 0.28270362 & 0.26012819 & 0.23537259 \\ 0.20960180 & 0.18375108 & 0.15855102 & 0.13455267 \end{bmatrix}$$

(5.16)

Table 5.2 presents the output of the equivalent open loop system using the convolution matrix with the help of equation (5.14), and is compared with the BPF coefficients of the direct expansion of the exact output. The percentage error is shown in the last column of the Table and the maximum error is found to be 4.24986821%.

5.2 Identification of an Open Loop System via "Deconvolution"

From equations (5.5) and (5.6), we have arrived at equation (5.14) with the support of the convolution process in BPF domain. If we know the input $e(t)$ and the output $c(t)$ of the system, then it is possible to determine $g(t)$.

Consider that $e(t)$ and $c(t)$ are known in BPF domain from equations (5.5) and (5.11). Then, using (5.14), we can write

$$\hat{\mathbf{G}}^T = \frac{2}{h}\mathbf{C}^T\mathbf{C}_c^{-1}$$

(5.17)

where $\mathbf{C} = \begin{bmatrix} c_0 & c_1 & \cdots & c_{(m-1)} \end{bmatrix}^T$ is the block pulse coefficient vector for $c(t)$, $\mathbf{C}_c$ is the convolution matrix as defined by equation (5.13), and $\hat{\mathbf{G}}$ is defined as

$$\hat{\mathbf{G}} \triangleq \begin{bmatrix} \hat{g}_0 & \hat{g}_1 & \cdots & \hat{g}_{(m-1)} \end{bmatrix}^T$$

(5.18)

TABLE 5.2

Output of the Equivalent Open Loop System of the Feedback System of Example 5.2 for $m = 16$ and $T = 1$ s in BPF Domain Using the Convolution Matrix and the Recursive Formula, and Its Comparison with the Direct BPF Expansion of the Output

t (sec)	Direct BPF Expansion of the Output	Output After Convolution in BPF Domain Using Equation (5.14) or (5.15)	Percentage Error
0.0000			
	0.05745426	0.05501253	4.24986821
0.0625			
	0.15347314	0.15135173	1.38226473
0.1250			
	0.22402022	0.22220679	0.80949440
0.1875			
	0.27278431	0.27126074	0.55852551
0.2500			
	0.30323895	0.30198315	0.41413172
0.3125			
	0.31859432	0.31758168	0.31784726
0.3750			
	0.32176688	0.32097154	0.24717869
0.4375			
	0.31536397	0.31475981	0.19157821
0.5000			
	0.30168026	0.30124164	0.14539337
0.5625			
	0.28270362	0.28240602	0.10526916
0.6250			
	0.26012819	0.25994862	0.06902881
0.6875			
	0.23537259	0.23528991	0.03512924
0.7500			
	0.20960180	0.20959685	0.00236403
0.8125			
	0.18375108	0.18380680	− 0.03032224
0.8750			
	0.15855102	0.15865246	− 0.06397477
0.9375			
	0.13455267	0.13468691	− 0.09976642
1.0000			

where, $\hat{g}_0, \hat{g}_1, \dots, \hat{g}_{(m-1)}$ are the BPF coefficients of the impulse response of the plant $G(s)$ as determined by equation (5.17). It should be noted that in equations (5.9) and (5.11), the like BPF coefficients $\hat{c}_i$'s and c_i's are not equal but very close. Similarly, comparing equations (5.8) and (5.18), we realize that the like BPF coefficients of $\mathbf{G}$ and $\hat{\mathbf{G}}$ are not equal but pretty close.

That is, the BPF coefficients of $\hat{\mathbf{G}}$, as determined by equation (5.17), are not equal to the corresponding BPF coefficients of the function $g(t)$ expanded directly in BPF domain.

Since $\mathbf{C}_c$ is an upper triangular matrix with diagonal elements e_0, $\mathbf{C}_c^{-1}$ always exists as long as $e_0 \neq 0$. As (5.14) signifies the process of convolution in BPF domain, (5.17) is the corresponding "deconvolution" process in BPF domain.

In the previous article we have presented an equation (5.15) through which the BPF coefficients for the output can be derived recursively.

Similarly, for determining the BPF coefficients of the impulse response of the plant, following equation (5.12), we have

$$c_0 = \frac{h}{2}\hat{g}_0 e_0$$

$$c_1 = \frac{h}{2}\left[\hat{g}_0(e_0 + e_1) + \hat{g}_1 e_0\right]$$

$$c_2 = \frac{h}{2}\left[\hat{g}_0(e_1 + e_2) + \hat{g}_1(e_0 + e_1) + \hat{g}_2 e_0\right]$$

$$\vdots$$

$$c_{(m-1)} = \frac{h}{2}\left[\hat{g}_0\left(e_{(m-2)} + e_{(m-1)}\right) + \cdots + \hat{g}_{(m-2)}(e_0 + e_1) + \hat{g}_{(m-1)}e_0\right]$$

Using these equations, we present one recursive equation below:

$$\hat{g}_k = \left[\frac{2}{h}c_k - \sum_{i=0}^{k-1} g_i\left(e_{(k-i-1)} + e_{(k-i)}\right)\right]e_0^{-1} \tag{5.19}$$

However, this equation yields results which are identical to that obtained via the "deconvolution" operation.

5.2.1 Numerical Example

Example 5.3 [2]

Consider the feedback system of Example 5.2 for identifying the equivalent open loop system via "deconvolution."

The response of the equivalent open loop system for a step input is given by

$$c(t) = \exp(-2t)\sin(2t), \quad t \geq 0$$

where, the equivalent impulse response of the system is

$$g_1(t) = 2\exp(-2t)[\cos(2t) - \sin(2t)]$$

Using equation (5.17), we have identified the equivalent open loop impulse response via "deconvolution" in BPF domain, and have compared the result with the direct BPF expansion of the equivalent impulse response in Table 5.3. It has been noted that the result obtained via recursive "deconvolution" operation using (5.19), is identical to that obtained using equation (5.17).

In Table 5.3, it is interesting to note that in the percentage error column there is a tendency of oscillation. Figure 5.5 shows that the BPF domain result of "deconvolution" oscillates with respect to the direct BPF expansion of the impulse response of $g_1(t)$. Finally, Figure 5.6 proves that these oscillations can be removed from these results by increasing the value of m, keeping T constant.

This numerical instability has been investigated further in the following.

5.3 Numerical Instability of the "Deconvolution" Operation: Its Mathematical Basis [5]

To study the instability further, we take up a very simple system with a step input.

Consider a first-order single-input, single output (SISO) open loop system shown in Figure 5.7.

Assume that the input to the system is $r(t) = u(t)$ and the impulse response of the plant is $g(t) = \exp(-at)$, where $\dfrac{1}{a}$ is the time constant of the system.

The output $c(t)$ of the system is obtained via convolution as

$$c(t) = u(t) * g(t) = [G(t) - G(0)] = \frac{1}{a}[1 - \exp(-at)], \text{ for } t \geq 0 \qquad (5.20)$$

TABLE 5.3

Identification of the Equivalent Open Loop System of Example 5.3 for $m = 16$ and $T = 1$ s in BPF Domain Using the "Deconvolution" Operation and the Result Is Compared with the Direct BPF Expansion of the Equivalent Impulse Response (Vide Appendix B, Program No. 5.2)

t(sec)	Direct Expansion of $g_1(t)$ in BPF Domain	"Deconvolution" Result in BPF Domain Using Equation (5.17) or (5.19)	Percentage Error
0.0000			
	1.76040106	1.83853643	– 4.43849855
0.0625			
	1.32245330	1.23406751	6.68347150
0.1250			
	0.94490856	1.02343919	– 8.31092313
0.1875			
	0.62481769	0.53701151	14. 05308843
0.2500			
	0.35829942	0.43753725	– 22.11497393
0.3125			
	0.14085359	0.05383447	61.77983746
0.3750			
	– 0.03237787	0.04768752	247.28430079
0.4375			
	– 0.16639774	– 0.25258058	– 51.79327580
0.5000			
	– 0.26618360	– 0.18529822	30.38706514
0.5625			
	– 0.33655608	– 0.42195421	– 25.37411644
0.6250			
	– 0.38208066	– 0.30045970	21.36223495
0.6875			
	– 0.40699821	– 0.49171930	– 20.81608448
0.7500			
	– 0.41517979	– 0.33294605	19.80677913
0.8125			
	– 0.41010177	– 0.49427704	– 20.52545742
0.8750			
	– 0.39483720	– 0.31212482	20.94847656
0.9375			
	– 0.37206030	– 0.45582246	– 22.51306081
1.0000			

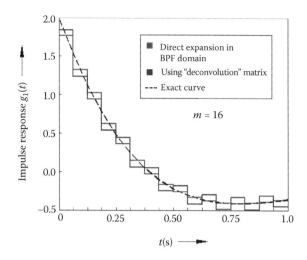

FIGURE 5.5
Identification of the equivalent open loop system of Example 5.3 using the "deconvolution" process and its comparison with the direct BPF expansion for $m = 16$ and $T = 1$ s (vide Appendix B, Program no. 5.2).

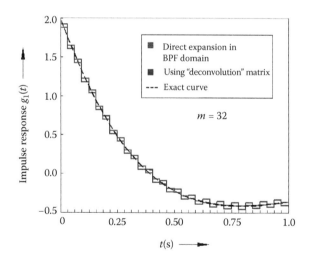

FIGURE 5.6
Identification of the equivalent open loop system of Example 5.3 using the "deconvolution" process and its comparison with the direct BPF expansion for $m = 32$ and $T = 1$ s. It is noted that with increasing m, the oscillation is reduced (vide Appendix B, Program no. 5.2).

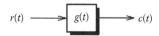

FIGURE 5.7
Plant modeled by its impulse response functions.

where $\int g(t)dt = G(t)$

Consider $T = 1$ s and $h = \dfrac{T}{m}$ s, where $m = $ number of BPF components. Then using equations (5.20) and (1.14), the $(k + 1)$th component c_k is given by

$$c_k = \frac{1}{a} - \frac{[1 - \exp(-ah)]}{a^2 h} \exp(-ahk) \qquad (5.21)$$

For $k = 0$, the coefficient is

$$c_0 = \frac{1}{a} - \frac{[1 - \exp(-ah)]}{a^2 h} \qquad (5.22)$$

Let $\mathbf{G}$ of equation (5.8) be determined from "deconvolution" operation, following equations (5.17) and (5.18). That is, $\mathbf{G} \approx \hat{\mathbf{G}}$.

Since $r_i = 1$ for all i's for a step input, $\mathbf{C}_c^{-1}$ may be written as

$$\mathbf{C}_c^{-1} = \frac{2}{h}
\begin{bmatrix}
1 & -2 & 2 & -2 & \cdots & (-1)^{m-1}.2 \\
 & 1 & -2 & 2 & & \\
 & & 1 & -2 & & \vdots \\
 & & & 1 & & \vdots \\
 & \mathbf{0} & & & \ddots & \\
 & & & & & 1
\end{bmatrix}_{(m \times m)} \qquad (5.23)$$

Then from equation (5.17), we can write

$$g_0 = \frac{2}{h}[c_0],$$

$$g_1 = \frac{2}{h}[c_1 - 2c_0],$$

$$g_2 = \frac{2}{h}[c_2 - 2c_1 + 2c_0],$$

$$\vdots$$

$$g_k = \frac{2}{h}\left[c_k - 2c_{(k-1)} + 2c_{(k-2)} - \cdots + (-1)^k 2c_0\right],$$

$$\vdots$$

and so on.
where $k = 1, 2, \ldots, (m - 1)$.

The above equations can be rearranged as

$$
\left.\begin{aligned}
g_0 &= \frac{2}{h}[c_0], \\
g_1 &= \frac{2}{h}[c_1 - c_0] - g_0, \\
g_2 &= \frac{2}{h}[c_2 - c_1] - g_1, \\
&\;\;\vdots \\
g_k &= \frac{2}{h}\left[c_k - c_{(k-1)}\right] - g_{(k-1)}, \\
&\;\;\vdots
\end{aligned}\right\}
\tag{5.24}
$$

and so on.

Putting the values of c_k and $c_{(k-1)}$ using equation (5.21), we get

$$
\begin{aligned}
g_k &= \frac{2}{h}\left[\frac{1}{a} - \frac{1-\exp(-ah)}{a^2 h}\exp(-ahk) - \frac{1}{a} + \frac{1-\exp(-ah)}{a^2 h}\exp\{-ah(k-1)\}\right] - g_{(k-1)} \\
&= A\exp\{-ah(k-1)\} - g_{(k-1)}
\end{aligned}
\tag{5.25}
$$

where,

$$
A \triangleq 2\left[\frac{1-\exp(-ah)}{ah}\right]^2
\tag{5.26}
$$

From equation (5.25), putting different values of $k = 1, 2, 3 \ldots, (m-1)$, we get

$$
\left.\begin{aligned}
g_1 &= A - g_0, \\
g_2 &= A\exp(-ah) - g_1, \\
&\;\;\vdots \\
&\;\;\vdots \\
g_k &= A\exp\{-ah(k-1)\} - g_{(k-1)}, \\
&\;\;\vdots \\
&\;\;\vdots
\end{aligned}\right\}
\tag{5.27}
$$

and so on.

Now, from equation (5.27)

$$\left(g_k + g_{(k-1)}\right) - \left(g_{(k-1)} + g_{(k-2)}\right) + \cdots + (g_2 + g_1) - (g_1 + g_0)$$

$$= A\exp\{-ah(k-1)\} - A\exp\{-ah(k-2)\} + \cdots + A\exp(-ah) - A$$

Then,

$$g_k - g_0 = A\exp\{-ah(k-1)\}[1 - \exp(ah) + \exp(2ah) - \ldots - (-1)^k \exp\{-ah(k-1)\}]$$

$$= A\exp\{-ah(k-1)\}\left[\frac{1-(-1)^k \exp(ahk)}{1+\exp(ah)}\right] \tag{5.28}$$

Case 1: When k is even, from equation (5.28), we write

$$g_k - g_0 = A\left[\frac{\exp\{-ah(k-1)\} - \exp(ah)}{1+\exp(ah)}\right]$$

$$= A\left[\frac{\exp(ah)}{1+\exp(ah)}\exp(-ahk)\right] - A\left[\frac{\exp(ah)}{1+\exp(ah)}\right] \tag{5.29}$$

Now,

$$g_k = A\left[\frac{\exp(ah)}{1+\exp(ah)}\exp(-ahk)\right] + \left[g_0 - A\frac{\exp(ah)}{1+\exp(ah)}\right] \tag{5.30}$$

Case 2: Similarly, when k is odd, we have

$$g_k = A\left[\frac{\exp(ah)}{1+\exp(ah)}\exp(-ahk)\right] - \left[g_0 - A\frac{\exp(ah)}{1+\exp(ah)}\right] \tag{5.31}$$

Hence, the general expression for g_k is

$$g_k = A\left[\frac{\exp(ah)}{1+\exp(ah)}\exp(-ahk)\right] + (-1)^k\left[g_0 - A\frac{\exp(ah)}{1+\exp(ah)}\right]$$

$$= \frac{A\exp(ah)}{1+\exp(ah)}\exp(-ahk) + (-1)^k B \tag{5.32}$$

$$= T_1 + T_2 \text{ (say)} \tag{5.33}$$

where,

$$\left.\begin{array}{l} T_1 = \dfrac{A\exp(ah)}{1+\exp(ah)}\exp(-ahk), \\[4mm] B = \left[g_0 - A\dfrac{\exp(ah)}{1+\exp(ah)} \right], \end{array}\right\} \tag{5.34}$$

and $T_2 = (-1)^k B.$

We know that $g_0 = \dfrac{2}{h}[c_0].$

Putting the value of c_0 from equation (5.22), we get

$$g_0 = \frac{2}{(ah)^2}[ah - 1 + \exp(-ah)] \tag{5.35}$$

Now using equations (5.26) and (5.35) in equation (5.34), we have

$$B = \frac{2}{(ah)^2}\left[ah - 1 + \exp(-ah)\right] - \left[\frac{2\{1-\exp(-ah)\}^2}{(ah)^2}\right]\left[\frac{1}{1+\exp(-ah)}\right]$$

$$= \frac{2}{(ah)^2\{1+\exp(-ah)\}}\left[ah\{1+\exp(-ah)\} + 2\exp(-ah) - 2\right]$$

Putting $ah \triangleq p$ and simplifying,

$$B = \left[\frac{2}{p^2\{1+\exp(p)\}}\right]\left[(p+2)+(p-2)\exp(p)\right] \tag{5.36}$$

Expanding the above equation, we get

$$B = \left[\frac{2}{p^2\{1+\exp(p)\}}\right]\left[\frac{p^3}{3!} + \frac{2p^4}{4!} + \frac{3p^5}{5!} + \cdots\right] \tag{5.37}$$

Now rewriting equation (5.32) and using equation (5.37),

$$g_k = A\left[\frac{\exp(p)}{1+\exp(p)}\exp(-kp)\right]+(-1)^k\left[\frac{2}{p^2\{1+\exp(p)\}}\left[\frac{p^3}{3!}+\frac{2p^4}{4!}+\frac{3p^5}{5!}+\cdots\right]\right]$$

(5.38)

Equation (5.37) proves beyond doubt the positivity of B in equation (5.32). Thus, the second term of equation (5.32) will oscillate between positive and negative values with increasing integral values of k, i.e., $k = 1, 2, \ldots, (m-1)$. This will lead to instability in the values of g_k as seen from equation (5.38).

In Table 5.4, we tabulate the values of two terms of equation (5.33), i.e., T_1 and T_2, to reveal the nature of variation of magnitudes of the two terms for a typical value of p, i.e., $p = \frac{1}{2}$.

Referring to equations (5.26) and (5.34), it is observed that, for a particular value of p, A and B are always constants. Since the term T_2 of equation (5.33) oscillates with the value of k, it is of interest to find the nature of variation of B with increasing p. Figure 5.8 shows the $(B-p)$ curve and it is seen that B attains a maximum value of 0.26497508 for $p = 3.21223060$. These values can also be obtained by differentiating B with respect to p in equation (5.36) so that

$$(4-p)\exp(2p)-6p\exp(p)-p-4=0$$

(5.39)

Equation (5.39) may be solved numerically to obtain the value of p for B_{max}, as indicated above.

Also, T_1 of equation (5.33) decreases with increasing k for the first-order plant considered. But magnitude of T_2 is constant as it is independent of k

TABLE 5.4

Values of the Two Terms T_1 and T_2 of Equation (5.33) for $p = ah = \frac{1}{2}$ and Different Values of k

k	First Term of g_k, T_1 Equation (5.33)	Second Term of g_k, T_2 Equation (5.33)	$g_k = T_1 + T_2$
0	0.77094388	0.08130140	0.85224528
1	0.46760110	− 0.08130140	0.38629970
2	0.28361440	0.08130140	0.36491580
3	0.17202083	− 0.08130140	0.09071943
4	0.10433591	0.08130140	0.18563731
5	0.06328293	− 0.08130140	− 0.01801847
6	0.03838304	0.08130140	0.11968444
7	0.02328048	− 0.08130140	− 0.05802091

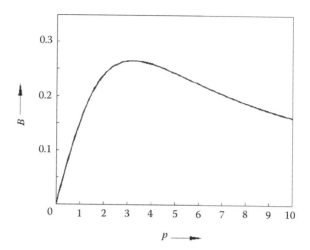

FIGURE 5.8
Variation of B with p (refer equation (5.36)).

because k only contributes the sign. This is why the tail end terms of g_k (i.e., for higher k) oscillate in a more pronounced manner.

5.4 Identification of a Closed Loop System [6]

In linear control theory, the problem of system identification is common and there are several approaches for solving such problems. The basic system identification problem is to determine the unknown plant from the input-output data available. The classical approach uses a nonparametric model involving the impulse response function to design a dynamic system.

Consider the linear SISO feedback control system shown in Figure 5.1. For this system the output is

$$c(t) = \int_0^t \left[r(\tau) - \int_0^\tau c(\sigma) h(\tau - \sigma) d\sigma \right] g(t - \tau) d\tau \tag{5.40}$$

It is required to solve $g(t)$, the unknown plant, from a knowledge of $r(t)$, $c(t)$, and $h(t)$.

In equation (5.40), it is not possible to decouple $g(t)$ from the convolution integral and hence $g(t)$ cannot be solved from equation (5.40). Hence, using equation (1.14), let all the time functions $r(t)$, $c(t)$, and $h(t)$ be transformed to their respective BPF vectors **R**, **C**, and **H**, each having m BPF components over a period T.

Following equation (5.17), it is possible to show that the BPF vector representing $g(t)$ is given by

$$\mathbf{G} \triangleq \begin{bmatrix} g_0 & g_1 & g_2 & \cdots & g_{(m-1)} \end{bmatrix}^{\mathrm{T}} \tag{5.41}$$

and the BPF vector of $g(t)$ to be identified through "deconvolution" operation is defined as

$$\hat{\mathbf{G}}^{\mathrm{T}} \triangleq \begin{bmatrix} \hat{g}_0 & \hat{g}_1 & \hat{g}_2 & \cdots & \hat{g}_{(m-1)} \end{bmatrix} = \frac{2}{h} \mathbf{C}^{\mathrm{T}} \hat{\mathbf{C}}_{\mathrm{e}}^{-1}$$

where $h = \dfrac{T}{m}$, and $\mathbf{C}_{\mathrm{e}}$ is the convolution matrix formed by the elements of the BPF vector for error signal $\hat{\mathbf{E}}$, determined from the following relation

$$\hat{\mathbf{E}} = \mathbf{R} - \hat{\mathbf{Z}} \tag{5.42}$$

where, $\hat{\mathbf{E}} \triangleq \begin{bmatrix} \hat{e}_0 & \hat{e}_1 & \cdots & \hat{e}_{(m-1)} \end{bmatrix}^{\mathrm{T}}$

Since the error signal $\hat{\mathbf{E}}$ is the algebraic sum of two signals, one from direct BPF expansion ($\mathbf{R}$) and another using the convolution operation ($\hat{\mathbf{Z}}$), we have used the "double hat" symbol to distinguish it.

In equation (5.42), $\hat{\mathbf{Z}}$ is given by

$$\hat{\mathbf{Z}} = \frac{h}{2} \mathbf{H} \mathbf{C}_{\mathrm{C}} \tag{5.43}$$

where $\mathbf{C}_{\mathrm{c}}$ is formed by the elements of $\mathbf{C}$,

$$\hat{\mathbf{Z}} \triangleq \begin{bmatrix} \hat{z}_0 & \hat{z}_1 & \cdots & \hat{z}_{(m-1)} \end{bmatrix}^{\mathrm{T}},$$

and

$$\mathbf{H} \triangleq \begin{bmatrix} h_0 & h_1 & \cdots & h_{(m-1)} \end{bmatrix}^{\mathrm{T}}.$$

Thus, equations (5.41), (5.42), and (5.43) together provide the solution for $\mathbf{G}$ and the plant is then identified in BPF domain.

In an alternative way, we can identify a feedback SISO system also in a recursive manner described below.

From the knowledge of $\mathbf{C}$, we can form the convolution matrix $\mathbf{C}_c$. Hence, knowing $\mathbf{H}$ and using a relation similar to equation (5.43), $\hat{\mathbf{Z}}$ can be determined whose elements are

$$\hat{z}_k = \sum_{i=0}^{k-1}(-1)^{k+i+1}\hat{z}_i + \frac{h}{2}\sum_{i=0}^{k}h_i c_{k-i} \tag{5.44}$$

where,

$\hat{z}_k$ is the $(k + 1)$th element of the BPF vector $\hat{\mathbf{Z}}$ of the feedback signal $z(t)$.
h_k is the $(k + 1)$th element of the BPF vector $\mathbf{H}$ of the impulse response of the feedback element $h(t)$.
and c_k is the $(k + 1)$th element of the BPF vector $\mathbf{C}$ of the output $c(t)$.

Then from equations (5.42) and (5.44), we can write

$$\hat{e}_k = r_k - \hat{z}_k = r_k - \left[\sum_{i=0}^{k-1}(-1)^{k+i+1}\hat{z}_i + \frac{h}{2}\sum_{i=0}^{k}h_i c_{k-i}\right] \tag{5.45}$$

where,

$\hat{e}_k$ is the $(k + 1)$th element of the BPF vector $\hat{\mathbf{E}}$ of the error signal $e(t)$.
and r_k is the $(k + 1)$th element of the BPF vector $\mathbf{R}$ of the input $r(t)$.

Now, $\hat{\mathbf{C}}_e$ can be formed from equation (5.45).
Then, we can write

$$\mathbf{C} = \frac{h}{2}\hat{\mathbf{G}}\hat{\mathbf{C}}_e \tag{5.46}$$

where,

$$\hat{\mathbf{C}}_e \triangleq \begin{bmatrix} \hat{e}_0 & \hat{e}_0 + \hat{e}_1 & \cdots & \cdots & \hat{e}_{(m-2)} + \hat{e}_{(m-1)} \\ & \hat{e}_0 & \cdots & \cdots & \hat{e}_{(m-3)} + \hat{e}_{(m-2)} \\ & & \ddots & & \vdots \\ & \mathbf{0} & & \ddots & \vdots \\ & & & & \hat{e}_0 \end{bmatrix}_{(m\times m)} \tag{5.47}$$

Using equations (5.46) and (5.47), we can arrive at the relationships between BPF coefficients of $\hat{\mathbf{E}}$, $\mathbf{C}$, and $\hat{\mathbf{G}}$ as shown by equation (5.12). The BPF coefficients of $\mathbf{C}$ are shown as $\hat{c}_i$ because these are obtained through the use of the convolution matrix.

For determining the BPF coefficients of the impulse response of the plant, following equation (5.12), we have

$$\hat{c}_0 = \frac{h}{2}\hat{g}_0\hat{e}_0$$

$$\hat{c}_1 = \frac{h}{2}\left[\hat{g}_0\left(\hat{e}_0 + \hat{e}_1\right) + \hat{g}_1\hat{e}_0\right]$$

$$\hat{c}_2 = \frac{h}{2}\left[\hat{g}_0\left(\hat{e}_1 + \hat{e}_2\right) + \hat{g}_1\left(\hat{e}_0 + \hat{e}_1\right) + \hat{g}_2\hat{e}_0\right] \qquad (5.48)$$

$$\vdots$$

$$\hat{c}_{(m-1)} = \frac{h}{2}\left[\hat{g}_0\left(\hat{e}_{(m-2)} + \hat{e}_{(m-1)}\right) + \cdots + \hat{g}_{(m-2)}\left(\hat{e}_0 + \hat{e}_1\right) + \hat{g}_{(m-1)}\hat{e}_0\right]$$

Using these equations, we can determine the $(k + 1)$th term of $\hat{\mathbf{G}}$ as

$$\hat{g}_k = \left[\frac{2}{h}c_k - \sum_{i=0}^{k-1}g_i\left(\hat{e}_{(k-i-1)} + \hat{e}_{(k-i)}\right)\right]\hat{e}_0^{-1} \qquad (5.49)$$

But to determine $\hat{g}_k$ from equation (5.48), we need to use the coefficients of $\mathbf{C}$ obtained via direct expansion, since $c(t)$ is known. Hence, replace $\hat{c}_i$ by c_i in (5.48) to write equation (5.49).

5.4.1 Numerical Example

Example 5.4

Consider the example treated by Kwong and Chen [2]. Referring to Figure 5.1, $r(t) = u(t)$, a unit step function, $g(t) = 2\exp(-4t)$ and $h(t) = 4u(t)$.
It is apparent that the output $c(t) = \exp(-2t)\sin(2t)$.
Following Reference [2], we take $m = 16$ and $T = 1$ s.
Then the BPF coefficients of the output $c(t)$ are

$$
\begin{aligned}
c(t) \approx \big[&0.05745426 \quad 0.15347314 \quad 0.22402022 \quad 0.27278431 \\
&0.30323895 \quad 0.31859432 \quad 0.32176688 \quad 0.31536397 \\
&0.30168026 \quad 0.28270362 \quad 0.26012819 \quad 0.23537259 \\
&0.20960180 \quad 0.18375108 \quad 0.15855102 \quad 0.13455267 \big]\Psi_{(16)}
\end{aligned}
$$

$$(5.50)$$

Using equations (5.41)–(5.43), or equivalently equation (5.49), we have identified the system $g(t)$ via the "deconvolution" operation in BPF domain. Table 5.5 presents the quantitative comparison of the results obtained for $g(t)$ after the "deconvolution" operation with the direct BPF expansion of the impulse response of the plant.

In Table 5.5, it is noted that in the percentage error column, there is an oscillation in the result, where the absolute minimum value of the percentage error is 4.64751759 while the maximum value is 202.24710652.

In comparison to the "deconvolution" results in Table 5.5, we discuss on the same that was published by Kwong and Chen in Reference [2], in the following article.

5.4.2 Discussion on the Reliability of Result

In Example 5.4, for identification of system $g(t)$, we have considered the direct expansion of the output in BPF domain, given by equation (5.50), as the reference result.

However, in their work [2], Kwong and Chen have considered the output vector as (for $m = 16$ and $T = 1$ s).

$$
\begin{aligned}
c(t) \approx \big[\, &0.05501253 \quad 0.15135173 \quad 0.22220679 \quad 0.27126074 \\
&0.30198315 \quad 0.31758168 \quad 0.32097154 \quad 0.31475981 \\
&0.30124163 \quad 0.28240602 \quad 0.25994862 \quad 0.23528991 \\
&0.20959685 \quad 0.18380679 \quad 0.15865246 \quad 0.13468691 \,\big] \mathbf{\Psi}_{(16)}
\end{aligned}
$$

$$(5.51)$$

and solved the system identification problem satisfactorily without any oscillation whatsoever. It may be observed that the elements of the output in equation (5.50) are somewhat different from those in equation (5.51). Further investigation revealed that the BPF coefficients of equation (5.51) were obtained via BPF domain analysis of the system using the convolution matrix. These coefficients were obtained when we treated Example 5.2 (vide the third column of Table 5.2).

This is the underlying reason that Kwong and Chen could identify the system and came up with satisfactory solution without any oscillation.

But the reality is the elements of the vector $\hat{\mathbf{G}}$ identified through "deconvolution" operation and the direct BPF expansion of $g(t)$ are not very close, and the elements of $\hat{\mathbf{G}}$ shows noticeable deviations from the desired result. This is due to the compounding errors in the BPF analysis while performing the convolution and "deconvolution" operations. However, Figure 5.9(a) and 5.9(b) show that with decreasing values of h, these discrepancies are removed.

TABLE 5.5

Identification of the System $g(t)$ of Example 5.4 via "Deconvolution" for $m = 16$ and $T = 1$ s in BPF Domain and the Result is Compared with the Direct Expansion of Impulse Response of the System in BPF Domain (Vide Appendix B, Program No. 5.3)

t (sec)	Direct Expansion of $g(t)$ in BPF Domain	"Deconvolution" Result in BPF Domain Using the Equations (5.41), (5.42), and (5.43) or Equation (5.49)	Percentage Error
0.0000			
	1.76959374	1.85183592	– 4.64751759
0.0625			
	1.37816099	1.29217298	6.23932949
0.1250			
	1.07331286	1.15317233	– 7.44046524
0.1875			
	0.83589689	0.74876684	10.42354014
0.2500			
	0.65099715	0.73052223	– 12.21588744
0.3125			
	0.50699709	0.42003832	17.15173099
0.3750			
	0.39484973	0.47484396	– 20.25941035
0.4375			
	0.30750928	0.22117703	28.07468281
0.5000			
	0.23948847	0.32017375	– 33.69067431
0.5625			
	0.18651381	0.10087772	45.91407178
0.6250			
	0.14525710	0.22660743	– 56.00437663
0.6875			
	0.11312634	0.02810377	75.15718213
0.7500			
	0.08810288	0.17000535	– 92.96230023
0.8125			
	0.06861460	– 0.01592016	123.20229367
0.8750			
	0.05343710	0.13576445	– 154.06402034
0.9375			
	0.04161686	– 0.04255203	202.24710652
1.0000			

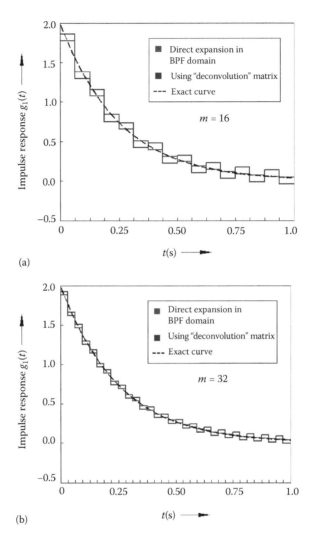

FIGURE 5.9
Identification of the closed loop system of Example 5.4 using the "deconvolution" process, and its comparison with the direct BPF expansion of the system impulse response for (a) $m = 16$ and $T = 1$ s, and (b) $m = 32$ and $T = 1$ s. It is noted that with decreasing h, the oscillation has reduced (vide Appendix B, Program no. 5.3).

5.5 Conclusion

For the analysis of linear open loop and closed loop systems, the BPF set has been used to form the convolution matrix. Using the convolution matrix, two illustrative examples have been presented with relevant tables and graphs.

In Tables 5.1 and 5.2, BPF domain convolution results have been compared with their equivalent direct BPF expansion coefficients. It is observed that the convolution matrix has been able to analyze successfully the open loop system of Example 5.1 and the equivalent open loop system of Example 5.2. Also, we have presented a simple recursive expression for the convolution operation leading to the same result without using any vector or matrix.

From the convolution matrix, we derived the "deconvolution" operation by inverting the same to obtain the "deconvolution" matrix. As was done in case of convolution, here also, we have presented one recursive formula for the "deconvolution" operation used effectively for the identification of an open loop system. However, the results obtained by the "deconvolution" matrix and the recursive formula are found to be identical. While identifying one feedback system using its equivalent open loop transfer function model, some oscillations have been noticed. These are reflected in Table 5.3 and Figure 5.5. Also, Figure 5.6 shows that with decreasing values of h, the oscillations are reduced.

Finally, this chapter deals with the identification of a closed loop system using both convolution and "deconvolution" and presents the same illustrative example treated by Kwong and Chen [2]. Table 5.5 tabulates the results of identification for this example. Results obtained via direct BPF expansion of the impulse response of the system have been compared with the BPF domain results of Table 5.5. The undesired oscillations predominant in the results can be reduced by decreasing the value of h. This fact is supported by Figure 5.9(a) and 5.9(b).

References

1. P. Eykhoff, *System identification*, Wiley, New York, 1974.
2. C. P. Kwong and C. F. Chen, Linear feedback system identification via block-pulse functions, Int. J. Syst. Sci., vol. **12**, no. 5, pp. 635–642, 1981.
3. K. Ogata, *Modern Control Engineering* (5th Ed.), Prentice Hall of India, 2011.
4. E. O. Brigham, *The fast Fourier transform and its applications*, Prentice Hall, New Jersey, 1988.
5. Anish Deb, Gautam Sarkar, Amitava Biswas, and Priyaranjan Mandal, Numerical instability of deconvolution operation via block pulse functions, J. Franklin Inst., vol. **345**, pp. 319–327, 2008.
6. Anish Deb, Gautam Sarkar, and Sunit K. Sen, Linearly pulse-width modulated block pulse functions and their application to linear SISO feedback control system identification, Proc. IEEE, Part—D, Control Theory and Appl., vol. **142**, no. 1, pp. 44–50, 1995.

Study Problems

5.1　Consider two block pulse functions of different heights c_0 and c_1 having the same width of h seconds. Then graphically determine the convolution result in time domain and subsequently convert it to BPF domain.

5.2　Consider an open loop system having a transfer function $G(s) = (s + 1)^{-1}$. Find its output $c(t)$ in BPF domain for a step input $u(t)$ using the convolution matrix. Consider $m = 4$ and $T = 1$ s. Compare the convolution results with direct BPF expansion of the output and determine the percentage errors of different coefficients.

5.3　Repeat Problem 5.2 for $m = 8$ and $T = 1$ s. Compare the two results and discuss.

5.4　Consider a closed loop system having the forward path transfer function $G(s) = (s + 1)^{-1}$ and the feedback path transfer function $H(s) = \dfrac{2}{s}$. Find its output $c(t)$ in BPF domain for a step input $u(t)$ using the convolution matrix. Consider $m = 4$ and $T = 1$ s. Finally, compare the result graphically with the direct BPF expansion of $c(t)$.

5.5　Consider the feedback system of Problem 5.4. Double the value of m for the same interval T and find its output $c(t)$ using the convolution matrix. Finally, comment on the convolution result.

5.6　Consider the open loop system of Problem 5.2. Knowing the input and output, identify the system using (a) the "deconvolution" matrix and (b) the recursive approach, for $m = 8$ and $T = 1$ s. Comment on the results obtained using these two approaches. Is there any oscillation in the result?

5.7　Consider the open loop system of Problem 5.6. After identifying the system using the "deconvolution" approach for $m = 8$ and $T = 1$ s, compare the result with the direct BPF expansion of the impulse response of the system. Compute percentage error for each segment and make your observations.

5.8　Consider the open loop system of Problem 5.6. Identify the system for double the value of m keeping the time interval T same. Compute percentage error for each segment and make your observation with respect to oscillation in the result.

5.9　Consider the closed loop system of Problem 5.4. Knowing the input, output, and feedback path transfer function, identify the system using (a) the "deconvolution" matrix and (b) the recursive approach for $m = 8$ and $T = 1$ s. Comment on the results obtained using these two approaches.

5.10 Consider the closed loop system of Problem 5.9. Identify the system for double the value of m keeping the time interval T same. Compute percentage error for each BPF coefficient and make your observation with respect to oscillation in the result.

6

Delayed Unit Step Functions (DUSF) for System Analysis and Fundamental Nature of the Block Pulse Function (BPF) Set

The applications of Walsh and block pulse functions (BPFs) to systems sciences have been extensively exploited by many investigators. This is mainly due to the development of various operational matrices for Walsh and BPF domain analysis. Chen and Hsiao [1] first derived Walsh integration matrix to solve state space equations. Chen and Shih [2,3] introduced the Walsh product matrix and Walsh delay matrix for solutions of time-varying systems and delay systems, respectively. The application of BPFs introduced by Chen et al. [4], Gopalsami and Deekshatulu [5] and Rao [6] enormously simplified the computation work via Walsh functions. This made BPFs most popular among the researchers. One major indicator of this popularity is the scores of research papers published [7] in the 1980s and 1990s of the last century and even in the first decade [8–10] of the twenty-first century.

The set of delayed unit step functions (DUSF) was added to the piecewise constant basis function family in 1983 by Hwang [11]. He derived the operational matrix for integration in DUSF domain and also analyzed a class of functional differential equations of the form

$$\left.\begin{aligned}\dot{\mathbf{x}}(t) &= a\mathbf{x}(\lambda t) + b\mathbf{x}(t)\\ \mathbf{x}(0)\ &\text{specified}\end{aligned}\right\} \tag{6.1}$$

Differential equations containing terms with a stretched argument play an important role in describing the dynamics of current collecting mechanism for electric locomotives [12] and in describing the particulate systems [13]. The solution of such type of differential equations has been presented by several authors. Hwang and Shih [14] obtained the solutions for both the cases of $\lambda > 1$ and $\lambda < 1$ by a Laguerre series expansion. Chen [15] obtained the block pulse infinite series solution for the case of $\lambda < 1$ by successively scaling the argument. Rao and Palanisamy [16] derived a Walsh stretch matrix to obtain Walsh series solutions of the equations.

In the following, the definition of the set of DUSFs is given and its operational matrix is derived. Finally, the application of DUSFs to analyse equation (6.1) has been illustrated with two examples.

6.1 The Set of DUSF and the Operational Matrix for Integration [11]

Let a set of m DUSF's $u_i(t)$ for $i = 0, 1, 2, \ldots, (m-1)$ be defined as follows:

$$u_i(t) = \begin{cases} 1 & t \geq ih \\ 0 & t < ih \end{cases} \tag{6.2}$$

In Figure 6.1 we reproduce Figure 1.12 showing an m-set DUSFs.

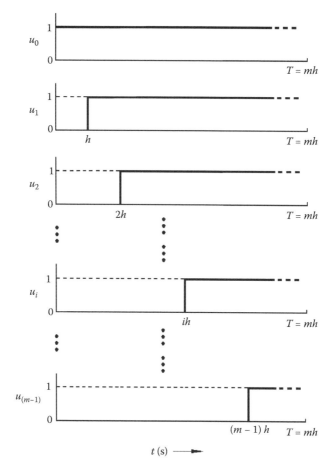

FIGURE 6.1
A set of delayed unit step function (DUSF) for m-component functions.

The Laplace transform of the ith component function $U_i(s)$ is given by

$$U_i(s) = \mathcal{L}\{u_i(t)\} = \frac{1}{s}\exp(-ihs) \tag{6.3}$$

Any function $f(t)$ that is both defined and absolutely integrable in the interval $0 \le t \le t_f$ ($t_f = mh$) and can be approximated in a minimum mean square error sense as

$$f(t) \approx \sum_{i=0}^{m-1} g_i u_i(t) \triangleq \mathbf{G}^{\mathrm{T}} \mathbf{U}_{(m)}(t) \tag{6.4}$$

where

$$g_i = \frac{1}{h}\left[\int_{ih}^{(i+1)h} f(t)\,\mathrm{d}t - \int_{(i-1)h}^{ih} f(t)\,\mathrm{d}t\right] \tag{6.5}$$

The DUSF coefficient vector $\mathbf{G}$ and the DUSF vector $\mathbf{U}(t)$ are respectively given by

$$\mathbf{G} = \begin{bmatrix} g_0 & g_1 & \cdots & g_{(m-1)} \end{bmatrix}^{\mathrm{T}} \tag{6.6}$$

and

$$\mathbf{U}(t) = \begin{bmatrix} u_0(t) & u_1(t) & \cdots & u_{(m-1)}(t) \end{bmatrix}^{\mathrm{T}} \tag{6.7}$$

where $[\ldots]^{\mathrm{T}}$ denotes transpose.

The first-order trapezoidal integrator is based upon the trapezoidal method of integration. Its z transfer function TR(z) is given by

$$\mathrm{TR}(z) = \frac{T}{2}\left[\frac{1+z^{-1}}{1-z^{-1}}\right] \quad \text{where } T \text{ is the sampling period.}$$

If we let $T = h$, then in s-domain, the transfer function becomes that of an integrator, expressed by

$$\mathrm{TR}(s) = \frac{1}{s} = \frac{h}{2}\left[\frac{1+\exp(-hs)}{1-\exp(-hs)}\right]$$

That is, $\dfrac{2}{hs} = \left[\dfrac{1+\exp(-hs)}{1-\exp(-hs)}\right]$

By componendo dividendo, we have the delay operator [17].

$$\exp(-hs) \approx \frac{(2-hs)}{(2+hs)} \tag{6.8}$$

Thus

$$\frac{1}{s} = \frac{h}{2}\Big[1 + 2\exp(-hs) + 2\exp(-2hs) + 2\exp(-3hs) + \ldots\Big] \tag{6.9}$$

and $\mathcal{L}\left\{\displaystyle\int_0^t u_i(\tau)\,d\tau\right\} = \dfrac{1}{s^2}\exp(-ihs)$

$$= \frac{h}{2}\frac{1}{s}\exp(-ihs) + \frac{h}{s}\exp\big(-(i+1)hs\big) + \frac{h}{s}\exp\big(-(i+2)hs\big) + \ldots \tag{6.10}$$

Using the above relation for different values of i, we may approximate the integrals of DUSFs over the interval $t \in [0, T)$, by a set of DUSFs as follows:

$$\int_0^t \mathbf{U}_{(m)}(\tau)\,d\tau = \begin{bmatrix} \displaystyle\int_0^t u_0(\tau)\,d\tau \\ \displaystyle\int_0^t u_1(\tau)\,d\tau \\ \displaystyle\int_0^t u_2(\tau)\,d\tau \\ \vdots \\ \displaystyle\int_0^t u_{(m-1)}(\tau)\,d\tau \end{bmatrix} = h\begin{bmatrix} \frac{1}{2} & 1 & 1 & 1 & \cdots & 1 \\ & \frac{1}{2} & 1 & 1 & \cdots & 1 \\ & & \frac{1}{2} & 1 & \cdots & 1 \\ & & & \ddots & & \vdots \\ \mathbf{0} & & & & \ddots & 1 \\ & & & & & \frac{1}{2} \end{bmatrix}\begin{bmatrix} u_0(t) \\ u_1(t) \\ u_2(t) \\ \vdots \\ u_{(m-1)}(t) \end{bmatrix}\quad \mathbf{P}_{\mathrm{D}(m)}\mathbf{U}_{(m)}(t)$$

$$\tag{6.11}$$

where $\mathbf{P}_{\mathrm{D}(m)}$ is a constant square matrix $\mathbf{P}_{\mathrm{D}(m)}[\in \mathbb{R}^{m\times m}]$ of order m, known as the operational matrix for integration in the DUSF domain [11]. It is of interest to note that this operational matrix for integration is identical with that for BPFs.

6.1.1 Alternative Way to Derive the Operational Matrix for Integration

The operational matrix for integration in DUSF domain may be derived in a more direct and straight forward manner if we follow the procedure adopted for deriving the integration matrix for BPF domain in Section 3.1. That is, we integrate a set of DUSFs and transform the results to DUSFs again. Figure 6.2 shows a set of DUSF for $m = 4$ and $T = 4h$ s, their subsequent integration and expansion of the results into DUSF domain again.

The equation of the $(i + 1)$th DUSF is given by

$$u_i(t) = u(t - ih)$$

Thus

$$\int u_i(t)\,dt = \int u(t - ih)\,dt = (t - ih)u(t - ih) \tag{6.12}$$

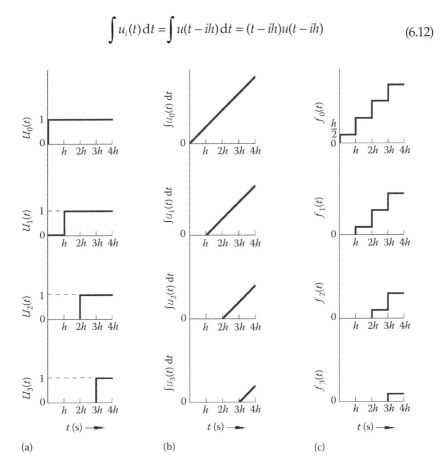

(a) (b) (c)

FIGURE 6.2
(a) First four delayed unit step functions, (b) their first integrations and (c) subsequent representation of the integrated functions by their delayed unit step function (DUSF) equivalents.

Hence, following equations (6.4) and (6.5), the general mathematical expression for (6.12) may be written as

$$\int u_i(t)\,dt \approx 0.u_0(t) + \ldots + 0.u_{(i-1)}(t) + \frac{h}{2} u_i(t) + hu_{i+1}(t) + \ldots + hu_{(m-1)}(t) \quad (6.13)$$

Putting proper values of i for the first four DUSFs in equation (6.13) we have

$$
\left.
\begin{aligned}
\int u_0(t)\,dt &= \frac{h}{2} u_0(t) + hu_1(t) + hu_2(t) + hu_3(t) \\
\int u_1(t)\,dt &= 0.u_0(t) + \frac{h}{2} u_1(t) + hu_2(t) + hu_3(t) \\
\int u_2(t)\,dt &= 0.u_0(t) + 0.u_1(t) + \frac{h}{2} u_2(t) + hu_3(t) \\
\int u_3(t)\,dt &= 0.u_0(t) + 0.u_1(t) + 0.u_2(t) + \frac{h}{2} u_3(t)
\end{aligned}
\right\} \quad (6.14)
$$

Writing the above set of equations in matrix form, we have

$$
\begin{bmatrix}
\int u_0(t)\,dt \\
\int u_1(t)\,dt \\
\int u_2(t)\,dt \\
\int u_3(t)\,dt
\end{bmatrix}
\approx h
\begin{bmatrix}
\frac{1}{2} & 1 & 1 & 1 \\
0 & \frac{1}{2} & 1 & 1 \\
0 & 0 & \frac{1}{2} & 1 \\
0 & 0 & 0 & \frac{1}{2}
\end{bmatrix}
\begin{bmatrix}
u_0(t) \\
u_1(t) \\
u_2(t) \\
u_3(t)
\end{bmatrix}
\quad (6.15)
$$

Writing in a compact form, we arrive at equation (6.11) for $m = 4$. That is

$$\int \mathbf{U}_{(4)}(t)\,dt \approx \mathbf{P}_{D(4)}\mathbf{U}_{(4)}(t)$$

Generalizing this fact for an m-set DUSF, we get the generalized relation (6.11).

6.1.2 Numerical Example

Example 6.1

Consider the function $f(t) = \dfrac{t^2}{2}$ for $t \geq 0$. Now, we use $\mathbf{P}_{D(m)}$ for $m = 10$ and $T = 1$ s. Integration results obtained using the integration operational matrices of both DUSF and BPF domains are compared with the direct expansion of the exact integration of $f(t)$. Table 6.1 presents a quantitative comparison of the results along with respective percentage errors.

From Table 6.1 it is observed that BPF coefficients are cumulative sums of the DUSF coefficients. This is because the DUSF set is related to the BPF set by some invertible transformation matrix $\mathbf{N}_{(m)}$ having a special pattern.

TABLE 6.1

Comparison of the Results of Integration Obtained via the Operational Matrix in DUSF Domain, the Operational Matrix in Block Pulse Function (BPF) Domain and Direct Delayed Unit Step Function (DUSF) and BPF Expansion along with Respective Percentage Errors for Example 6.1, for $m = 10$ and $T = 1$ s (Vide Appendix B, Program No. 6.1)

$t(s)$	Coefficients of Integrated Function in DUSF Domain		Percentage Error in DUSF Domain	Coefficients of Integrated Function in BPF Domain		Percentage Error in BPF Domain
	Using Operational Matrix	Direct Expansion		Using Operational Matrix	Direct Expansion	
0.0						
	0.00008333	0.00004167	– 1.0000 e02	0.00008333	0.00004167	– 1.0000 e02
0.1						
	0.00066667	0.00058333	– 0.1429 e02	0.00075000	0.00062500	– 0.2000 e02
0.2						
	0.00216667	0.00208333	– 0.0400 e02	0.00291667	0.00270833	– 0.0769 e02
0.3						
	0.00466667	0.00458333	– 0.0182 e02	0.00758333	0.00729167	– 0.0400 e02
0.4						
	0.00816667	0.00808333	– 0.0103 e02	0.01575000	0.01537500	– 0.0244 e02
0.5						
	0.01266667	0.01258333	– 0.0066 e02	0.02841667	0.02795833	– 0.0163 e02
0.6						
	0.01816667	0.01808333	– 0.0046 e02	0.04658333	0.04604167	– 0.0118 e02
0.7						
	0.02466667	0.02458333	– 0.0034 e02	0.07125000	0.07062500	– 0.0088 e02
0.8						
	0.03216667	0.03208333	– 0.0026 e02	0.10341667	0.10270833	– 0.0069 e02
0.9						
	0.04066667	0.04058333	– 0.0021 e02	0.14408333	0.14329167	– 0.0055 e02
1.0						

From Figure 6.1, it is apparent that the set of DUSFs truncated at $T = mh$ is related to the set of BPFs by the following relation:

$$
\begin{bmatrix} u_0(t) \\ u_1(t) \\ \vdots \\ u_{(m-1)}(t) \end{bmatrix} = \begin{bmatrix} 1 & 1 & 1 & \cdots & 1 \\ & 1 & 1 & \cdots & 1 \\ & & 1 & \cdots & 1 \\ & 0 & & \ddots & \vdots \\ & & & & 1 \end{bmatrix} \begin{bmatrix} \boldsymbol{\psi}_0(t) \\ \boldsymbol{\psi}_1(t) \\ \vdots \\ \boldsymbol{\psi}_{(m-1)}(t) \end{bmatrix} \tag{6.16}
$$

or,

$$
\mathbf{U}_{(m)}(t) \triangleq \mathbf{N}_{(m)} \boldsymbol{\Psi}_{(m)}(t) \tag{6.17}
$$

where $\mathbf{N}_{(m)}$ is the upper triangular Toeplitz matrix of order m.

Thus, though the coefficients of DUSF and BPF look different, they essentially end up with an *identical approximation*.

6.2 Block Pulse Function versus Delayed Unit Step Function: A Comparative Study [18]

With the necessary discussion of the DUSFs, now the qualitative attributes of the BPF set and the DUSF set may be explored.

The basic properties of BPFs and DUSFs have already been tabulated in Table 1.1. These properties are apparent from Figures 1.6 and 6.1. rom Table 1.1 it is clear that the strong bias in favor of DUSFs must include some inherent fallacies. That this is so is very clear from Section 6.2.2.

6.2.1 Function Approximation: BPF versus DUSF

For a comparison of function approximation in BPF and DUSF domain, we take up a unit ramp function and expand the same in both the domains for $m = 4$ and $T = 4h$ s. The results are shown below:

$$
f(t)\big|_{\text{BPF}} \approx h \begin{bmatrix} \dfrac{1}{2} & \dfrac{3}{2} & \dfrac{5}{2} & \dfrac{7}{2} \end{bmatrix} \boldsymbol{\Psi}_{(4)} \tag{6.18}
$$

and

$$
f(t)\big|_{\text{DUSF}} \approx h \begin{bmatrix} \dfrac{1}{2} & 1 & 1 & 1 \end{bmatrix} \mathbf{U}_{(4)} \tag{6.19}
$$

Figure 6.3 shows the resulting approximated ramp function along with the component functions drawn below, for both the BPF as well as DUSF domains. It is noted that equations (6.18) and (6.19) eventually lead to the same result shown in Figure 6.3. This proves the fact that BPF domain and DUSF domain approximations are identical. Also, using the transformation

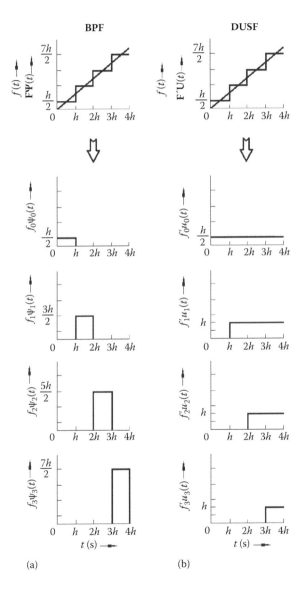

FIGURE 6.3
Approximation of a unit ramp function via (a) block pulse function (BPF) and (b) delayed unit step function (DUSF) for $m = 4$ and $T = 4h$ s.

matrix of equation (6.17) we can transform equation (6.19) into equation (6.18). That is

$$f(t)\big|_{\text{DUSF}} \approx h \begin{bmatrix} \dfrac{1}{2} & 1 & 1 & 1 \end{bmatrix} \mathbf{U}_{(4)}$$

$$= h \begin{bmatrix} \dfrac{1}{2} & 1 & 1 & 1 \end{bmatrix} \mathbf{N}_{(4)} \boldsymbol{\Psi}_{(4)}(t)$$

$$= h \begin{bmatrix} \dfrac{1}{2} & \dfrac{3}{2} & \dfrac{5}{2} & \dfrac{7}{2} \end{bmatrix} \boldsymbol{\Psi}_{(4)}$$

6.2.2 Analytical Assessment

6.2.2.1 Identification of the Last Member of the DUSF Set

Considering the similarities of equations (1.12)–(1.14) with equations (6.2), (6.4), and (6.5), it may be of interest to establish a relation between f_i and g_i. If the function $f(t)$ is expanded both in a BPF as well as a DUSF series, we must have

$$f(t) \approx \mathbf{F}^{\mathrm{T}} \boldsymbol{\Psi}_{(m)}(t) = \mathbf{G}^{\mathrm{T}} \mathbf{U}_{(m)}(t) \tag{6.20}$$

Then, from equation (6.5), we obtain

$$g_0 = f_0 \text{ and } g_i = f_i - f_{i-1}$$

Hence,

$$\mathbf{G} = \begin{bmatrix} f_0 \left(f_1 - f_0 \right) \left(f_2 - f_1 \right) \cdots \left(f_{(m-1)} - f_{(m-2)} \right) \end{bmatrix}^{\mathrm{T}}$$

Substituting $\mathbf{G}$ in (6.4), we have

$$f(t) \approx \begin{bmatrix} f_0 \left(f_1 - f_0 \right) \left(f_2 - f_1 \right) \cdots \left(f_{(m-1)} - f_{(m-2)} \right) \end{bmatrix} \begin{bmatrix} u_0(t) \\ u_1(t) \\ u_2(t) \\ \vdots \\ u_{(m-1)}(t) \end{bmatrix}$$

or, $f(t) \approx f_0 u_0(t) + (f_1 - f_0) u_1(t) + \cdots + \left(f_{(m-1)} - f_{(m-2)} \right) u_{(m-1)}(t)$

Rearranging, we have

$$f(t) \approx f_0 \left[u_0(t) - u_1(t) \right] + f_1 \left[u_1(t) - u_2(t) \right]$$
$$+ f_{(m-2)} \left[u_{(m-2)}(t) - u_{(m-1)}(t) \right] + f_{(m-1)} u_{(m-1)}(t) \qquad (6.21)$$

Comparing the function sets of Figures 1.6 and 6.1, we can write

$$\left. \begin{aligned} u_0(t) - u_1(t) &= \psi_0(t) \\ u_1(t) - u_2(t) &= \psi_1(t) \\ &\vdots \\ &\vdots \\ u_{(m-2)}(t) - u_{(m-1)}(t) &= \psi_{(m-2)}(t) \end{aligned} \right\} \qquad (6.22)$$

Using equation (6.22) in equation (6.21),

$$f(t) \approx f_0 \psi_0(t) + f_1 \psi_1(t) + \cdots + f_{(m-2)} \psi_{(m-2)}(t) + f_{(m-1)} \psi_{(m-1)}(t) \qquad (6.23)$$

Comparing equation (6.23) with equation (1.13), we are rather forced to write

$$\psi_{(m-1)}(t) = u_{(m-1)}(t) \qquad (6.24)$$

Equation (6.24) is of paramount importance to prove the dependency of DUSFs upon BPFs. We know that, mathematically, equation (6.24) can never be true. But yet it seems to be so from the above derivation. This is because, for a finite number m, the DUSF $u_{(m-1)}(t)$ is truncated at $mh(= T)$ to make itself equal to $\psi_{(m-1)}(t)$. This is also obvious from Figures 1.6 and 6.1.

Remark:

For any large positive integer m, if m is finite, the last member of the DUSF set is identical to the last member of a compatible BPF set.

6.2.2.2 Operational Matrix for Integration

As pointed out by Hwang [11], derivation of the operational matrices for integration in both the DUSF and BPF domains is based upon the approximation [17]

$$\exp(-hs) \approx \frac{(2 - hs)}{(2 + hs)}$$

In the DUSF domain, the vector-matrix relation (6.11) holds.

Similarly, in BPF domain, we have the following relation:

$$\int_0^t \boldsymbol{\Psi}_{(m)}(\tau)d\tau \triangleq \mathbf{P}_{B(m)}\boldsymbol{\Psi}_{(m)}(t) \tag{6.25}$$

where $\mathbf{P}_{B(m)}$ is a constant square matrix $[\mathbf{P}_{B(m)} \in \mathbb{R}^{m\times m}]$ of order m, already introduced in equation (3.9) as $\mathbf{P}_{(m)}$, the operational matrix for integration in the BPF domain [4,7].

Hwang [11] found it of interest that

$$\mathbf{P}_{D(m)} = \mathbf{P}_{B(m)} = h \begin{bmatrix} \frac{1}{2} & 1 & 1 & 1 & \cdots & 1 \\ & \frac{1}{2} & 1 & 1 & \cdots & 1 \\ & & \frac{1}{2} & 1 & \cdots & 1 \\ & & & \ddots & & \vdots \\ \mathbf{0} & & & & \ddots & 1 \\ & & & & & \frac{1}{2} \end{bmatrix} \triangleq \mathbf{P}_{(m)}(\text{say}) \tag{6.26}$$

We shall now prove that the nature of $\mathbf{P}_{D(m)}$ is a direct consequence of the nature of $\mathbf{P}_{B(m)}$. That is, $\mathbf{P}_{D(m)}$ directly follows from $\mathbf{P}_{B(m)}$ to establish the equality given in equation (6.26), i.e.,

$$\left\langle p_{Dij} \right\rangle = \left\langle p_{Bij} \right\rangle$$

Theorem 6.2.2.2.1

For a very large but finite $[m \in I : \text{integer}]$, the matrices $\mathbf{P}_{D(m)} \in \mathbb{R}^{m\times m}$ and $\mathbf{P}_{B(m)} \in \mathbb{R}^{m\times m}$ are always identical.

Proof

Considering the general nature of equation (6.25), for an m-set BPF, we can write

$$\begin{bmatrix} \int_0^t \psi_i(\tau)d\tau \\ \int_0^t \psi_{(i+1)}(\tau)d\tau \\ \vdots \\ \int_0^t \psi_{(i+m-1)}(\tau)d\tau \end{bmatrix} = \mathbf{P}_{B(m)} \begin{bmatrix} \psi_i(t) \\ \psi_{(i+1)}(t) \\ \vdots \\ \psi_{(i+m-1)}(t) \end{bmatrix} \tag{6.27}$$

Now, expressing each BPF in terms of two DUSFs as

$$\psi_i(t) = u_i(t) - u_{(i+1)}(t) \tag{6.28}$$

we can write equation (6.27) as

$$\left[\begin{array}{c} \int_0^t u_i(\tau)d\tau \\ \int_0^t u_{(i+1)}(\tau)d\tau \\ \vdots \\ \int_0^t u_{(i+m-1)}(\tau)d\tau \end{array}\right] \left[\begin{array}{c} \int_0^t u_{(i+1)}(\tau)d\tau \\ \int_0^t u_{(i+2)}(\tau)d\tau \\ \vdots \\ \int_0^t u_{(i+m)}(\tau)d\tau \end{array}\right] = \mathbf{P}_{B(m)} \left[\begin{array}{c} u_i(t) \\ u_{(i+1)}(t) \\ \vdots \\ u_{(i+m-1)}(t) \end{array}\right] - \mathbf{P}_{B(m)} \left[\begin{array}{c} u_{(i+1)}(t) \\ u_{(i+2)}(t) \\ \vdots \\ u_{(i+m)}(t) \end{array}\right] \tag{6.29}$$

For $i = 0, 1, 2, \ldots, (m-1)$, we can write a series of equations from equation (6.29) as follows:

$$\left[\begin{array}{c} \int_0^t u_0(\tau)d\tau \\ \int_0^t u_1(\tau)d\tau \\ \vdots \\ \int_0^t u_{(m-1)}(\tau)d\tau \end{array}\right] \left[\begin{array}{c} \int_0^t u_1(\tau)d\tau \\ \int_0^t u_2(\tau)d\tau \\ \vdots \\ \int_0^t u_m(\tau)d\tau \end{array}\right] = \mathbf{P}_{B(m)} \left[\begin{array}{c} u_0(t) \\ u_1(t) \\ \vdots \\ u_{(m-1)}(t) \end{array}\right] - \mathbf{P}_{B(m)} \left[\begin{array}{c} u_1(t) \\ u_2(t) \\ \vdots \\ u_m(t) \end{array}\right] \tag{6.29.1}$$

$$\left[\begin{array}{c} \int_0^t u_1(\tau)d\tau \\ \int_0^t u_2(\tau)d\tau \\ \vdots \\ \int_0^t u_m(\tau)d\tau \end{array}\right] \left[\begin{array}{c} \int_0^t u_2(\tau)d\tau \\ \int_0^t u_3(\tau)d\tau \\ \vdots \\ \int_0^t u_{(m+1)}(\tau)d\tau \end{array}\right] = \mathbf{P}_{B(m)} \left[\begin{array}{c} u_1(t) \\ u_2(t) \\ \vdots \\ u_m(t) \end{array}\right] - \mathbf{P}_{B(m)} \left[\begin{array}{c} u_2(t) \\ u_3(t) \\ \vdots \\ u_{(m+1)}(t) \end{array}\right] \tag{6.29.2}$$

$$
\begin{bmatrix} \vdots \\ \int_0^t u_i(\tau)d\tau \\ \int_0^t u_{(i+1)}(\tau)d\tau \\ \vdots \\ \int_0^t u_{(i+m-1)}(\tau)d\tau \end{bmatrix} - \begin{bmatrix} \vdots \\ \int_0^t u_{(i+1)}(\tau)d\tau \\ \int_0^t u_{i+2}(\tau)d\tau \\ \vdots \\ \int_0^t u_{(i+m)}(\tau)d\tau \end{bmatrix} = \mathbf{P}_{B(m)} \begin{bmatrix} \vdots \\ u_i(t) \\ u_{(i+1)}(t) \\ \vdots \\ u_{(i+m-1)}(t) \end{bmatrix} - \mathbf{P}_{B(m)} \begin{bmatrix} \vdots \\ u_{(i+1)}(t) \\ u_{(i+2)}(t) \\ \vdots \\ u_{(i+m)}(t) \end{bmatrix}
$$

$$(6.29.(i+1))$$

$$
\begin{bmatrix} \vdots \\ \int_0^t u_{(m-1)}(\tau)d\tau \\ \int_0^t u_m(\tau)d\tau \\ \vdots \\ \int_0^t u_{(2m-2)}(\tau)d\tau \end{bmatrix} - \begin{bmatrix} \vdots \\ \int_0^t u_m(\tau)d\tau \\ \int_0^t u_{(m+1)}(\tau)d\tau \\ \vdots \\ \int_0^t u_{(2m-1)}(\tau)d\tau \end{bmatrix} = \mathbf{P}_{B(m)} \begin{bmatrix} \vdots \\ u_{(m-1)}(t) \\ u_m(t) \\ \vdots \\ u_{(2m-2)}(t) \end{bmatrix} - \mathbf{P}_{B(m)} \begin{bmatrix} \vdots \\ u_m(t) \\ u_{(m+1)}(t) \\ \vdots \\ u_{(2m-1)}(t) \end{bmatrix}
$$

$$(6.29.m)$$

Hence, $i \le (m-1)$.
Adding all the above m equations of (6.29), we have

$$
\begin{bmatrix} \int_0^t u_0(\tau)d\tau \\ \int_0^t u_1(\tau)d\tau \\ \vdots \\ \int_0^t u_{(m-1)}(\tau)d\tau \end{bmatrix} - \begin{bmatrix} \int_0^t u_m(\tau)d\tau \\ \int_0^t u_{(m+1)}(\tau)d\tau \\ \vdots \\ \int_0^t u_{(2m-1)}(\tau)d\tau \end{bmatrix} = \mathbf{P}_{B(m)} \begin{bmatrix} u_0(t) \\ u_1(t) \\ \vdots \\ u_{(m-1)}(t) \end{bmatrix} - \mathbf{P}_{B(m)} \begin{bmatrix} u_m(t) \\ u_{(m+1)}(t) \\ \vdots \\ u_{(2m-1)}(t) \end{bmatrix}
$$

$$(6.30)$$

In equation (6.30), the second term on the right- or left-hand side contains a set of DUSFs starting from $t = mh$. Since our domain of interest is $0 < t \le mh$, these components of the DUSF, starting from $u_{(m)}(t)$, i.e., $u_{(m)}(t)$, $u_{(m+1)}(t)$,..., $u_{(2m-1)}(t)$ should be ignored.

Thus, we have

$$\int_0^t \mathbf{U}_{(m)}(\tau)d\tau = \mathbf{P}_{B(m)}\mathbf{U}_{(m)}(\tau) \tag{6.31}$$

Comparing equations (6.11) and (6.31), we have

$$\mathbf{P}_{D(m)} = \mathbf{P}_{B(m)} \qquad\qquad \square \tag{}$$

6.2.2.3 Operational Matrix for Integration and Related Transformation Matrices

Proof of Theorem 6.2.2.2.1 is dependent upon the truncation of the DUSFs at $T = mh$. However, in this section, we present a more direct proof involving an invertible transformation matrix $\mathbf{N}_{(m)}$ described in equation (6.17)

Substituting equation (6.17) in equation (6.11), we write

$$\int_0^t \mathbf{\Psi}_{(m)}(\tau)d\tau = \mathbf{N}_{(m)}^{-1}\mathbf{P}_{D(m)}\mathbf{N}_{(m)}\mathbf{\Psi}_{(m)}(t) \tag{6.32}$$

Comparing equations (6.25) and (6.32),

$$\mathbf{P}_{B(m)} = \mathbf{N}_{(m)}^{-1}\mathbf{P}_{D(m)}\mathbf{N}_{(m)} \tag{6.33}$$

For simplicity, we drop the subscript (m) and write

$$\mathbf{P}_B = \mathbf{N}^{-1}\mathbf{P}_D\mathbf{N} \tag{6.34}$$

Equation (6.34) is really interesting because for equation (6.34) to be true, it appears that $\mathbf{P}_D$ should remain invariant under the transformation. This is possible only if either of the matrix products on the right-hand side of equation (6.34) is commutative. However, it is well-known that, in general, a matrix product is not commutative [19].

Lemma 6.2.2.3.1

The product of two non-null, non-unity, and non-circulant square matrices $\mathbf{A}, \mathbf{B} \in \mathbb{R}^{m \times m}$, where $\mathbf{A}^{-1} \neq \mathbf{B}$ is commutative *if and only if* both the matrices can be expressed in the form of finite matrix polynomials involving any invertible square matrix $\mathbf{S} \in \mathbb{R}^{m \times m}$ and the unit matrix $\mathbf{I} \in \mathbb{R}^{m \times m}$ with constant coefficients.

Proof

Let

$$\mathbf{A} = \sum_{i=-M}^{+N} a_i \mathbf{S}^i \tag{6.35}$$

and

$$\mathbf{B} = \sum_{j=-P}^{+Q} b_j \mathbf{S}^j \tag{6.36}$$

where M, N, P, and Q are finite integers and a_i and b_j are constant real coefficients having positive or negative values.
Then the product $\mathbf{AB}$ is given by

$$\mathbf{AB} = \left[\sum_{i=-M}^{+N} a_i \mathbf{S}^i\right]\left[\sum_{j=-P}^{+Q} b_j \mathbf{S}^j\right]$$

$$\mathbf{AB} = \sum_{i=-M}^{+N}\sum_{j=-P}^{+Q} a_i b_j \mathbf{S}^{i+j} \tag{6.37}$$

All the product terms on the right-hand side of equation (6.37) involve a matrix product only of the following type:

$$\mathbf{S}^m.\mathbf{S}^n, \text{ where} - M \le m \le N \text{ and} - P \le n \le Q.$$

Since this matrix product is always commutative, it follows that

$$\mathbf{A}.\mathbf{B} = \mathbf{B}.\mathbf{A} \qquad\qquad \square$$

Now the integral operational matrix $\mathbf{P}_\mathrm{D}$ can be expressed in the following form [18]:

$$\mathbf{P}_\mathrm{D} = h\left[\frac{1}{2}\mathbf{Q}^0 + \mathbf{Q}^1 + \mathbf{Q}^2 + \mathbf{Q}^3 + \cdots + \mathbf{Q}^{(m-1)}\right] \tag{6.38}$$

where

$$\mathbf{Q}_{(m)} = \left[\begin{array}{c|c} \mathbf{0}_{(m-1)\times 1} & \mathbf{I}_{(m-1)} \\ \hline 0 & \mathbf{0}_{1\times(m-1)} \end{array}\right]$$

is the delay matrix and it has the properties [4]

$$Q^i_{(m)} = \left[\begin{array}{c|c} 0_{(m-i) \times i} & I_{(m-i)} \\ \hline 0_{(i)} & 0_{i \times (m-i)} \end{array} \right], \quad \text{for} \quad i < m$$

and $Q^i_{(m)} = 0_{(m)}$ for $i \geq m$.

Using the above property,

$$P_D = \frac{h}{2}(I + Q)(I - Q)^{-1} \tag{6.39}$$

From (6.16) and (6.17), N is given by

$$N = Q^0 + Q^1 + Q^2 + \cdots + Q^{(m-1)} \tag{6.40}$$

or,

$$N = (I - Q)^{-1} \tag{6.41}$$

which prompts us to find N^{-1} as

$$N^{-1} = (I - Q) \tag{6.42}$$

It is to be noted that Q is not invertible, but since equations (6.38) and (6.40) do not involve any negative powers of Q, use of Lemma 6.2.2.3.1 is possible. Hence, from equations (6.38), (6.40), and (6.42), by virtue of Lemma 6.2.2.3.1, we see that it is only natural that the upper triangular Toeplitz matrices P_D, N, and N^{-1} mutually commute. Hence, in equation (6.33), P_D remains invariant under the transformation. Thus,

$$P_B = N^{-1}P_D \, N = N^{-1} \, NP_D = P_D$$

or,

$$P_B = P_D \tag{6.43}$$

As an alternative proof, we can substitute the values of P_D, N, and N^{-1} from equations (6.39), (6.41), and (6.42) in equation (6.34) to obtain

$$\mathbf{P}_B = (\mathbf{I} - \mathbf{Q})\frac{h}{2}(\mathbf{I} - \mathbf{Q})^{-1}(\mathbf{I} + \mathbf{Q})(\mathbf{I} - \mathbf{Q})^{-1} = \mathbf{P}_D$$

Thus, the operational matrix for integration in the DUSF domain follows directly from the operational matrix for integration in the BPF domain. Since $\mathbf{P}_D$ (or $\mathbf{P}_B$ for that matter) remains invariant under the transformation described by equation (6.34), the transformation is transparent leading to apparent astonishment in the first place.

Finally, if we are given the elements of the BPF set as jigsaw puzzle pieces, we can form the DUSF set by putting those pieces according to relation (6.16). This also proves the elementary nature of the BPF set. Associating proper weightage to an individual BPF component, we can even form all the members of PCBF family by putting weighted BPF components piece by piece.

6.3 Stretch Matrix in DUSF Domain [11]

For the purpose of functions with stretched argument we form a stretch matrix for DUSFs. To do this for $f(t/\lambda)$, we write

$$f(t/\lambda) \approx \mathbf{G}^T \mathbf{U}(t/\lambda) \tag{6.44}$$

and

$$\mathbf{U}(t/\lambda) \approx \mathbf{S}_D \mathbf{U}(t) \tag{6.45}$$

where $\mathbf{G}$ and $\mathbf{U}$ are defined by equations (6.6) and (6.7), and $\mathbf{S}_D$ is the stretch matrix in DUSF domain.

Then

$$f(t/\lambda) \approx \mathbf{G}^T \mathbf{S}_D \mathbf{U}(t) \tag{6.46}$$

Now, the development of $\mathbf{S}_D$ is presented in the following.

The development is based on the approximation of $\exp(-\alpha hs)$ as

$$\exp(-\alpha hs) \approx (1 - \alpha) + \alpha \exp(-hs), \quad \text{where,} \quad 0 \le \alpha \le 1 \tag{6.47}$$

Since, $\mathcal{L}\{u_i(t/\lambda)\} = \lambda U_i(\lambda s)$, from equation (6.3) we get

$$\lambda U_i(\lambda s) = \frac{1}{s}\exp(-i\lambda hs) \tag{6.48}$$

Consider

$$n_i = [i\lambda] \tag{6.49}$$

and

$$\alpha_i = i\lambda - [i\lambda] \tag{6.50}$$

where $[r]$ denotes the greatest integer in r.

By using equation (6.47), equation (6.48) can be written as

$$\mathcal{L}\{u_i(t/\lambda)\} = \frac{1}{s}\left[(1-\alpha_i) + \alpha_i \exp(-hs)\right]\exp(-n_i hs) \tag{6.51}$$

In equation (6.51), taking the inverse Laplace transform of the RHS, we write

$$u_i(t/\lambda) = (1-\alpha_i)u_{ni}(t) + \alpha_i u_{ni+1}(t) \tag{6.52}$$

Using equation (6.52) in equation (6.45), we have the stretch matrix **S** of the form

$$\mathbf{S}_D = \begin{bmatrix} \mathbf{s}_0^T \\ \mathbf{s}_1^T \\ \vdots \\ \mathbf{s}_i^T \\ \vdots \\ \mathbf{s}_{(m-1)}^T \end{bmatrix} \tag{6.53}$$

where $\mathbf{s}_i^T$ is the $(i+1)$ th row of $\mathbf{S}_D$ given by

$$\mathbf{s}_i^T = \begin{cases} \begin{bmatrix} 0 & \cdots & 0 & (1-\alpha_i) & \alpha_i & 0 & \cdots & 0 \end{bmatrix}, & n_i+1 \leq m \\ \qquad\qquad\qquad \uparrow \\ \qquad\quad (n_i+1)\text{th element} \\ \begin{bmatrix} 0 & 0 & \cdots & 0 \end{bmatrix}, & n_i+1 > m \end{cases} \tag{6.54}$$

For example, the stretch matrix $\mathbf{S}_D$ for $\lambda = 0.5$ and $m = 8$ is [11]

$$
\mathbf{S}_{D,0.5} =
\begin{bmatrix}
1 & 0 & 0 & 0 & 0 & 0 & 0 & 0 \\
0.5 & 0.5 & 0 & 0 & 0 & 0 & 0 & 0 \\
0 & 1 & 0 & 0 & 0 & 0 & 0 & 0 \\
0 & 0.5 & 0.5 & 0 & 0 & 0 & 0 & 0 \\
0 & 0 & 1 & 0 & 0 & 0 & 0 & 0 \\
0 & 0 & 0.5 & 0.5 & 0 & 0 & 0 & 0 \\
0 & 0 & 0 & 1 & 0 & 0 & 0 & 0 \\
0 & 0 & 0 & 0.5 & 0.5 & 0 & 0 & 0
\end{bmatrix}
\tag{6.55}
$$

and for $\lambda = 1.25$ and $m = 8$, the stretch matrix $\mathbf{S}_D$ is

$$
\mathbf{S}_{D1.25} =
\begin{bmatrix}
1 & 0 & 0 & 0 & 0 & 0 & 0 & 0 \\
0 & 0.75 & 0.25 & 0 & 0 & 0 & 0 & 0 \\
0 & 0 & 0.5 & 0.5 & 0 & 0 & 0 & 0 \\
0 & 0 & 0 & 0.25 & 0.75 & 0 & 0 & 0 \\
0 & 0 & 0 & 0 & 0 & 1 & 0 & 0 \\
0 & 0 & 0 & 0 & 0 & 0 & 0.75 & 0.25 \\
0 & 0 & 0 & 0 & 0 & 0 & 0 & 0.5 \\
0 & 0 & 0 & 0 & 0 & 0 & 0 & 0
\end{bmatrix}
\tag{6.56}
$$

6.3.1 Stretch Matrices in Walsh and BPF Domain

It is well known that

$$
\boldsymbol{\Phi}_{(m)}(t) = \mathbf{W}_{(m)}\boldsymbol{\Psi}_{(m)}(t)
\tag{6.57}
$$

where $\mathbf{W}_{(m)}$ is the Walsh matrix [4] of order m.

From equations (6.17) and (6.57), the relationship between the Walsh function and the DUSF may easily be found as

$$
\boldsymbol{\Phi}_{(m)}(t) = \mathbf{W}_{(m)}\mathbf{N}_{(m)}^{-1}\mathbf{U}_{(m)}(t) \triangleq \mathbf{M}_{(m)}\mathbf{U}_{(m)}(t)
\tag{6.58}
$$

Using this relation, the stretch matrix $\mathbf{S}_W$ for $\lambda = 1.25$ and $m = 8$ in Walsh domain may be formed as

$$
\mathbf{S}_{W,1.25} =
\begin{bmatrix}
1 & 0 & 0 & 0 & 0 & 0 & 0 & 0 \\
0.25 & 0.75 & 0.25 & -0.25 & 0.25 & -0.25 & 0.25 & -0.25 \\
0.25 & 0 & 0.25 & 0.5 & 0 & 0.25 & -0.5 & 0.25 \\
-0.25 & 0.5 & 0.25 & 0.5 & 0 & 0.25 & 0 & -0.25 \\
0.0625 & 0.1875 & 0.0625 & -0.0625 & -0.1875 & 0.4375 & 0.0625 & 0.4375 \\
-0.0625 & 0.3125 & -0.3125 & 0.3125 & 0.1875 & 0.0625 & 0.1875 & 0.3125 \\
-0.0625 & -0.1875 & 0.6875 & -0.1875 & 0.1875 & 0.0625 & 0.1875 & 0.3125 \\
0.0625 & -0.3125 & 0.0625 & 0.4375 & 0.3125 & -0.0625 & 0.5625 & -0.0625
\end{bmatrix}
$$

$$(6.59)$$

Similarly, using equation (6.57) and the stretch matrix $\mathbf{S}_B$ for $\lambda = 1.25$ and $m = 8$, in BPF domain is

$$
\mathbf{S}_{B,1.25} =
\begin{bmatrix}
1 & 1 & 1 & 1 & 1 & 1 & 1 & 1 \\
1 & 1 & 1 & 1 & 1 & -1 & -1 & -1 \\
1 & 1 & 0 & -1 & -1 & 1 & 1 & 0 \\
1 & 1 & 0 & -1 & -1 & -1 & -1 & 0 \\
1 & -0.5 & 0 & 0.5 & -1 & 1 & -0.5 & 0 \\
1 & -0.5 & 0 & 0.5 & -1 & -1 & 0.5 & 0 \\
1 & -0.5 & -1 & -0.5 & 1 & 1 & -0.5 & -1 \\
1 & -0.5 & -1 & -0.5 & 1 & -1 & 0.5 & 1
\end{bmatrix}
$$

$$(6.60)$$

6.3.2 Numerical Example

Example 6.2

Let us consider a function

$$
f(t) = \begin{cases} \sin(\pi t) & \text{for } 0 \le t < 1\,\text{s} \\ 0 & \text{for } 1 \le t \le 1.5\,\text{s} \end{cases}
$$

to illustrate the application of stretch matrix $\mathbf{S}_D$ in equation (6.56) to obtain the function with stretched argument (t/λ) in DUSF domain, taking $\lambda = 1.25$, $m = 12$, and $T = 1.5$ s.

The DUSF coefficients of the original function are

$$
f(t) \approx \big[0.19383918 \;\; 0.35816810 \;\; 0.27413000 \;\; 0.14835808
$$
$$
0.00000000 \;\; -0.14835808 \;\; -0.27412999 \;\; -0.35816810
$$
$$
-0.19383918 \;\; 0.00000000 \;\; 0.00000000 \;\; 0.00000000 \big] \mathbf{U}_{(12)}(t)
$$

$$(6.61)$$

To get the approximation of the original function in BPF domain, apply the transformation matrix **N** in equation (6.61) and obtain the BPF coefficients of the original function as

$$f(t) \approx \big[0.19383918 \ 0.55200728 \ 0.82613727 \ 0.97449536$$

$$0.97449536 \ 0.82613727 \ 0.55200728 \ 0.19383918$$

$$0.00000000 \ 0.00000000 \ 0.00000000 \ 0.00000000 \big] \mathbf{\Psi}_{(12)}(t)$$

(6.62)

Using the $\mathbf{S}_D$ matrix of equation (6.56) for $\lambda = 1.25$ in equation (6.61), we directly obtain the result of approximation of stretched function in DUSF domain as

$$f(t/\lambda) \approx \big[0.19383918 \ 0.26862607 \ 0.22660702 \ 0.17415452$$

$$0.11126856 \ 0.00000000 \ -0.11126856 \ -0.17415452$$

$$-0.22660702 \ -0.26862607 \ -0.19383918 \ 0.00000000 \big] \mathbf{U}_{(12)}(t)$$

(6.63)

Similar to equation (6.62), we obtain the approximated stretched function using similarity transformation matrix **N** as

$$f(t/\lambda) \approx \big[0.19383918 \ 0.46246525 \ 0.68907228 \ 0.86322680$$

$$0.97449536 \ 0.97449536 \ 0.86322680 \ 0.68907228$$

$$0.46246525 \ 0.19383918 \ 0.00000000 \ 0.00000000 \big] \mathbf{\Psi}_{(12)}(t)$$

(6.64)

Figure 6.4 shows the exact function $f(t)$ along with the exact curve for the stretched function $f(t/\lambda)$, and Figure 6.5 shows the equivalent BPF or DUSF approximation of the function $f(t)$ along with the coefficients for the stretched function using the stretch matrix for $\lambda = 1.25$.

6.4 Solution of a Functional Differential Equation Using DUSF [11,18]

Consider the following functional differential equation

$$\left. \begin{aligned} \dot{\mathbf{x}}(t) &= \mathbf{A}\mathbf{x}(t/\lambda) + \mathbf{B}\mathbf{x}(t) \\ \mathbf{x}(0) \text{ specified} \end{aligned} \right\}$$

(6.65)

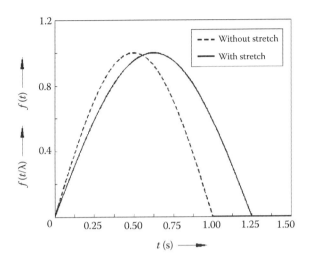

FIGURE 6.4

Exact curves for $f(t) = \sin(\pi t)$ and $f(t/\lambda) = \sin(\pi t/\lambda)$ of Example 6.2 for $\lambda = 1.25$ (vide Appendix B, Program no. 6.2).

where $\mathbf{x}$ is an n-vector and $\mathbf{A}$ and $\mathbf{B}$ are matrices of appropriate dimensions. First, $\mathbf{x}(t)$ and $\dot{\mathbf{x}}(t)$ are expanded into DUSF series as

$$\mathbf{x}(t) \approx \sum_{i=0}^{m-1} x_i u_i(t) \triangleq \mathbf{XU}(t) \tag{6.66}$$

$$\dot{\mathbf{x}}(t) \approx \sum_{i=0}^{m-1} d_i u_i(t) \triangleq \mathbf{DU}(t) \tag{6.67}$$

Then, from equation (6.45) we can write

$$\mathbf{x}(t/\lambda) = \mathbf{XS}_{\mathrm{D}}\mathbf{U}(t) \tag{6.68}$$

For the purpose of computation, the initial condition $\mathbf{x}(0)$ can be expressed in terms of a DUSF series as

$$\mathbf{x}(0) = \begin{bmatrix} \mathbf{x}(0) & \underbrace{0 \quad 0 \quad \cdots \quad 0}_{(m-1)\,\text{columns}} \end{bmatrix} \mathbf{U}(t) \tag{6.69}$$

Integration of the both sides of equation (6.67) from $t = 0$ to t and use of operational matrix for integration of DUSFs give

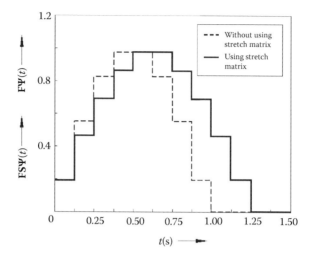

FIGURE 6.5
Equivalent block pulse function (BPF) or delayed unit step function (DUSF) approximation obtained (a) without using the stretch matrix and (b) using the stretch matrix for $\lambda = 1.25$, $m = 8$ and $T = 1$ s (vide Appendix B, Program no. 6.2).

$$\mathbf{x}(t) = \left| \mathbf{DP} + \left[\mathbf{x}(0) \quad \underbrace{0 \quad 0 \quad \cdots \quad 0}_{(m-1)\ \text{columns}} \right] \right| \mathbf{U}(t) \tag{6.70}$$

where equation (6.69) has been used.

Comparison of Equation (6.70) with (6.66) yields

$$\mathbf{X} = \mathbf{DP} + \left[\mathbf{x}(0) \quad \underbrace{0 \quad 0 \quad \cdots \quad 0}_{(m-1)\ \text{columns}} \right] \tag{6.71}$$

Substituting equations (6.67), (6.68), and (6.71) into equation (6.65), we have

$$\mathbf{DU}(t) = \mathbf{A} \left| \mathbf{DPS}_D + \left[\mathbf{x}(0) \quad \underbrace{0 \quad 0 \quad \cdots \quad 0}_{(m-1)\ \text{columns}} \right] \mathbf{S}_D \right| \mathbf{U}(t)$$
$$+ \mathbf{B} \left| \mathbf{DP} + \left[\mathbf{x}(0) \quad \underbrace{0 \quad 0 \quad \cdots \quad 0}_{(m-1)\ \text{columns}} \right] \right| \mathbf{U}(t) \tag{6.72}$$

or,

$$\mathbf{D} = \mathbf{ADPS}_D + \mathbf{BDP} + \mathbf{C} \tag{6.73}$$

where

$$C = A \begin{bmatrix} x(0) & \underbrace{0 \quad 0 \quad \cdots \quad 0}_{(m-1)\text{ columns}} \end{bmatrix} S_D + B \begin{bmatrix} x(0) & \underbrace{0 \quad 0 \quad \cdots \quad 0}_{(m-1)\text{ columns}} \end{bmatrix}$$

$$= \begin{bmatrix} c_1 & c_2 & \cdots & c_{(m-1)} \end{bmatrix} \tag{6.74}$$

equation (6.73) can be solved to give

$$d = \left[I - A \otimes (PS_D)^T - B \otimes P^T \right]^{-1} c \tag{6.75}$$

where

$$d = \begin{bmatrix} d_0^T & d_1^T & \cdots & d_{(m-1)}^T \end{bmatrix}^T \tag{6.76}$$

$$c = \begin{bmatrix} c_0^T & c_1^T & \cdots & c_{(m-1)}^T \end{bmatrix}^T \tag{6.77}$$

and the Kronecker product of two matrices A and B is defined as

$$A \otimes B = \begin{bmatrix} a_{11}B & a_{12}B & \cdots & a_{1n}B \\ a_{21}B & a_{22}B & \cdots & a_{2n}B \\ \vdots & \vdots & \ddots & \vdots \\ a_{m1}B & a_{m2}B & \cdots & a_{mn}B \end{bmatrix} \tag{6.78}$$

Once D is computed, the DUSF coefficients of $x(t)$ can be obtained from equation (6.71).

The solution of $x(t)$ obtained from equations (6.71) and (6.73) are piecewise constant.

In order to obtain the solution at discrete points $t = ih, i = 1, 2, \ldots,$ we may integrate equation (6.67) from $t = (j-1)h$ to $t = jh$ to give

$$x(jh) = x\left((j-1)h\right) + \sum_{i=0}^{m-1} d_i \int_{(j-1)h}^{jh} u_i(t)\,dt$$

$$= x\left((j-1)h\right) + h\sum_{i=0}^{j-1} d_i \tag{6.79}$$

6.4.1 Numerical Example

Example 6.3 [11,12]

Consider the example given by

$$\dot{x}(t) = -x\left(t/1.25\right) - x(t) \left. \right\}$$
$$x(0) = 1 \qquad\qquad\qquad\quad$$

(6.80)

By choosing $h = 0.1$ and $m = 10$ and expanding equation (6.80) into DUSF domain we obtain the coefficient vector of $\dot{x}(t)$

$$\mathbf{d} = \left[\mathbf{I} + \mathbf{S}_D^T \mathbf{P}^T + \mathbf{P}^T \right]^{-1} \mathbf{c}$$

(6.81)

$$\mathbf{c} = \mathbf{S}^T \left[-1 \quad \underbrace{0 \quad 0 \quad \cdots \quad 0}_{(m-1)\ \text{elements}} \right]^T + \left[-1 \quad \underbrace{0 \quad 0 \quad \cdots \quad 0}_{(m-1)\ \text{elements}} \right]^T$$

(6.82)

and since $\lambda = 1.25$, equation (6.56) is used for **S**. By using equations (6.66) and (6.71) we get the piecewise constant solution of $x(t)$. The discrete-time values of $x(t)$ can be obtained using equation (6.79). Table 6.2 shows the discrete-time solutions of equation (6.80) with $m = 10$ and $h = 0.1$. The results of Rao and Palanisamy [16] by Walsh series approach and those of Fox et al. [20] are also shown in Table 6.2 for comparison.

The computed result for $m = 8$ and $T = 1$ s is shown in Figure 6.6, and the results at discrete-time values are tabulated in Table 6.3 for better clarity.

In Table 6.2, we find that the results obtained using Walsh series method provide us the solution of equation (6.80) for $m = 10$ and $T = 1$ s. Using both DUSF and BPF methods for same m and T we arrive at the same results. But in Reference [11] Hwang claimed that the DUSF series method (using equation (6.79)) provides a better result compared to the Walsh series approach [6], for $m = 100$ and $h = 0.01$ s.

This apparently "improved" result is due to the increased value of m. While Rao and Palanisamy [16] considered $m = 10$, Hwang [11] considered $m = 100$ to compute his results. After computation, Hwang considered the time instants presented in the first column of Table 6.2 and fished out 11 values out of his 101 values to compare with the results obtained by Rao and Palanisamy. In this manner, Hwang could achieve the apparent improvement in his DUSF results presented in Table 6.2.

6.5 Conclusion

It has been established that the BPFs are superior to the DUSF proposed by Hwang [11]. This has been presented through qualitative comparison and quantitative analysis. The superiority of BPF is mainly due to its fundamental nature in comparison to any other PCBF.

It has also been proved that the operational matrices for integration in DUSF domain ($\mathbf{P_D}$) and BPF domain ($\mathbf{P_B}$) is connected by simple linear transformation involving an invertible Toeplitz matrix $\mathbf{N}$. The transformation is transparent because the integration operational matrices $\mathbf{P_D}$ or $\mathbf{P_B}$ is invariant under this transformation. The reason for such invariance has also been studied in detail.

Hwang [11] claimed superiority of DUSFs compared with Walsh functions in obtaining solution of functional differential equations using a stretch matrix in the DUSF domain. Here, it has been proved that the stretch matrices $\mathbf{S_W}$ in Walsh domain, $\mathbf{S_B}$ in BPF domain and $\mathbf{S_D}$ in DUSF domain are related to one another by similarity transformations [21]. It has also been proved that

TABLE 6.2

Solution of the Scaled System of Example 6.3, for $\lambda = 1.25$, $m = 10$, and $T = 1$ s (Vide Appendix B, Program No. 6.3)

	Fox et al. (1971)		Rao et al. (1982)	Hwang (1983)	
$t(s)$	Improved Finite Difference Method $x(t)$	Deferred Correction Method $x(t)$	Walsh Series Method $x(t)$ ($m = 10$)	DUSF Series Method $x(t)$ ($m = 100$)	Using Walsh or DUSF or BPF Method ($m = 10$)
0.0	1.000000	1.000000	1.00000000	1.00000000	
					0.90909091
0.1	0.816636	0.817054	0.81818182	0.81704360	
					0.74190178
0.2	0.663989	0.664695	0.66562173	0.66467664	
					0.60192948
0.3	0.537317	0.538210	0.53823722	0.53818757	
					0.48533164
0.4	0.432562	0.433565	0.43242606	0.43354010	
					0.38872331
0.5	0.346255	0.347307	0.34502056	0.34728046	
					0.31008038
0.6	0.275434	0.276486	0.27514019	0.27646027	
					0.24633632
0.7	0.217572	0.218592	0.21753245	0.21856697	
					0.19392575
0.8	0.172459	0.171487	0.17031904	0.17146348	
					0.15110224
0.9	0.132459	0.133360	0.13188545	0.13333699	
					0.11637063
1.0	0.101849	0.102673	0.10085581	0.10265233	

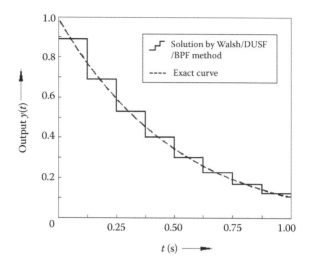

FIGURE 6.6
Solution of Example 6.3 (vide Table 6.3) for $\lambda = 1.25$, $m = 8$ and $T = 1$ s and its comparison with exact curve (vide Appendix B, Program no. 6.3).

TABLE 6.3

Solution of the Scaled System of Example 6.3 for $\lambda = 1.25$, $m = 8$, and $T = 1$ s (Vide Appendix B, Program No. 6.3)

t(s)	Rao et al. (1982) Walsh Series Method $x(t)(m = 8)$	Using Walsh or DUSF or BPF Method ($m = 8$)
0.000	1.00000000	
		0.88888889
0.125	0.77777778	
		0.68857590
0.250	0.59937402	
		0.52832551
0.375	0.45727700	
		0.40117031
0.500	0.34506362	
		0.30116751
0.625	0.25727139	
		0.22442205
0.750	0.19157270	
		0.16597381
0.875	0.14037492	
		0.12063534
1.000	0.10089576	

use of any of these three matrices leads to exactly the same result. This fact has been supported by an example. It may be noted that this eventuality is true for any analysis approach involving any of the function sets from the PCBF family which are related to one another by similarity transformation.

References

1. C. F. Chen and C. H. Hsiao, A state-space approach to Walsh series solution of linear systems, Int. J. Syst. Sci., vol. **6**, no. 9, pp. 833–858, 1975.
2. W. L. Chen and Y. P. Shih, Analysis and optimal control of time-varying linear systems via Walsh functions, Int. J. Control, vol. **27**, no. 6, pp. 917–932, 1978.
3. W. L. Chen and Y. P. Shih, Shift Walsh matrix and delay-differential equations, IEEE Trans. Autom. Control, vol. **23**, no. 6, pp. 1023–1028, 1978.
4. C. F. Chen, Y. T. Tsay, and T. T. Wu, Walsh operational matrices for fractional calculus and their application to distributed systems, J. Franklin Inst., vol. **303**, no. 3, pp. 267–284, 1977.
5. N. Gopalsami and B. L. Deekshatulu, Comments on "Design of piecewise constant gains for optimal control via Walsh functions", IEEE Trans. Autom. Control, vol. **21**, no. 1, pp. 634–636, 1976.
6. G. P. Rao, *Piecewise constant orthogonal functions and their application to system and control*, LNC1S, vol. **55**, Springer-Verlag, Berlin, 1983.
7. Z. H. Jiang and W. Schaufelberger, *Block pulse functions and their applications in control systems*, LNCIS, vol. **179**, Springer-Verlag, Berlin, 1992.
8. K. Maleknejad and Y. Mahmoudi, Numerical solution of linear Fredholm integral equation by using hybrid Taylor and block-pulse functions, Appl. Math. and Comp., vol. **149**, no. 3, pp. 799–806, 2004.
9. X. T. Wang, Numerical solutions of optimal control for time delay systems by hybrid of block-pulse functions and Legendre polynomials, Appl. Math. and Comp., vol. **184**, no. 2, pp. 849–856, 2007.
10. E. Babolian and Z. Masouri, Direct method to solve Volterra integral equation of the first kind using operational matrix with block-pulse functions, J. Comp. and Appl. Math., vol. **220**, no. 1, pp. 51–57, 2008.
11. C. Hwang, Solution of a functional differential equation via delayed unit step functions, Int. J. Syst. Sci., vol. **14**, no. 9, pp. 1065–1073, 1983.
12. J. Ockendon and A. B. Tayler, The dynamics of a current collection system for an electric locomotive, Proc. Royal. Soc. London A: Math. Phys. and Engg. Sci., vol. **322**, no. 1551, pp. 447–468, 1971.
13. A. D. Randolph, Effect of crystal breakage on crystal size distribution in mixed suspension crystallizer, Inds. Engg. Chem. Fundls., vol. **8**, no. 1, pp. 58–63, 1969.
14. C. Hwang and Y. P. Shih, Laguerre series solution of a functional differential equation, Int. J. Syst. Sci., vol. **13**, no. 7, pp. 783–788, 1982.
15. W. L. Chen, Block pulse series analysis of scaled systems, Int. J. Syst. Sci., vol. **12**, no. 7, pp. 885–891, 1981.

16. G. P. Rao and K. R. Palanisamy, Walsh stretch matrices for functional differential equations, IEEE Trans. Autom. Control, vol. **27**, no. 1, pp. 272–276, 1982.
17. L. S. Shieh, W. P. Schneider, and D. R. Williams, A chain of factored matrices for Routh array inversion and continued fraction inversion, Int. J. Control, vol. **13**, no. 4, pp. 691–703, 1971.
18. Anish Deb, Gautam Sarkar, and Sunit K. Sen, Block pulse functions, the most fundamental of all piecewise constant basis functions, Int. J. Syst. Sci., vol. **25**, no. 2, pp. 351–363, 1994.
19. K. B. Dutta, *Matrix and Linear Algebra*, Prentice-Hall of India, New Delhi, 1991.
20. L. Fox, D. F. Mayers, J. R. Ockendon, and A. B. Tayler, On a functional differential equation, IMA J. Appl. Math., vol. **8**, no. 3, pp. 271–307, 1971.
21. A. Deb, S. Roychoudhury, and G. Sarkar, *Analysis and identification of time-invariant systems, time-varying systems and multi-delay systems using orthogonal hybrid functions: theory and algorithms with MATLAB®*, vol. **46**, Springer, Switzerland, 2016.

Study Problems

6.1 Derive the delayed unit step function (DUSF) coefficients for the function $f(t) = \sin(\pi t)$, $0 \le t < 1$s, with $m = 10$. Also show that the results in DUSF domain are effectively same as that obtained using block pulse function (BPF) domain approximation.

6.2 Compare qualitatively the basic properties of the two piecewise constant basis function sets: BPF and DUSF.

6.3 Derive the operational matrix for integration P_D in DUSF domain. Explain why the integration matrix is identical with P_B, the operational matrix for integration in BPF domain ?

6.4 Expand the function $f(t) = 1 - \exp(-t)$ in BPF domain with $m = 8$ and $T = 1$ s and plot the result along with the exact curve of $f(t)$.

Also, obtain the approximation of the scaled function $f(t/\lambda)$ using the stretch matrix S_B in BPF domain, where $\lambda = 0.5$. Draw the BPF expanded forms of $f(t)$ and $f(t/\lambda)$ in a same plot.

6.5 Expand the function $f(t) = \cos(\pi t)$ in BPF domain for $m = 10$ and $T = 1$ s and then integrate it using the operational matrix for integration P_B.

Also, using the BPF expanded form of $f(t)$, the operational matrix for integration P_B and the stretch matrix S_B in BPF domain, integrate the function $f(t/\lambda)$ when $\lambda = 0.5$. Draw the BPF expanded forms of both the integrations of $f(t)$ and $f(t/\lambda)$ in the same plot.

6.6 Consider the equation $\dot{x}(t) = -x(t/\lambda)$, $x(0) = 1$. For $m = 16$, $T = 2$ s, and $\lambda = 2$, determine the solution in DUSF domain. Convert the DUSF result into BPF domain as well as Walsh domain. Represent the results pictorially along the time scale. Does the pictorial representations differ in these three domains?

6.7 Consider the equation $\dot{x}(t) = -x(0.99t) - 0.95x(t)$, $x(0) = 1$. For $m = 8$ and $T = 1$ s determine the solution in DUSF domain. Convert the DUSF result into BPF domain as well as Walsh domain. Represent the results pictorially along the time scale. Do the pictorial representations differ in these three domains?

7

Sample-and-Hold Functions (SHFs) for System Analysis

This chapter presents a suitable set of orthogonal functions for the analysis of control systems with sample-and-hold (S/H). The quest for such a set starts with the applicability of the well-known block pulse function (BPF) [1] set and uses an operational technique by defining a block pulse operational transfer function (**BPOTF**) [2] to analyze a few control systems, as discussed in Chapter 4. The results obtained are found to be fairly accurate. But this method failed to distinguish between an input sampled system and an error sampled system.

To remove these limitations, another improved approach was followed using an S/H matrix **S**, but it also failed to come up with accurate results. Further, the method needed a large number of component BPFs leading to a much larger amount of storage as well as computational time.

To search for a more efficient technique, the set of piecewise constant orthogonal functions, termed sample-and-hold functions (SHFs), is used. The analysis technique, based upon a similar operational technique used for BPFs in Chapter 4, has been adopted for, SHF domain investigation. Surprisingly, the results obtained have the same accuracy as the conventional z-transform analysis.

Here, the input signal is expressed as a linear combination of SHFs; the plant having a Laplace transfer function $G(s)$ is represented by an equivalent sample-and-hold operational transfer function (**SHOTF**), and the output in the SHF domain is obtained by means of simple matrix multiplication. For higher order systems, the one-shot operational technique, introduced in Chapter 3, has been utilized. The presented technique is able to do away with the laborious algebraic manipulations associated with the z-transform technique without sacrificing accuracy. Also, the accuracy does not depend upon m and the presented method does not need any kind of inverse transformation.

A few linear S/H SISO control systems, open loop as well as closed loop, are analyzed as illustrative examples. Finally, an error analysis has been carried out to estimate the upper bound of the mean integral square error (MISE) of the SHF approximation of a function $f(t)$ of Lebesgue measure and the same has been compared with that of BPF-based approximation.

7.1 Brief Review of Sample-and-Hold Functions (SHF) [3]

Any square integrable function $f(t)$ may be represented by a SHF set in the semi-open interval $[0, T)$ by considering the $(i+1)$th component of the expansion to be

$$f_i(t) \approx f(ih) = f_i s_i(t) \text{ where } ih \le t \le (i+1)h$$

and $i = 0, 1, 2, \ldots , (m{-}1)$
   h is the sampling period,
   f_i is the $(i + 1)$th coefficient of the $(i + 1)$th SHF component $s_i(t)$ which is given by

$$s_i(t) = \begin{cases} 1 & \text{for} \quad ih \le t < (i+1)h \\ 0 & \text{elsewhere} \end{cases}$$

   As mentioned in Chapter 1, the basic functions of the SHF set are look-alikes of the members of the BPF set shown in Figure 1.6. Only the method of computation of the coefficients differs in the two cases. That is, the expansion coefficients in SHF domain do not depend upon the traditional integration formula (1.14). Rather the coefficients are determined by the following relation:

$$f_i = \int_0^T f(t)\delta(t - ih)dt \quad ih < T$$

where $\delta(t)$ is the well-known Dirac delta function.

7.2 Analysis of Control Systems with Sample-and-Hold Using the Operational Transfer Function Approach [3]

Consider the systems shown in Figure 7.1(a)–7.1(c).
   Using the conventional z-transform approach, the exact solution with step input for the system shown in Figure 7.1(a) is given by

$$C1(z) = \frac{0.63212056z}{z^2 - 1.36787944z + 0.36787944} \tag{7.1}$$

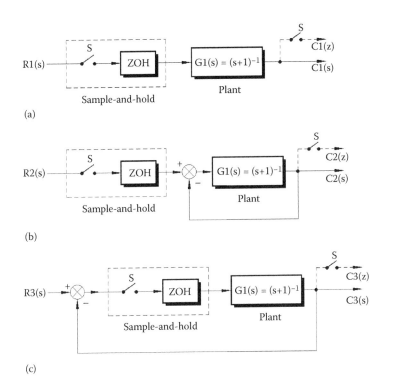

FIGURE 7.1
(a) An open loop control system with sample-and-hold (S/H) having first order transfer function. (b) A simple feedback control system with a S/H device in the input, (c) Error sampled feedback control system having a S/H device.

For the system shown in Figure 7.1(b), the exact output is given by

$$C2\left(z\right) = \frac{0.43233236z}{z^2 - 1.13533528z + 0.13533528} \tag{7.2}$$

and for the system of Figure 7.1(c), the output is

$$C3\left(z\right) = \frac{0.63212056z}{z^2 - 0.73575888z - 0.26424112} \tag{7.3}$$

If we assume that a single block pulse *approximately* represents the S/H operation, the results for the system of Figure 7.1(a) using the **BPOTF** approach (as obtained from equation (4.10) for Example 4.1 of Chapter 4) and $C1(z)$ of equation (7.1) may be compared, because in effect, they analyze the same system. It is apparent from Table 7.1 that these do not match satisfactorily. This is graphically compared in Figure 7.2. Also, when we try to compare the results of the system of Figure 7.1(b) using the **BPOTF** approach

TABLE 7.1

Comparison of the Exact Solutions of the Sampled Data Systems of Figure 7.1 with Corresponding Results of BPF Domain Analysis ($m = 10$, $T = 10$ s, $h = 1$ s)

t (sec)	C1(z)	C1	C2(z)	C2
0				
	0.00000000	0.33333333	0.00000000	0.25000000
1				
	0.63212056	0.77777778	0.43233236	0.50000000
2				
	0.86466472	0.92592593	0.49084218	0.50000000
3				
	0.95021293	0.97530864	0.49876062	0.50000000
4				
	0.98168436	0.99176955	0.49983227	0.50000000
5				
	0.99326205	0.99725652	0.49997730	0.50000000
6				
	0.99752125	0.99908551	0.49999693	0.50000000
7				
	0.99908812	0.99969517	0.49999958	0.50000000
8				
	0.99966454	0.99989839	0.49999994	0.50000000
9				
	0.99987659	0.99996613	0.49999999	0.50000000
10				

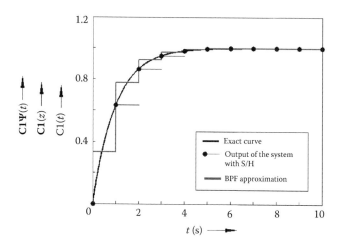

FIGURE 7.2

Graphical comparison of the exact solution of the sampled data system of Figure 7.1(a) with results obtained using **BPOTF** approach for $m = 10$, $T = 10$ s and $h = 1$ s.

(obtained from equation (4.14) for Example 4.2 of Chapter 4) and $C2(z)$ of equation (7.2) in Table 7.1, we find similar unsatisfactory matching of the two sets of results. Further, it is interesting to note that the **BPOTF** approach *cannot distinguish* between the systems of Figure 7.1(b) and 7.1(c). Hence, while the results $C2(z)$ and $C3(z)$ are quite different, the **BPOTF** approach is only able to come up with the output **C2**, obtained in equation (4.14). Table 7.1 compares the outputs obtained from equations (4.10) and (4.14) with those obtained from equations (7.1) and (7.2), respectively.

The apparent failure of the **BPOTF** approach to distinguish between the systems of Figure 7.1(b) and 7.1(c) is due to the fact that in BPF analysis, each and every signal is approximated by its BPF equivalent. Hence, the location of the S/H device cannot be identified by the presented BPF domain approach.

7.2.1 Sample-and-Hold Matrix for SHF-Based Analysis

At this point, it is felt necessary to introduce an S/H matrix [4] which should be able to give results with less deviation from the exact solution. Also, a more important point, it should be able to distinguish between the systems of Figures 7.1(b) and 7.1(c).

As derived by Chen and Wu [4], we consider an m-member BPF set over an interval $[0, T)$. The sampling period $T_s = qh$ and n sampling periods are considered in the entire interval. Then, $T = nT_s = nqh = mh$.

The S/H matrix is given by

$$
\mathbf{S} = \begin{bmatrix}
\underbrace{1\ 1\ \cdots\ 1\ 1}_{q} & 0\ 0\ \cdots\ 0\ \cdots & 0\ 0\ \cdots\ 0 & \cdots & 0 \\
0\ 0\ \cdots\ 0\ 0 & \cdots & \cdots\ 0 & \cdots & 0 \\
\cdots & \cdots & \cdots\ 0 & \cdots & 0 \\
0\ 0\ \cdots\ 0\ 0 & \cdots & \cdots\ 0 & \cdots & 0 \\
\underbrace{0\ 0\ \cdots\ 0\ 0}_{q} & \underbrace{1\ 1\ \cdots\ 1\ 1}_{q}\ 0\ 0\ \cdots\ 0 & \cdots & 0 \\
0\ 0\ \cdots\ 0\ 0 & \cdots & \cdots\ 0 & \cdots & 0 \\
\cdots & \cdots & \cdots\ 0 & \cdots & 0 \\
0\ 0\ \cdots\ 0\ 0 & \cdots & \cdots\ 0 & \cdots & 0 \\
\underbrace{0\ 0\ \cdots\ 0\ \cdots\ 0\ 0}_{2q}\ \underbrace{1\ 1\ \cdots\ 1\ 1}_{q}\ 0\ 0\ \cdots\ 0 & \cdots & 0 \\
\cdots\ \cdots & \cdots\ \cdots & \cdots\ \cdots & \cdots \\
\cdots\ \cdots & \cdots\ \cdots & \cdots\ \cdots & \cdots \\
\underbrace{0\ 0\ \cdots\ 0\ 0 \qquad 0\ 0\ \cdots\ 0\ 0}_{(n-1)q} & \cdots & \cdots\ \underbrace{1\ 1\ \cdots\ 1\ 1}_{q}
\end{bmatrix}_{(m \times m)}
$$

$$(7.4)$$

Let us expand the function $f(t)$ in BPF domain

$$f(t) \approx \mathbf{C}^{\mathrm{T}} \mathbf{\Psi}_{(m)}(t)$$ (7.5)

where $\mathbf{C} \triangleq \begin{bmatrix} c_0 & c_1 & \cdots & c_{m-1} \end{bmatrix}^{\mathrm{T}}$

When $f(t)$ is modified by an S/H device, let us call it $f_s(t)$. Then using equation (7.4),

$$f_s(t) \approx \mathbf{C}^{\mathrm{T}} \mathbf{S} \mathbf{\Psi}_{(m)}(t) = \left[\underbrace{\begin{matrix} c_0 & c_0 & \cdots & c_0 \end{matrix}}_{q \text{ terms}} \underbrace{\begin{matrix} c_q & c_q & \cdots & c_q \end{matrix}}_{q \text{ terms}} \cdots \right.$$

$$\left. \underbrace{\begin{matrix} c_{(n-1)q} & c_{(n-1)q} & \cdots & c_{(n-1)q} \end{matrix}}_{q \text{ terms}} \right] \mathbf{\Psi}_{(m)}(t)$$

For the operational transfer function analysis, in Figure 7.1(a), the plant transfer function $G1(s)$ is equivalently represented by **BPOTF1** [5] and the S/H device is represented by the matrix **S**. Hence the output **C1** is

$$\mathbf{C1}^{\mathrm{T}} = \mathbf{R1}^{\mathrm{T}} \begin{bmatrix} \mathbf{S} \ \mathbf{BPOTF1} \end{bmatrix}$$ (7.6)

Also, for the system of Figure 7.1(b), the output is

$$\mathbf{C2}^{\mathrm{T}} = \mathbf{R2}^{\mathrm{T}} \begin{bmatrix} \mathbf{S} \ \mathbf{BPOTF2} \end{bmatrix}$$ (7.7)

And for the system of Figure 7.1(c), the output will be

$$\mathbf{C3}^{\mathrm{T}} = \mathbf{R3}^{\mathrm{T}} \begin{bmatrix} \mathbf{S} \ \mathbf{BPOTF1} \end{bmatrix} \begin{bmatrix} \mathbf{I} + \mathbf{S} \ \mathbf{BPOTF1} \end{bmatrix}^{-1}$$ (7.8)

It is noted that, contrary to the ordinary **BPOTF** approach, introduction of the S/H matrix **S** has been able to distinguish between the systems of Figure 7.1(b) and 7.1(c), because equations (7.7) and (7.8) are not identical. Also, the **S** matrix, in effect, modifies the **BPOTF** like a cascaded plant producing an equivalent operational transfer function. If $\mathbf{S} = \mathbf{I}$, then we see that, as expected, equation (7.8) reverts back to equation (7.7).

Table 7.2 tabulates the outputs obtained from equations (7.6), (7.7), and (7.8) for a step input and compared them with corresponding exact solutions obtained from equations (7.1), (7.2), and (7.3).

It is observed that the results obtained from the S/H **BPOTF** analysis are not reasonably close to the exact solutions. The reason for this deviation may be found in equation (1.14), from which it is apparent that a block pulse function $\psi_i(t)$ can never truly represent an S/H device. If proper modifications

TABLE 7.2

Comparison of the Exact Solutions of the Sample-Data Systems of Figure 7.1 with Block Pulse Function (BPF) Domain Analysis Using the Sample-and-Hold (S/H) Matrix ($m = 40$, $T = 10$ s, $h = 0.25$ s, $q = 4$, $n = 10$) (Vide Appendix B, Program No. 7.2)

$t(s)$	C1(z)	C1	C2(z)	C2	C3(z)	C3
0						
	0.00000000	0.11111111	0.00000000	0.10000000	0.00000000	0.10000000
1						
	0.63212056	0.67471083	0.43233236	0.44816000	0.63212056	0.55651578
2						
	0.86466472	0.88096033	0.49084218	0.49328154	0.46508832	0.49201492
3						
	0.95021293	0.95643739	0.49876062	0.49912929	0.50922510	0.50112821
4						
	0.98168436	0.98405825	0.49983227	0.49988716	0.49756235	0.49984060
5						
	0.99326205	0.99416611	0.49997730	0.49998538	0.50064413	0.50002252
6						
	0.99752125	0.99786508	0.49999693	0.49999810	0.49982979	0.49999682
7						
	0.99908812	0.99921873	0.49999958	0.49999975	0.50004498	0.50000045
8						
	0.99966454	0.99971409	0.49999994	0.49999997	0.49998812	0.49999994
9						
	0.99987659	0.99989537	0.49999999	0.50000000	0.50000314	0.50000001
10						

to this effect could be made, it is expected that more accurate results could be arrived at with less number of component functions. This need calls for a new set of orthogonal functions known as "sample-and-hold functions."

7.3 Operational Matrix for Integration in SHF Domain [3]

As with the BPFs, we derive the operational matrix for integration in the SHF domain in a similar manner.

We know that any square integrable function $f(t)$ may be represented by a SHF set in the semi-open interval $[0, T)$ by considering the $(i + 1)$th component of the expansion to be

$$f_i(t) \approx f(ih), \quad ih \leq t < (i+1)h, \quad i = 0, 1, 2, \ldots, (m-1)$$

where h is the sampling period ($= T/m$), $f_i(t)$ is the part of the function $f(t)$ in the $(i + 1)$th interval, and $f(ih)$ is the first term of the Taylor series expansion of the function $f(t)$ around the point $t = ih$. Thus,

$$f_i(t) \approx f(ih) = f_i S_i(t) \tag{7.9}$$

If a time function $f(t)$ is fed to an S/H device, as shown in Figure 7.3, the output of the device approximates $f(t)$ following equation (7.9). Thus,

$$f(t) \approx \sum_{i=0}^{m-1} f_i S_i(t), \quad i = 0, 1, 2, \dots, (m-1)$$

$$= \begin{bmatrix} f_0 & f_1 & \cdots & f_{(m-1)} \end{bmatrix} \mathbf{S}_{(m)}(t) \triangleq \mathbf{F}_{(m)}^T \mathbf{S}_{(m)}(t) \tag{7.10}$$

where we have chosen an m-set SHFs.

Figure 7.4 shows (a) an m-set sample-and-hold functions (SHFs), (b) their first integrations and (c) subsequent representation of the integrated functions by their respective SHF equivalents.

It is easy to see the relationship

$$\int \mathbf{S}_{(m)}(t) dt \approx \mathbf{P1}_{S(m)} \mathbf{S}_{(m)}(t) \tag{7.11}$$

where $\mathbf{S}_{(m)}(t)$ is the m-term SHF vector and $\mathbf{P1}_{s(m)}$ is the operational matrix for integration of order m in the SHF domain given by

$$\mathbf{P1}_{S(m)} = h \begin{bmatrix} 0 & 1 & 1 & \cdots & 1 \\ & 0 & 1 & \cdots & 1 \\ & & 0 & \ddots & \vdots \\ & \mathbf{0} & & \ddots & 1 \\ & & & & 0 \end{bmatrix}_{(m \times m)} \tag{7.12}$$

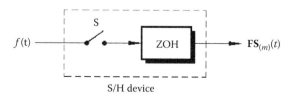

FIGURE 7.3
Input-output of a sample-and-hold (S/H) device.

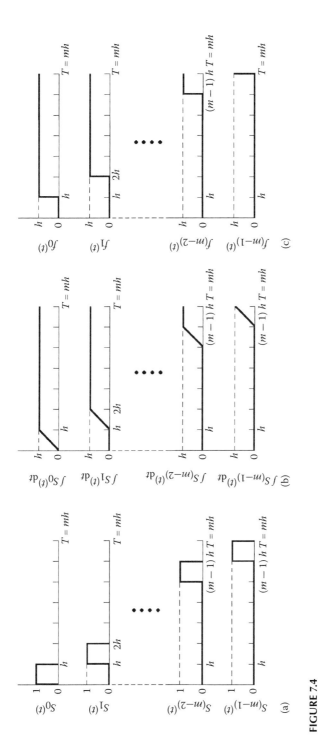

FIGURE 7.4
(a) An *m*-set sample-and-hold functions, (b) their first integrations and (c) subsequent representation of the integrated functions by their respective SHF equivalents.

It is noted that, like equation (3.12), $\mathbf{P1}_{S(m)}$ can be expressed as

$$\mathbf{P1}_{S(m)} = h\mathbf{Q}_{(m)}\left[\mathbf{I}_{(m)} - \mathbf{Q}_{(m)}\right]^{-1} \tag{7.13}$$

$\mathbf{P1}_{S(m)}$ is defective in the sense that it is not invertible like $\mathbf{P}_{(m)}$, and it has an eigenvalue 0 repeated m times.

However, we can integrate the functions shown in Figure 7.4(c) further and again express them in terms of component SHFs. Figure 7.5 shows the stages graphically.

7.3.1 Numerical Example

Example 7.1

Let us consider the function $f(t) = \dfrac{t^2}{2}$ for $t \geq 0$ and use $\mathbf{P1}_{S(m)}$ to integrate the same for $m = 10$ and $T = 1$ s.

In Figure 7.6, the integration results obtained in SHF domain are compared with the samples of the exact integration of $f(t)$. From the figure, it is noted that the SHF set is not suitable for integrating a time function. Whereas it is to be mentioned that the SHF sets are mainly used for analyzing the systems with S/H devices, where BPF set fails to do so.

7.4 One-Shot Operational Matrices for Repeated Integration

For double integration in SHF domain, the integration operational matrix can be obtained using equation (7.11), as

$$\iint \mathbf{S}_{(m)}(t)\,\mathrm{d}t \approx \mathbf{P1}_{S(m)} \int \mathbf{S}_{(m)}(t)\,\mathrm{d}t \approx \mathbf{P1}_{S(m)}^2 \mathbf{S}_{(m)}(t) \tag{7.14}$$

From equations (7.11) and (7.14), we note that at each stage of integration, $\mathbf{P1}_{S(m)}$ introduces some error. If we repeat the integration n times in this manner, the resulting $\mathbf{P1}_{S(m)}^n$ is obtained, using equation (7.13), as

$$\mathbf{P1}_{S(m)}^n = h^n\mathbf{Q}_{(m)}^n\left[\mathbf{I}_{(m)} - \mathbf{Q}_{(m)}\right]^{-n}, \quad n < m \tag{7.15}$$

It is obvious that if we work with $\mathbf{P1}_{S(m)}^n$, $n = 1, 2, 3, \ldots, (m-1)$, for system analysis, introduction of error at each of the n stages will cripple the usefulness of the SHF domain analysis.

To avoid such a situation, we seek resort to one-shot integration operational matrices like we have done for BPF domain integration, vide Section 4.3.

So, in case of SHF domain also, we integrate the SHF set directly n times and then expand the result of integration in terms of the basis SHFs.

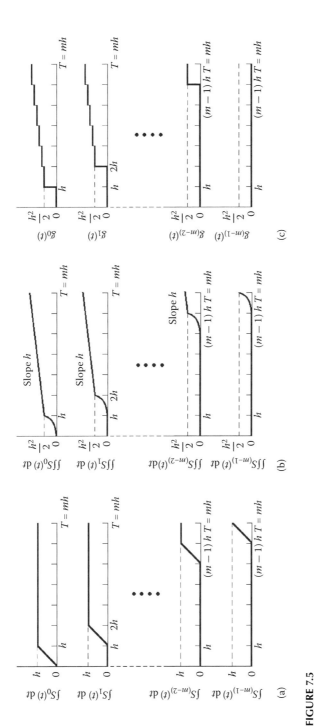

FIGURE 7.5

(a) First integration of an *m*-set sample-and-hold functions, (b) second integration of an *m*-set sample-and-hold functions, and (c) their subsequent representation by respective SHF equivalent.

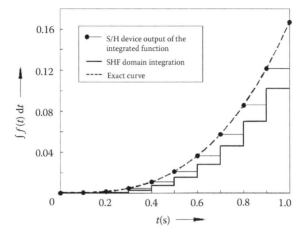

FIGURE 7.6
Integration of the function $f(t)$ of Example 7.1 using the operational matrix $\mathbf{P1}_{S(10)}$ and the result compared with the samples obtained from the exact integration for $m = 10$ and $T = 1$ s (vide Appendix B, Program no. 7.1).

The method is similar to the one-shot operational matrices for repeated integration (OSOMRI) introduced by Rao et al. [6] for Walsh function analysis.

When $n = 1$, the operational matrix is obviously $\mathbf{P1}_{s(m)}$ and we call it $\mathbf{P1(1)}_{s(m)}$.

When $n = 2$, the resulting one-shot operational matrix $\mathbf{P1(2)}_{s(m)}$ is obtained as

$$\mathbf{P1(2)}_{S(m)} = h \begin{bmatrix} 0 & 1 & 3 & 5 & \cdots & (2m-3) \\ & 0 & 1 & 3 & \cdots & (2m-5) \\ & & \ddots & & \ddots & \vdots \\ & & & \ddots & & \vdots \\ \mathbf{0} & & & & \ddots & 1 \\ & & & & & 0 \end{bmatrix}_{(m \times m),\, m \geq 2} \tag{7.16}$$

Upon detailed investigation, the operational matrix for n times repeated integration $\mathbf{P1(n)}_{s(m)}$ is obtained as

$$\mathbf{P1(n)}_{S(m)} = \frac{h^n}{n!} \begin{bmatrix} 0 & \left(1^n - 0^n\right) & \left(2^n - 1^n\right) & \cdots & \left[(m-1)^n - (m-2)^n\right] \\ & 0 & \left(1^n - 0^n\right) & \cdots & \left[(m-2)^n - (m-3)^n\right] \\ & & 0 & & \vdots \\ \mathbf{0} & & & \ddots & \vdots \\ & & & & 0 \end{bmatrix}_{(m \times m),\, m \geq 2}$$

$$\tag{7.17}$$

It is observed that we cannot find any operational matrix for differentiation in SHF domain like that for block pulse or Walsh function analysis. However, the integration operational matrices of all orders may be helpful in analyzing control systems with S/H which is discussed in the next section.

7.5 System Analysis Using One-Shot Operational Matrices and Operational Transfer Function [3]

In this section, the **SHOTFs** for several standard plants are presented. these **SHOTFs** are SHF domain equivalents of the Laplace domain transfer functions. Since the set of SHF is described in time domain, no inverse transformations are necessary, as with **BPOTF**, to determine the outputs. it will be evident from the following that any system having any type of complex transfer function can be analyzed using the results obtained for the simple plants with the help of the partial fraction technique.

7.5.1 First Order Plant

Consider a first order plant having a transfer function given by

$$G1(s) = (s+a)^{-1}$$

where a is real and greater than zero.

G1(s) can be expressed as a power series involving negative powers of the Laplace operator s. That is

$$G1(s) = (s+a)^{-1} = a^{-1}\left[as^{-1} - a^2 s^{-2} + a^3 s^{-3} - a^4 s^{-4} + \cdots \text{ to } \infty\right]$$

Since the terms within the square bracket involve integrators of different orders, it is logical that these can be replaced by one-shot operational matrices of respective orders to obtain an equivalent operational transfer function in the S/H domain. Calling this operational transfer function **SHOTF1** and using equations (7.12), (7.16), and (7.17), we can write

$$\textbf{SHOTF1}_{(m)} = a^{-1}\left[a\textbf{P1}(1)_{S(m)} - a^2\textbf{P1}(2)_{S(m)} + a^3\textbf{P1}(3)_{S(m)} - a^4\textbf{P1}(4)_{S(m)} + \cdots \text{ to } \infty\right]$$

where m basis functions have been considered.

Upon simplification, we have

$$\mathbf{SHOTF1}_{(m)} = \frac{1-\exp(-ah)}{a}\begin{bmatrix} 0 & 1 & \exp(-ah) & \exp(-2ah) & \cdots & \exp\big[-(m-2)ah\big] \\ & 0 & 1 & \exp(-ah) & \cdots & \exp\big[-(m-3)ah\big] \\ & & \ddots & 1 & \cdots & \vdots \\ & & & & \ddots & \exp(-ah) \\ & \mathbf{0} & & & & 1 \\ & & & & & 0 \end{bmatrix}_{(m\times m)}$$

(7.18)

If we consider $a = 1$, then from equation (7.18), the SHOTF for $G1(s) = (s + 1)^{-1}$ is obtained as

$$\mathbf{SHOTF1}_{(m)} = \big[1-\exp(-h)\big]\begin{bmatrix} 0 & 1 & \exp(-h) & \exp(-2h) & \cdots & \exp\big[-(m-2)h\big] \\ & 0 & 1 & \exp(-h) & \cdots & \exp\big[-(m-3)h\big] \\ & & \ddots & 1 & \cdots & \vdots \\ & & & & \ddots & \exp(-h) \\ & \mathbf{0} & & & & 1 \\ & & & & & 0 \end{bmatrix}_{(m\times m)}$$

(7.19)

Now we analyze systems shown in Figure 7.1(a)–7.1(c) for unit step inputs, considering $m = 10$, $T = 10$ s and $h = 1$ s.

Following equation (7.10), the unit step input is given by

$$u(t) = \mathbf{R1}^{\mathrm{T}}\mathbf{S}_{(10)}(t) = [1\ 1\ 1\ 1\ 1\ 1\ 1\ 1\ 1\ 1]\mathbf{S}_{(10)}(t)$$

For the system shown in Figure 7.1(a), following equation (7.6), the output $\mathbf{C1'}$ in SHF domain is given by

$$\mathbf{C1'}^{\mathrm{T}} = \mathbf{R1}^{\mathrm{T}}\,\mathbf{SHOTF1}$$

(7.20)

For the system of Figure 7.1(b), the equivalent open loop transfer function is

$$G2(s) = G1(s)\left[1 + G1(s)\right]^{-1}$$

and in SHF domain, the operational transfer function **SHOTF2** may be derived from equation (7.18) by substituting $a = 2$. That is

$$\mathbf{SHOTF2}_{(m)} = \frac{1 - \exp(-2h)}{2}\begin{bmatrix} 0 & 1 & \exp(-2h) & \exp(-4h) & \cdots & \exp\left[-2(m-2)h\right] \\ 0 & 1 & \exp(-2h) & \cdots & \exp\left[-2(m-3)h\right] \\ & & \ddots & 1 & \cdots & \vdots \\ & & & \ddots & \exp(-2h) \\ \mathbf{0} & & & & 1 \\ & & & & & 0 \end{bmatrix}_{(m \times m)}$$

(7.21)

Then, following equation (7.7), the output **C2**′ in SHF is

$$\mathbf{C2}'^{\mathrm{T}} = \mathbf{R1}^{\mathrm{T}}\mathbf{SHOTF2} \qquad (7.22)$$

In a similar fashion, as we have derived equation (7.8), the output **C3**′ in SHF domain will be

$$\mathbf{C3}'^{\mathrm{T}} = \mathbf{R1}^{\mathrm{T}}\mathbf{SHOTF1}\left[\mathbf{I} + \mathbf{SHOTF1}\right]^{-1} \qquad (7.23)$$

Equations (7.20), (7.22), and (7.23) are solved and the results are compared with the exact solutions given by equations (7.1), (7.2), and (7.3). These are tabulated in Table 7.3. It is observed that the results of SHF analysis match exactly with the results obtained from z-transform analysis.

TABLE 7.3

Comparison of the Exact Solutions with SHF Domain Analysis for the Sampled-Data Systems Shown in Figure 7.1(a)–7.1(c), with Unit Step Input ($m = 10$, $T = 10$ s, $h = 1$ s) (Vide Appendix B, Program No. 7.3)

t(s)	C1(z)	C1′	C2(z)	C2′	C3(z)	C3′
0						
	0.00000000	0.00000000	0.00000000	0.00000000	0.00000000	0.00000000
1						
	0.63212056	0.63212056	0.43233236	0.43233236	0.63212056	0.63212056
2						
	0.86466472	0.86466472	0.49084218	0.49084218	0.46508832	0.46508832
3						
	0.95021293	0.95021293	0.49876062	0.49876062	0.50922510	0.50922510
4						
	0.98168436	0.98168436	0.49983227	0.49983227	0.49756235	0.49756235
5						
	0.99326205	0.99326205	0.49997730	0.49997730	0.50064413	0.50064413
6						
	0.99752125	0.99752125	0.49999693	0.49999693	0.49982979	0.49982979
7						
	0.99908812	0.99908812	0.49999958	0.49999958	0.50004498	0.50004498
8						
	0.99966454	0.99966454	0.49999994	0.49999994	0.49998812	0.49998812
9						
	0.99987659	0.99987659	0.49999999	0.49999999	0.50000314	0.50000314
10						

7.5.2 nth Order Plant with Single Pole of Multiplicity n

Consider a plant having a transfer function

$$G2(s) = (s+a)^{-n}$$

The equivalent **SHOTF** of this plant is given by

$$\mathbf{SHOTF3}_{(m)} = a^{-n}\left[a^n \mathbf{P1}(\mathbf{n})_{S(m)} - na^{n+1}\mathbf{P1}(\mathbf{n+1})_{S(m)} + \frac{n(n+1)}{2!}a^{n+2}\mathbf{P1}(\mathbf{n+2})_{S(m)} \right.$$
$$\left. - \frac{n(n+1)(n+2)}{3!}a^{n+3}\mathbf{P1}(\mathbf{n+3})_{S(m)} + \cdots \text{ to } \infty \right]$$

It can be shown that

$$\mathbf{SHOTF3}_{(m)} = \begin{bmatrix} 0 & u_{12} & u_{13} & u_{14} & \cdots & u_{1m} \\ & 0 & u_{12} & u_{13} & \cdots & u_{1(m-1)} \\ & & \ddots & u_{12} & \cdots & u_{1(m-2)} \\ & & & & \ddots & \vdots \\ & \mathbf{0} & & & & u_{12} \\ & & & & & 0 \end{bmatrix}_{(m \times m)} \tag{7.24}$$

where

$$u_{ij} = a^{-n} \exp\left[-(j-2)ah\right]\left[\sum_{p=1}^{n} \frac{(ah)^{p-1}}{(p-1)!}\left\{(j-2)^{p-1} - \left[\exp(-ah)\right](j-1)^{p-1}\right\}\right]$$

for $2 \le j \le m$

Consider an open loop control system having a plant transfer function $G4(s)$ with $a = 1$ and $n = 2$, and an S/H device at the input. For a unit step input and a sampling period of 1 s, the output response of the system $C4(z)$ is determined by the conventional z-transform analysis as

$$C4(z) = \frac{0.26424112z^2 + 0.13533528z}{z^3 - 1.73575888z^2 + 0.87109417z - 0.13533528} \tag{7.25}$$

Using equation (7.24), we can find out SHF domain responses with a unit step input for $m = 10$, $T = 10$ s and $h = 1$ s. The results obtained are compared with the exact solutions in Table 7.4. It is only apparent that the results match exactly.

7.5.3 Second-Order Plant with Imaginary Roots

Consider a system having the transfer function $G3(s) = \left(s^2 + a^2\right)^{-1}$.
    This can be decomposed as

$$G3(s) = \frac{1}{2ja}\left[\frac{1}{s - ja} - \frac{1}{s + ja}\right], \quad \text{where } j = \sqrt{-1}$$

TABLE 7.4

Comparison of the Exact Solutions with SHF Domain Analysis for the Sampled-Data System Having Plant Transfer Function $(s + 1)^{-2}$, with Unit Step Input ($m = 10$, $T = 10$ s, $h = 1$ s)

t(s)	z-Transform Analysis, C4(z)	SHF Domain Analysis C4′
0		
	0.00000000	0.00000000
1		
	0.26424112	0.26424112
2		
	0.59399415	0.59399415
3		
	0.80085173	0.80085173
4		
	0.90842181	0.90842181
5		
	0.95957232	0.95957232
6		
	0.98264873	0.98264873
7		
	0.99270494	0.99270494
8		
	0.99698084	0.99698084
9		
	0.99876590	0.99876590
10		

Using equation (7.18), the **SHOTF** is given by

$$\textbf{SHOTF4}_{(m)} = \begin{bmatrix} 0 & v_{12} & v_{13} & v_{14} & \cdots & v_{1m} \\ & 0 & v_{12} & v_{13} & \cdots & v_{1(m-1)} \\ & & \ddots & & v_{12} & \cdots & v_{1(m-2)} \\ & & & & \ddots & \vdots \\ & \textbf{0} & & & & v_{12} \\ & & & & & 0 \end{bmatrix}_{(m \times m)} \tag{7.26}$$

where

$$v_{ij} = \frac{2}{a^2} \sin\left(\frac{ah}{2}\right) \sin\left[\left(\frac{ah}{2}\right)(2j - 3)\right], \quad \text{for } 2 \le j \le m$$

Consider an open loop control system having a plant transfer function $G3(s)$ and an S/H device at the input. For a unit step input and a sampling period of 1 s, the output response **C5′** of the system is determined for $a = \sqrt{2}$ by the conventional z-transform analysis. Calling this output $C5(z)$, we have

$$C5(z) = \frac{0.42202815z^2 + 0.42202815z}{z^3 - 1.31188738z^2 + 1.31188738z - 1} \qquad (7.27)$$

Using equation (7.26), the equivalent SHF domain responses for $m = 10$, $T = 10$ s, and $h = 1$ s are determined. The results obtained are compared in Table 7.5. As before, the results match exactly.

TABLE 7.5

Comparison of the Exact Solutions with SHF Domain Analysis for the Sampled-Data System Having Plant Transfer Function $(s^2 + 2)^{-1}$, with Unit Step Input ($m = 10$, $T = 10$ s, $h = 1$ s)

t(s)	z-Transform Analysis, $C5(z)$	SHF Domain Analysis $C5'$
0		
	0.00000000	0.00000000
1		
	0.42202815	0.42202815
2		
	0.97568156	0.97568156
3		
	0.72633093	0.72633093
4		
	0.09490820	0.09490820
5		
	0.14732605	0.14732605
6		
	0.79509724	0.79509724
7		
	0.94471106	0.94471106
8		
	0.34360253	0.34360253
9		
	0.00651054	0.00651054
10		

7.5.4 Second-Order Plant with Complex Roots

Let us consider a system with complex conjugate poles having the transfer function

$$G6(s) = \left(s^2 + \alpha s + \beta\right)^{-1}, \quad \alpha, \beta > 0.$$

This can be decomposed as

$$G6(s) = \frac{1}{2jb}\left[\frac{1}{s + a - jb} - \frac{1}{s + a + jb}\right]$$

where

$$a = \frac{1}{2}\alpha \quad \text{and} \quad b = \frac{1}{2}\sqrt{\left(4\beta - \alpha^2\right)}$$

Again, using equation (7.18), the **SHOTF** is given by

$$\mathbf{SHOTF5}_{(m)} = \begin{bmatrix} 0 & w_{12} & w_{13} & w_{14} & \cdots & w_{1m} \\ & 0 & w_{12} & w_{13} & \cdots & w_{1(m-1)} \\ & & \ddots & w_{12} & \cdots & w_{1(m-2)} \\ & & & \ddots & & \vdots \\ & \mathbf{0} & & & & w_{12} \\ & & & & & 0 \end{bmatrix}_{(m \times m)} \tag{7.28}$$

where

$$w_{ij} = \frac{\exp\left[-(j-2)ah\right]}{b\left(a^2 + b^2\right)}\left[\left\{a\sin\left(j-2\right)bh + b\cos\left(j-2\right)bh\right\}\right.$$
$$\left. - \exp(-ah)\left\{a\sin\left(j-1\right)bh + b\cos\left(j-1\right)bh\right\}\right], \quad \text{for } 2 \leq j \leq m$$

Let us consider an open loop control system having plant transfer function $G4(s)$ with an S/H device. For a unit step input, the output **C6′** of this system is determined for $m = 10$, $T = 10$ s, $h = 1$ s and $\alpha = \beta = 1$ using equation (7.28).

TABLE 7.6

Comparison of the Exact Solutions with SHF Domain Analysis for the Sampled-Data System Having Plant Transfer Function $(s^2 + s + 1)^{-1}$, with Unit Step Input ($m = 10$, $T = 10$ s, $h = 1$ s)

t(s)	z-Transform Analysis, C6(z)	SHF Domain Analysis C6$'$
0		
	0.00000000	0.00000000
1		
	0.34029985	0.34029985
2		
	0.84942563	0.84942563
3		
	1.12435477	1.12435477
4		
	1.15312277	1.15312277
5		
	1.07459057	1.07459057
6		
	1.00228949	1.00228949
7		
	0.97435896	0.97435896
8		
	0.97900663	0.97900663
9		
	0.99293426	0.99293426
10		

Also, by the conventional z-transform analysis, the output is determined as

$$C6(z) = \frac{0.34029985z^2 + 0.24168648z}{z^3 - 0.78589311z^2 + 1.1537725z - 0.36787944} \tag{7.29}$$

These results are compared in Table 7.6 and as expected, the results match exactly.

7.6 Error Analysis: A Comparison between SHF and BPF

To estimate the error of approximation of a function $f(t)$ of Lebesgue measure, we focus our attention on a single subinterval of duration h, $t \in (ih, (i+1)h)$, as shown in Figure 7.7(a).

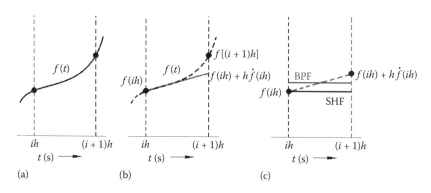

FIGURE 7.7
(a) A time function $f(t)$ over the interval $(ih, (i + 1)h)$, along with its (b) first order Taylor series representation and (c) its equivalent BPF and SHF representation over the same interval.

The function $f(t)$ is approximated via first order Taylor series in the specified interval, shown in Figure 7.7(b), knowing $f(ih)$ and $\dot{f}(ih)$, assuming h to be small. Hence

$$f(t) \approx f(ih) + \dot{f}(ih)(t - ih) \tag{7.30}$$

We consider that for estimation of error of approximation, the function $f(t)$ is represented by its equivalent first order Taylor approximation given by equation (7.30).

7.6.1 Error Estimate for Sample-and-Hold Function Domain Approximation

In SHF domain, the integral square error (ISE) in the $(i + 1)$th interval is

$$
\begin{aligned}
\left[\varepsilon_i\right]^2_{\text{SHF}} &= \int_{ih}^{(i+1)h} \left[\left\{f(ih) + \dot{f}(ih)(t - ih)\right\} - f(ih)\right]^2 dt \\
&= \int_{ih}^{(i+1)h} \left[\dot{f}(ih)(t - ih)\right]^2 dt \\
&= \frac{1}{3}\left[\dot{f}(ih)\right]^2 h^3
\end{aligned}
$$

So, the MISE in the $(i + 1)$th interval is,

$$\text{MISE}_{\text{SHF}} = \frac{1}{h} \cdot \frac{1}{3}\left[\dot{f}(ih)\right]^2 h^3 = \frac{mh^3}{3T}\left[\dot{f}(ih)\right]^2 \tag{7.31}$$

where, $T (= mh)$, the time period under consideration.

7.6.2 Error Estimate for Block Pulse Function Domain Approximation

As discussed in Section 2.3, the ISE in the $(i + 1)$th interval is

$$
\begin{aligned}
\left[\varepsilon_i\right]_{\text{BPF}}^2 &= \int_{ih}^{(i+1)h}\left[f(t)-c_i\psi_i(t)\right]^2 dt \\
&= \int_{ih}^{(i+1)h}\left[f(t)\right]^2 dt - hc_i^2
\end{aligned}
\tag{7.32}
$$

Following equation (7.30), the BPF coefficient is

$$
c_i = \frac{f(ih)+\left\{f(ih)+h\dot{f}(ih)\right\}}{2} = f(ih)+\frac{h}{2}\dot{f}(ih)
\tag{7.33}
$$

The MISE in the $(i+1)$th interval, using equations (7.30), (7.32), and (7.33), is

$$
\text{MISE}_{\text{BPF}} = \frac{mh^3}{12T}\left[\dot{f}(ih)\right]^2
\tag{7.34}
$$

7.6.3 A Comparative Study

Now define

$$
\begin{aligned}
\Delta_{\text{MISE}} &= \text{MISE}_{\text{SHF}} - \text{MISE}_{\text{BPF}} = \frac{mh^3}{4T}\left[\dot{f}(ih)\right]^2 \\
&= \frac{h^2}{4}\left[\dot{f}(ih)\right]^2 = \left[\frac{h}{2}\dot{f}(ih)\right]^2
\end{aligned}
\tag{7.35}
$$

Referring to Figure 7.7(c), it is observed that the difference d_i between SHF and BPF coefficient in the specified interval is

$$
d_i \triangleq \left[f(ih)+\frac{h}{2}\dot{f}(ih)\right]-f(ih) = \frac{h}{2}\dot{f}(ih)
\tag{7.36}
$$

From equation (7.35), using equation (7.36), we can write

$$
\Delta_{\text{MISE}} = d_i^2
\tag{7.37}
$$

By studying equation (7.37), we find that if $h \to 0$, $\Delta_{\text{MISE}} \to 0$, meaning, both the SHF and BPF approximations incur equal MISE. Also, from equations

(7.31) and (7.34), we see that when $h \to 0$, both the MISEs, namely MISE_{BPF} and MISE_{SHF}, tend to zero.

Also, if the slope of the working function is zero at $t = ih$, i.e. $\dot{f}(ih) = 0$, then both the MISEs are zero.

Then, the upper bound of MISEs is

$$\text{MISE}_{\text{BPF}} = \frac{h^2}{12}\left[\dot{f}_{\text{max}}\right]^2$$

and

$$\text{MISE}_{\text{SHF}} = \frac{h^2}{3}\left[\dot{f}_{\text{max}}\right]^2$$

where $\dot{f}_{\text{max}}$ is largest among all $\dot{f}(ih)$.

Therefore, finally the upper bound of MISEs over $[0, T)$ is

$$\text{MISE}_{\text{BPF}}\Big|_T = \frac{mh^2}{12}\left[\dot{f}_{\text{max}}\right]^2 \tag{7.38}$$

and

$$\text{MISE}_{\text{SHF}}\Big|_T = \frac{mh^2}{3}\left[\dot{f}_{\text{max}}\right]^2 \tag{7.39}$$

7.7 Conclusion

This chapter started with the main task of finding a suitable technique for the analysis of control systems with S/H to avoid the labors of transformations and/or manipulations associated with the conventional z-transform analysis.

As a first attempt, the well established BPFs were chosen for S/H system analysis because of the visible similarity between a BPF and the output of an S/H device. Block pulse domain operational transfer function **BPOTF** (defined and discussed in detail in Chapter 4) was utilized for such analysis in BPF domain and the results obtained were compared with the conventional z-transform analysis. But the results were not very close to the exact solution, and also, the technique failed to distinguish between the feedback control system of Figures 7.1(b) and 7.1(c).

To remove these shortcomings, another approach was followed using an S/H matrix introduced by Chen and Wu [4]. Though this method was able to distinguish between the control systems of Figure 7.1(b) and 7.1(c) properly, it failed to come up with results having desired accuracy. Further, the method needed a large number of component BPFs leading to a much larger amount of storage as well as computational time.

Finally, to achieve our objective, a set of piecewise constant orthogonal functions was employed. These functions, known as SHFs, have been introduced in Chapter 1. Using this set, a SHOTF is defined. This SHOTF has been used to analyse control systems with S/H in the SHF domain. Several open loop as well as closed loop control systems are analysed with different forms of transfer functions, e.g., $(s + a)^{-1}$, $(s + a)^{-n}$, $(s^2 + a^2)^{-1}$, and $(s^2 + \alpha s + \beta)^{-1}$ using a unit step input.

The results are found to match exactly with the solutions obtained by the conventional z-transform analysis and these are compared in various tables.

Another interesting point is, the accuracy of the method is entirely independent of m—the number of component SHFs.

In the conventional z-transform method, evaluation of m number of coefficients by long division requires computation of m number of determinants [7] of orders (2×2) to $(m \times m)$. But in the SHF domain, the **SHOTF**s are always upper triangular circulant matrix and hence provide economy in storage [8]. Also, the SHF analysis needs only one multiplication involving a $(1 \times m)$ and a $(m \times m)$ matrix, and thus requires much less computation time.

Considering these aspects and accuracy of the computed results, it is no wonder that the SHF is much more effective than the BPF for the analysis of control systems with S/H. Since the SHF set is defined in time domain, no inverse transformations are necessary for obtaining the solution. Simple matrix multiplications were effective to yield desired results.

For SHF and BPF domain approximations of a function $f(t)$ of Lebesgue measure, an error analysis has been carried out to provide an estimate of the upper bound of the MISEs and the MISE of SHF has been compared with that of BPF.

References

1. A. Deb, G. Sarkar, and S. K. Sen, Block pulse functions, the most fundamental of all piecewise constant basis functions, Int. J. Syst. Sci., vol. **25**, no. 2, pp. 351–363, 1994.

2. A. Deb, G. Sarkar, M. Bhattacharjee, and S. K. Sen, All-integrator approach to linear SISO control system analysis using block pulse functions (BPF), J. Franklin Inst., vol. **334B**, no. 2, pp. 319–335, 1997.

3. A. Deb, G. Sarkar, M. Bhattacharjee, and S. K. Sen, A new set of piecewise constant orthogonal functions for the analysis of linear SISO systems with sample-and-hold, J. Franklin Inst., vol. **335B**, no. 2, pp. 333–358, 1998.
4. W. L. Chen and S. G. Wu, Analysis of sampled-data systems by block-pulse functions, Int. J. Syst. Sci., vol. **16**, no. 6, pp. 747–752, 1985.
5. A. Deb, G. Sarkar, M. Bhattacharjee, and S. K. Sen, Analysis of linear SISO (single input single output) control systems with sample-and-hold using block pulse function (BPF) operational technique, Indian J. Engg. & Materials Sc., vol. **6**, pp. 5–8, February, 1999.
6. G. P. Rao and K. R. Palanisamy, Improved algorithms for parameter identification in lumped continuous systems via Walsh functions, Proceedings IEE, Part – D, Control Theory and Applications, vol. **130**, no. 1, pp. 9–16, 1983.
7. E. I. Jury, *Theory and application of the z-transform method*, Wiley, New York, 1964.
8. S. Pissanetzky, *Sparse matrix technology*, Academic Press, London, 1984.

Study Problems

7.1 What are the advantages of using sample-and-hold operational transfer function (**SHOTF**) over the block pulse operational transfer function (**BPOTF**) in analyzing control systems involving sample-and-hold device?

7.2 To implement an S/H device in practice what components do we need?

Draw the circuit diagram of an S/H device.

7.3 Consider an open loop system having a transfer function $G(s) = (s + 4)^{-1}$ with an S/H device at the input. For a unit step input $u(t)$, find the output $c(t)$ of the system by the conventional z-transform analysis, and also using **BPOTF** and **SHOTF** approaches. Consider $m = 8$, $T = 1$ s and sampling period $h = 0.125$ s.

7.4 Compare the results of **Problem 7.3** obtained using **BPOTF**, **SHOTF**, and conventional z-transform analysis by estimating the percentage errors at the sampling instants and discuss the effectiveness of the **SHOTF** approach.

7.5 Consider an open loop system having a transfer function $G(s) = (s^2 + s + 4)^{-1}$ and an S/H device at the input. For a unit step input $u(t)$, determine its output $c(t)$ by the conventional z-transform analysis, and also using **BPOTF** and **SHOTF** approaches. Consider $m = 10$ and $T = 1$ s and sampling period $h = 0.1$ s.

7.6 Compare the results of **Problem 7.5** obtained using **BPOTF**, **SHOTF**, and conventional z-transform analysis by estimating the percentage errors at the sampling instants and discuss the effectiveness of the **SHOTF** approach.

8

Discrete Time System Analysis Using a Set of Delta Functions (DFs)

Piecewise constant basis functions (PCBFs), like Haar functions, Rademacher functions, Walsh functions, block pulse functions (BPFs), etc. [1,2], have been around in the literature for about nine decades. Of all PCBFs, the BPFs turned out to be the most fundamental and its qualitative as well as quantitative appraisal were presented in Chapter 6.

BPFs and their variants have been used successfully for the analysis, synthesis [3], and design of control systems, and other related problems [4,5].

It is apparent that any set of PCBF is unsuitable for analyzing discrete systems because a discrete system always deals with impulses, namely, Dirac delta functions (DFs), which can never be handled by the PCBFs mentioned above. However, Chen and Wu [6] made a lone attempt to analyze a discrete control system using the set of BPF and came up with approximate results. The main reason for such inexact results was their assumption that a block pulse could approximate an impulse.

In this Chapter, discrete system analysis is carried out using a set of mutually disjoint DFs instead of the conventional z-domain. This DF set forms the δ-domain and is used to develop a special type of integration operational matrices of different orders. Then an operational technique, using delta operational transfer function (**DOTF**), is used to analyze various linear discrete control systems via this set of DFs and δ-domain operational matrices. Results are compared with the actual solutions and found to match exactly.

The present analysis does not need any inverse transformation, or long division, like the z-transform method and algebraic manipulations associated with the z-transform analysis are entirely avoided. Further, once the **DOTF** of any plant is computed, this technique is able to analyze the plant for different input functions in one mathematical operation.

8.1 A Set of Mutually Disjoint Delta Functions [7]

DF has a long as well as rich history and heritage in the area of mathematical physics. It was introduced by Paul A. M. Dirac [8] and, from the very moment of its appearance in the literature, it never stopped fascinating scientists with its enigmatic appeal.

δ function is not a function in the ordinary sense because, although $\delta(t) = 0$ if $t \neq 0$, $\delta(0) = \infty$. Also, the function does not have a definite value for each point in its infinitesimal domain. Thus, δ function is considered to be a generalized function [9], and can be regarded as a distribution in the physical sense.

The DF set is comprised of m (say) component functions of which the $(i + 1)$ th member is defined as:

$$\delta_i(t) = \delta(t - ih) \quad \text{where} \quad i = 0, 1, 2, \ldots, (m-1) \tag{8.1}$$

It is easy to see that h is the time delay between any two consecutive members of the set and may be regarded as the *sampling period*. In line with the definitions given in references [8] and [9], we can write

$$\delta_i(t) = \begin{cases} 0 & \text{for } t \neq ih \\ \infty & \text{for } t = ih \end{cases} \tag{8.2}$$

A set of first eight δ functions in the semi-open interval $t \in [0, T)$ is shown in Figure 8.1.

The *value* of $\delta_i(t)$ at $t = ih$ has no meaning but it attains numerical character if we use $\delta_i(t)$ only within an interval.

$\delta_i(t)$ satisfies the condition:

$$\int_0^\infty \delta_i(t)\, dt = 1$$

With these basic ideas, the set of DFs is formed in line with the set of BPFs. But it should be kept in mind that, unlike BPFs, the set of DFs does not form an orthogonal basis because the inner product of DFs is undefined.

From the nature of BPFs as defined by equation (1.12), it is easy to note that the component DFs are actually samples of BPFs with a sampling interval equal to the width of each component block pulse.

This set of impulse functions is used to represent a function $f^*(t)$ which is the sampled output of a square integrable function $f(t)$.

Thus, in the interval $[0, T)$, the sampled function $f^*(t)$ is given by:

$$f^*(t) = \sum_{i=0}^{m-1} f(t)\delta(t - ih) \tag{8.3}$$

Considering the nature of DF and comparing equations (1.12) and (8.2), we observe that any two members of the set of DF is mutually disjoint in $t \in [0, T)$, like the BPFs. But, other properties of orthogonal/orthonormal basis functions like completeness, approximations, etc. are not defined for this set of DF.

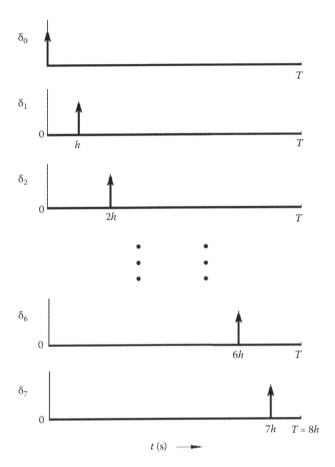

FIGURE 8.1
A set of first eight δ functions.

If $f(t)$ is fed to a sampling device, the output of the device modulates $f(t)$ as in equation (8.3). Thus,

$$f^*(t) = \sum_{i=0}^{m-1} f_i \delta_i(t), \quad \text{where } i = 0, 1, 2, \dots, (m-1)$$

$$= \begin{bmatrix} f_0 & f_1 & \cdots & f_{(m-1)} \end{bmatrix} \Delta_{(m)}(t) \triangleq \mathbf{F}_{(m)}^T \Delta_{(m)}(t) \tag{8.4}$$

where, we have chosen an m-set DFs given by:

$$\Delta_{(m)}(t) = \begin{bmatrix} \delta_0(t) & \delta_1(t) & \cdots & \delta_i(t) & \cdots & \delta_{(m-1)}(t) \end{bmatrix}^T$$

Unlike the coefficients of BPFs given by equation (1.13), the coefficients f_i's of equation (8.4) are given by:

$$f_i = \int_0^T f(t)\delta(t - ih)\, dt, \quad ih < T \tag{8.5}$$

8.2 Delta Function Domain Operational Matrices for Integration [7]

In this Section we derive the operational matrix for integration in DF domain.

It is noted that, if we integrate an m-set DFs, the result is a set of delayed unit step functions [2]. This integrated version may be represented by means of the same set of DFs (i.e., the output step functions are sampled) to obtain their DF equivalent.

Figure 8.2 shows (a) an m-set DFs, (b) their first integrations, and (c) subsequent representation of the integrated functions by their respective DF equivalents.

It is easy to see the following relationship involving integration of an m-set DF:

$$\int \Delta_{(m)}(t)\, dt = \mathbf{U}_{(m)}(t) \tag{8.6}$$

and

$$\mathbf{U}_{(m)}^{*}(t) = \mathbf{P1}_{\mathrm{DEL}(m)}\Delta_{(m)}(t) \tag{8.7}$$

Equation (8.7) needs some explaining. Integration of $\Delta_{(m)}(t)$ produces a set of delayed unit step functions $\mathbf{U}_{(m)}(t)$. We sample the vector $\mathbf{U}_{(m)}(t)$ to obtain a set involving impulse functions only. This is represented by the vector $\mathbf{U}_{(m)}^{*}(t)$. This sampled function set may be represented by the product $\mathbf{P1}_{\mathrm{DEL}(m)}\Delta_{(m)}(t)$ as shown in equation (8.7). We call the upper triangular matrix $\mathbf{P1}_{\mathrm{DEL}(m)}$ the operational matrix for integration of order m in the δ-domain. Thus,

$$\mathbf{P1}_{\mathrm{DEL}(m)} = \begin{bmatrix} 1 & 1 & 1 & \cdots & \cdots & 1 \\ & 1 & 1 & \cdots & \cdots & 1 \\ & & 1 & 1 & \cdots & 1 \\ & & & \ddots & \ddots & \vdots \\ \mathbf{0} & & & & 1 & 1 \\ & & & & & 1 \end{bmatrix}_{(m \times m)} \tag{8.8}$$

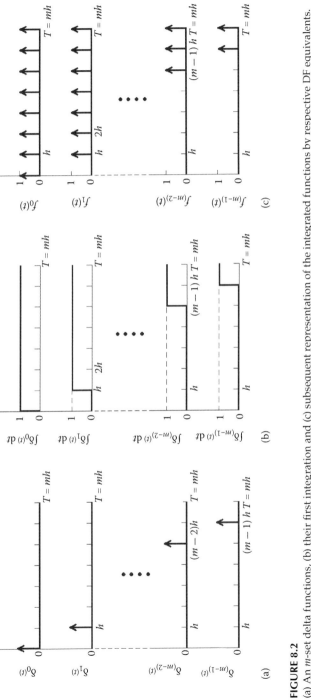

FIGURE 8.2

(a) An m-set delta functions, (b) their first integration and (c) subsequent representation of the integrated functions by respective DF equivalents.

$\mathbf{P1}_{\text{DEL}(m)}$ can also be expressed as:

$$\mathbf{P1}_{\text{DEL}(m)} = \left[\mathbf{I}_{(m)} - \mathbf{Q}_{(m)} \right]^{-1} \tag{8.9}$$

where $\mathbf{I}_{(m)}$ is a unit matrix and $\mathbf{Q}_{(m)}$ is the delay matrix [10], both of orders m. It is noted that $\mathbf{P1}_{\text{DEL}(m)}$ has an eigenvalue 1 repeated m times.

The character of $\mathbf{P1}_{\text{DEL}(m)}$ is slightly different from $\mathbf{P}_{(m)}$, the conventional operational matrix for integration in BPF domain. While the matrix $\mathbf{P}_{(m)}$ operates upon a BPF vector to obtain the BPF equivalent of the integrated time function, the matrix $\mathbf{P1}_{\text{DEL}(m)}$ operates upon a DF vector not only to integrate it, but also to *sample* the integrated output. Thus, the operational matrix $\mathbf{P1}_{\text{DEL}(m)}$ may be called an integrator-modulator which relates two impulse trains as input-output with the integration operation in the sense of Euler.

8.2.1 Numerical Example

Example 8.1

Let us consider the function $f(t) = \dfrac{t^2}{2}$ for $t \geq 0$ and use the operational matrix for integration in DF domain for $m = 10$, $T = 1$ s and $h = 0.1$ s.

The sampled function $f^*(t)$ may be expressed as:

$$f^*(t) = \left[\frac{0^2}{2} \quad \frac{0.1^2}{2} \quad \frac{0.2^2}{2} \quad \cdots \quad \cdots \quad \frac{0.9^2}{2} \right] \Delta_{(10)}(t) \quad \triangleq \mathbf{F}_{(10)}^{\mathsf{T}} \Delta_{(10)}(t)$$

$$\int f^*(t)\,dt = \mathbf{F}_{(10)}^{\mathsf{T}} \int \Delta_{(10)}(t)\,dt \approx \mathbf{F}_{(10)}^{\mathsf{T}} \mathbf{P1}_{\text{DEL}(10)} \Delta_{(10)}(t)$$

$$= \left[0.00000000 \quad 0.00500000 \quad 0.02500000 \quad 0.07000000 \quad 0.15000000 \right.$$
$$\left. 0.27500000 \quad 0.45500000 \quad 0.70000000 \quad 1.02000000 \quad 1.42500000 \right] \Delta_{(10)}(t)$$

In Figure 8.3, the integration results obtained in DF domain using the operational matrix $\mathbf{P1}_{\text{DEL}}$, are graphically compared with the samples of the exact integration of $f^*(t)$. Here we find that the resulting samples obtained after integration using $\mathbf{P1}_{\text{DEL}}$ is exactly same as the samples of the exact integration of $f^*(t)$.

From the figure, it is noted that this DF set is not particularly suitable for integrating a discrete time function. But it will be shown in the following that the set of DF is very useful for analyzing open loop and closed loop sampled-data systems, where as BPF set fails to do so.

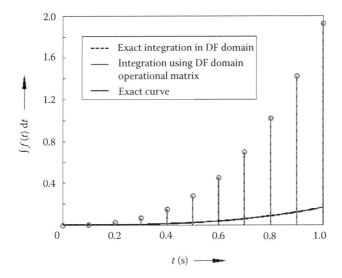

FIGURE 8.3

Integration of the function $f^*(t)$ of Example 8.1 using the operational matrix $\mathbf{P1}_{DEL}$ in delta function (DF) domain and compared with the samples of the exact Integration of $f^*(t)$, for $m = 10$ and $T = 1$ s. It is observed that the corresponding samples overlap (vide Appendix B, Program no. 8.3).

8.3 One-Shot Operational Matrices for Repeated Integration [7]

This Section presents the operational matrices for repeated integration in DF domain. These matrices are called one-shot operational matrices for integration.

Following equation (8.7), we can integrate $\mathbf{U}^*_{(m)}(t)$ as:

$$\int \mathbf{U}^*_{(m)}(t)\, dt = \mathbf{G}_{(m)}(t)$$

and

$$\mathbf{G}^*_{(m)}(t) = \mathbf{P1}^2_{DEL(m)} \mathbf{\Delta}_{(m)}(t) \tag{8.10}$$

It is obvious that, at each stage of integration, $\mathbf{P1}_{DEL(m)}$ introduces some error. If we repeat the integration n times in this manner, the resulting $\mathbf{P1}^n_{DEL(m)}$ is obtained, using equation (8.9), as:

$$\mathbf{P1}^n_{DEL(m)} = \left[\mathbf{I}_{(m)} - \mathbf{Q}_{(m)} \right]^{-n}, \quad n < m \tag{8.11}$$

If we work with $\mathbf{P1}^n_{\mathrm{DEL}(m)}$, $n = 1, 2, 3, \ldots, (m - 1)$, for discrete system analysis, introduction of error at each of the n stages will cripple the usefulness of the DF domain analysis.

To avoid such a situation, we integrate the DF set directly n times and then expand the result of integration in terms of the set of DFs. The method is similar to the one-shot operational matrices for repeated integration (OSOMRI) introduced by Rao and Palanisamy [11,12] for Walsh function analysis.

For $n = 2$, as shown in Figure 8.4, the resulting one-shot operational matrix $\mathbf{P1(2)}_{\mathrm{DEL}(m)}$ is obtained as:

$$\mathbf{P1(2)}_{\mathrm{DEL}(m)} = h \begin{bmatrix} 0 & 1 & 2 & 3 & \cdots & \cdots & (m-1) \end{bmatrix}_{(m \times m)} \qquad (8.12)$$

Upon a detailed investigation, the operational matrix for n times repeated integration $\mathbf{P1}(n)_{\mathrm{DEL}(m)}$ is obtained as:

$$\mathbf{P1(n)}_{\mathrm{DEL}(m)} = \frac{h^{n-1}}{(n-1)!} \begin{bmatrix} 0 & 1^{n-1} & 2^{n-1} & \cdots & (m-1)^{n-1} \end{bmatrix}_{(m \times m)} \qquad (8.13)$$

where $\begin{bmatrix} a & b & c \end{bmatrix} \triangleq \begin{bmatrix} a & b & c \\ 0 & a & b \\ 0 & 0 & a \end{bmatrix}$

We cannot find any operational matrix for differentiation for DF like that for block pulse or Walsh function analysis. However, the operational matrices for integration of all orders may be helpful in analyzing discrete control systems as discussed in the next Section.

8.4 Analysis of Discrete SISO Systems Using One-Shot Operational Matrices and Delta Operational Transfer Function [7]

In this Section we present the principle of the operational transfer function technique for linear discrete system analysis and develop **DOTFs** for several standard plants. These **DOTFs** are the equivalents of the Laplace domain transfer functions.

It will be evident from the following that any system having any type of complex transfer function can be analyzed using the results obtained for the simple plants with the help of the partial fraction technique.

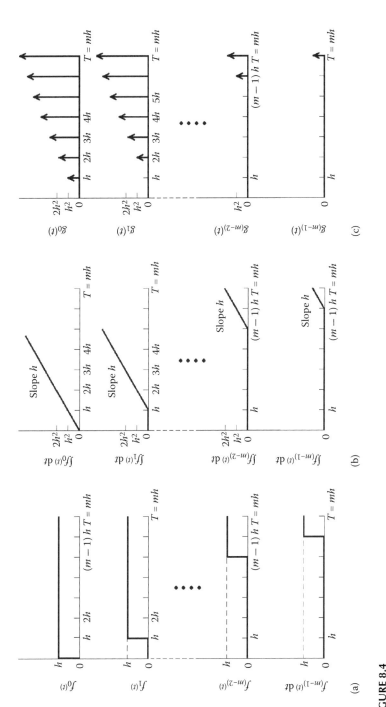

FIGURE 8.4

Second integration of the set of functions shown in Figure 8.2(b) and their subsequent representation by respective DF equivalents.

For a linear SISO control system, with the usual notation, we know

$$C(s) = G(s)R(s) \tag{8.14}$$

We can express the sampled versions of the input $r(t)$ and output $c(t)$ in terms of an m-set DFs as:

$$\left. \begin{array}{l} r^*(t) \triangleq \mathbf{R}^\mathrm{T}\mathbf{\Delta}_{(m)} \\ c^*(t) \triangleq \mathbf{C}^\mathrm{T}\mathbf{\Delta}_{(m)} \end{array} \right\} \tag{8.15}$$

Now the Laplace domain transfer function $G(s)$ of (8.14) is expanded in a power series containing negative powers of the Laplace operator s. Since the terms of this series involve integrators of different orders, it is logical that these can be replaced by one-shot operational matrices of respective orders to obtain an equivalent operational transfer function in the discrete δ-domain. Hence, equations (8.8), (8.12), and (8.13) can now be used to determine the delta domain operational transfer function $\mathbf{DOTF}_{(m)}$. It should be noted that $\mathbf{DOTF}_{(m)}$ is always a square matrix and is operative upon DFs only.

Using $\mathbf{DOTF}_{(m)}$ and equation (8.15), we can write an equation similar to equation (8.14) to obtain

$$\mathbf{C}^\mathrm{T} = \mathbf{R}^\mathrm{T}\mathbf{DOTF}_{(m)} \tag{8.16}$$

which is the key equation.

We can test this system for n number of inputs by making $\mathbf{R}$ a rectangular matrix of dimension $(n \times m)$ and the simple multiplication operation on the RHS of equation (8.16) will yield $\mathbf{C}_{(n \times m)}$ giving the n number of outputs.

That is, the input-output correspondence exists between each row of $\mathbf{R}_{(n \times m)}$ and each row of $\mathbf{C}_{(n \times m)}$.

8.4.1 First-Order Plant

Consider a first-order plant having a transfer function
$G1(s) = (s + a)^{-1}$, where, a is real and greater than zero.

Expressing $G1(s)$ as a power series involving negative powers of the Laplace operator s, we have

$$G1(s) = \left(s + a\right)^{-1} = a^{-1}\left[as^{-1} - a^2 s^{-2} + a^3 s^{-3} - a^4 s^{-4} + \cdots \quad \text{to } \infty \right]$$

Since the terms within the square bracket involve integrators of different orders, it is logical that these can be replaced by one-shot operational matrices of respective orders to obtain an equivalent operational transfer function in the delta domain. Using equations (8.8), (8.12), and 8.13, we can write

$$\mathbf{DOTF1}_{(m)} = a^{-1}\left[a\mathbf{P1(1)}_{D(m)} - a^2\mathbf{P1(2)}_{D(m)} + a^3\mathbf{P1(3)}_{D(m)} - a^4\mathbf{P1(4)}_{D(m)} + \cdots \quad \text{to } \infty\right]$$

where m basis functions have been considered.

Upon simplification, we have

$$\mathbf{DOTF1}_{(m)} =
\begin{bmatrix}
1 & \exp(-ah) & \exp(-2ah) & \exp(-3ah) & \cdots & \exp\left[-(m-1)ah\right] \\
 & 1 & \exp(-ah) & \exp(-2ah) & \cdots & \exp\left[-(m-2)ah\right] \\
 & & \ddots & \exp(-ah) & \cdots & \vdots \\
 & & & \ddots & \ddots & \exp(-2ah) \\
 & & 0 & & \ddots & \exp(-ah) \\
 & & & & & 1
\end{bmatrix}_{(m \times m)}$$

$$= \begin{bmatrix} 1 & \exp(-ah) & \exp(-2ah) & \cdots & \exp\left[-(m-1)ah\right] \end{bmatrix}_{(m \times m)}$$

(8.17)

where $h = \dfrac{T}{m}$ is the sampling period.

Equation (8.17) will now be used to analyze the sampled-data systems, shown in Figures 8.5(a), 8.5(b), and 8.5(c), for unit step inputs considering $m = 10$, $T = 10$ s, and $h = 1$ s.

Following equation (8.4), a unit step input in δ-domain is given by

$$u^*(t) \triangleq \mathbf{R1}^T \Delta_{(10)}(t) = \begin{bmatrix} 1 & 1 & \cdots & 1 & 1 \end{bmatrix} \Delta_{(10)}(t)$$

We consider $a = 1$ for the plant G1(s) shown in Figures 8.5(a), 8.5(b), and 8.5(c). For the system shown in Figure 8.5(a), following equation (8.16), the output **C1** in the DF domain is given by

$$\mathbf{C1}^T = \mathbf{R1}^T \mathbf{DOTF1} \tag{8.18}$$

where **DOTF1** is obtained from equation (8.17) putting $a = 1$.

For the system shown in Figure 8.5(b), the output **C2** in the DF domain is

$$\mathbf{C2}^T = \mathbf{R1}^T \mathbf{DOTF1}' \tag{8.19}$$

where **DOTF1**′ is obtained considering $a = 2$ in equation (8.17).

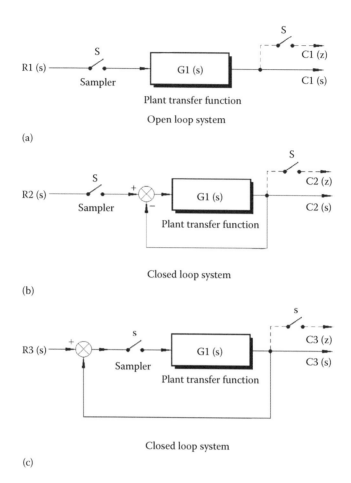

FIGURE 8.5
(a) An open loop discrete control system having a first-order transfer function, (b) Simple feedback control system with sampler at the input and (c) Error sampled feedback control system.

In a similar fashion the DF domain output **C3** for the system shown in Figure 8.5(c) is:

$$\mathbf{C3}^{\mathrm{T}} = \mathbf{R1}^{\mathrm{T}}\,\mathbf{DOTF1}\left[\mathbf{I} + \mathbf{DOTF1}\right]^{-1} \tag{8.20}$$

Using the conventional z-transform approach, the exact solutions with unit step input, for the systems shown in Figures 8.5(a), 8.5(b), and 8.5(c), are given by:

$$C1(z) = \frac{z^2}{z^2 - 1.36787944z + 0.36787944} \tag{8.21}$$

TABLE 8.1

Comparison of the Exact Solutions of the Systems of Figures 8.5(a), 8.5(b), and 8.5(c) Obtained via z-Transform Analysis with δ-Domain Analysis Results Obtained Using **DOTF**s ($m = 10$, $T = 10$ s, $h = 1$ s) (Vide Appendix B, Program No. 8.2)

$t(s)$	C1(z)	C1	C2(z)	C2	C3(z)	C3
0	1.00000000	1.00000000	1.00000000	1.00000000	0.50000000	0.50000000
1	1.36787944	1.36787944	1.13533528	1.13533528	0.59196986	0.59196986
2	1.50321472	1.50321472	1.15365092	1.15365092	0.60888677	0.60888677
3	1.55300179	1.55300179	1.15612967	1.15612967	0.61199846	0.61199846
4	1.57131743	1.57131743	1.15646514	1.15646514	0.61257083	0.61257083
5	1.57805538	1.57805538	1.15651054	1.15651054	0.61267611	0.61267611
6	1.58053413	1.58053413	1.15651668	1.15651668	0.61269547	0.61269547
7	1.58144601	1.58144601	1.15651751	1.15651751	0.61269903	0.61269903
8	1.58178148	1.58178148	1.15651763	1.15651763	0.61269968	0.61269968
9	1.58190489	1.58190489	1.15651764	1.15651764	0.61269981	0.61269981

$$C2(z) = \frac{z^2}{z^2 - 1.13533528z + 0.13533528} \tag{8.22}$$

$$C3(z) = \frac{0.5z^2}{z^2 - 1.18393972z + 0.18393972} \tag{8.23}$$

Equations (8.18), (8.19), and (8.20) are solved and the results are compared with the exact solutions given by equations (8.21), (8.22), and (8.23). From Table 8.1, it is observed that the results of the DF analysis match exactly with the results obtained from z-transform analysis.

8.4.2 *n*th-Order Plant with Single Pole of Multiplicity *n*

Consider a plant having a transfer function
$G2(s) = (s + a)^{-n}$, *a* being real and greater than zero.

As before, *G2(s)* can be expanded in a power series given by:

$$G2(s) = a^{-n}\left[a^n s^{-n} - na^{n+1}s^{-(n+1)} + \frac{n(n+1)}{2!}a^{n+2}s^{-(n+2)}\right.$$

$$\left. -\frac{n(n+1)(n+2)}{3!}a^{n+3}s^{-(n+3)} + \cdots \text{ to } \infty\right]$$

The equivalent **DOTF** is given by:

$$\mathbf{DOTF2}_{(m)} = \begin{bmatrix} 0 & u_{12} & u_{13} & u_{14} & \cdots & u_{1j} & \cdots & u_{1m} \end{bmatrix}_{(m \times m)} \quad (8.24)$$

where

$$u_{1j} = \frac{(j-1)^{n-1}}{(n-1)!}\exp\left[-(j-1)a\right], \quad \text{for } 1 \le j \le m$$

Now, consider an open loop control system having a plant transfer function *G2(s)* with $a = 1$ and $n = 2$, and sampler at the input. For a unit step input, the output response of the system is determined by both the conventional *z*-transform analysis and the δ-domain analysis for $m = 10$, $T = 10\,\text{s}$, and $h = 1\,\text{s}$. From Table 8.2, it is noted that the results are the same as those given by the *z*-transform analysis.

8.4.3 Second-Order Plant with Imaginary Roots

Consider a system having the transfer function $G3(s) = (s^2 + a^2)^{-1}$, where, *a* is real and greater than zero.

The transfer function may be decomposed as:

$$G3(s) = \frac{1}{2ja}\left[\frac{1}{s-ja} - \frac{1}{s+ja}\right], \quad \text{where } j = \sqrt{-1}$$

Using equation (8.17), the **DOTF** is given by:

$$\mathbf{DOTF3}_{(m)} = \begin{bmatrix} 0 & v_{12} & v_{13} & v_{14} & \cdots & v_{1j} & \cdots & v_{1m} \end{bmatrix}_{(m \times m)} \quad (8.25)$$

TABLE 8.2

Exact Output of the Input Sampled System Having the Transfer Function $(s + 1)^{-2}$, with Unit Step Input, Obtained via z-Transform Analysis, Compared with δ-Domain Results Obtained Using **DOTFs** ($m = 10$, $T = 10$ s, $h = 1$ s)

t(s)	z-Transform Analysis, C4(z)	δ-Domain Analysis
0	0.00000000	0.00000000
1	0.36787944	0.36787944
2	0.63855001	0.63855001
3	0.78791121	0.78791121
4	0.86117377	0.86117377
5	0.89486350	0.89486350
6	0.90973601	0.90973601
7	0.91611919	0.91611919
8	0.91880289	0.91880289
9	0.91991358	0.91991358

where

$$v_{1j} = \frac{1}{a}\sin\left[(j-1)a\right], \quad \text{for } 1 \le j \le m$$

Now, consider an open loop sampled-data system having a plant transfer function $G3(s)$ with $a = \sqrt{2}$. For a unit step input, the output response of the system is determined by both the conventional z-transform analysis and the δ-domain **DOTF** approach for $m = 10$, $T = 10$ s, and $h = 1$ s. From Table 8.3, it is noted that the results are the same as those obtained by z-transform analysis.

TABLE 8.3

Exact Output of the Input Sampled System Having
the Transfer Function $(s^2 + 2)^{-1}$, with Unit Step Input,
Obtained via z-Transform Analysis, Compared with
δ-Domain Results Obtained Using **DOTF**s ($m = 10$,
$T = 10$ s, $h = 1$ s)

t(s)	z-Transform Analysis, C5(z)	δ-Domain Analysis
0	0.00000000	0.00000000
1	0.69845600	0.69845600
2	0.91629562	0.91629562
3	0.28578105	0.28578105
4	– 0.12870811	– 0.12870811
5	0.37253251	0.37253251
6	0.94335230	0.94335230
7	0.62014317	0.62014317
8	– 0.05148147	– 0.05148147
9	0.06225640	0.06225640

8.4.4 Second-Order Plant with Complex Roots

Let us consider a system with complex conjugate poles having a transfer function:

$$G4(s) = \left(s^2 + \alpha s + \beta\right)^{-1}, \quad \alpha, \beta \text{ real and greater than zero.}$$

This can be decomposed as:

$$G4(s) = \frac{1}{2jb}\left[\frac{1}{s + a - jb} - \frac{1}{s + a + jb}\right]$$

where

$$a = \frac{1}{2}\alpha \quad \text{and} \quad b = \frac{1}{2}\sqrt{\left(4\beta - \alpha^2\right)}$$

Again using equation (8.17), the **DOTF** is given by

$$\textbf{DOTF4}_{(m)} = \begin{bmatrix} 0 & w_{12} & w_{13} & w_{14} & \cdots & w_{1j} & \cdots & w_{1m} \end{bmatrix}_{(m \times m)} \qquad (8.26)$$

where

$$w_{1j} = \frac{1}{b} \exp\left[-(j-1)a\right] \sin\left[(j-1)b\right], \quad \text{for } 1 \le j \le m$$

Consider an open loop sampled-data system having a plant transfer function $G4(s)$ with $\alpha = \beta = 1$. For a step input, the output response of the system is computed via both the conventional z-transform method and the δ-domain technique for $m = 10$, $T = 10\,\text{s}$ and $h = 1\,\text{s}$. From Table 8.4, it is noted that the results obtained are the same as those given by the z-transform method.

TABLE 8.4

Exact Output of the Input Sampled System Having the Transfer Function $(s^2 + s + 1)^{-1}$, with Unit Step Input, Obtained via z-Transform Analysis, Compared with δ-Domain Results Obtained Using **DOTFs** ($m = 10$, $T = 10$ s, $h = 1$ s)

t(s)	z-Transform Analysis, C6(z)	δ-Domain Analysis
0	0.00000000	0.00000000
1	0.53350720	0.53350720
2	0.95278682	0.95278682
3	1.08602947	1.08602947
4	1.03649959	1.03649959
5	0.94855717	0.94855717
6	0.89766485	0.89766485
7	0.89002114	0.89002114
8	0.90273623	0.90273623
9	0.91554090	0.91554090

8.5 Conclusion

This chapter used a set of mutually Disjoint Dirac DFs for discrete control system analysis. The principle of delta domain operational technique is presented and using this set, a **DOTF** is defined. This new **DOTF** has been used to analyze linear discrete SISO control systems. Several open loop and closed loop control systems are investigated with different forms of transfer functions using a unit step input. The results are found to match exactly with the solutions obtained by the conventional z-transform analysis. Also, the accuracy of sample values derived by this method is entirely independent of m, or the value of h, for a fixed time period T. So, if we have unequal sampling intervals for any input signal, the output sequence is also spaced in a similar fashion, provided the **DOTF**s are deduced considering the same pattern of unequal sampling intervals. The presented technique can analyze any plant having a specific **DOTF** for any number of inputs in one mathematical operation using equation (8.16). Since DFs are defined in time domain, no inverse transformations are necessary to obtain the solution. Several numerical examples have been treated to establish the effectiveness of the δ-domain operational technique and the results are compared with that of conventional z-transform analysis. Several tables are presented to this effect. It was found that the results match exactly.

References

1. K. G. Beauchamp, *Applications of Walsh and related functions: With an introduction to sequency theory*, Academic Press, London, 1984.
2. Anish Deb, Gautam Sarkar, and Sunit K. Sen, Block pulse functions, the most fundamental of all piecewise constant basis functions, Int. J. Syst. Sci., vol. **25**, no. 2, pp. 351–363, 1994.
3. Anish Deb, Gautam Sarkar, and Sunit K. Sen, Linearly pulse-width modulated block pulse functions and their application to linear SISO feedback control system identification, Proc. IEE, Part D, Control Theory and Appl., vol. **142**, no. 1, pp. 44–50, 1995.
4. Anish Deb, Gautam Sarkar, Sunit K. Sen, and A. K. Dutta, A new set of pulse-width modulated generalized block pulse functions (PWM-GBPF) and its application to cross-/auto-correlation of time-varying functions, Int. J. Syst. Sci., vol. **26**, no. 1, pp. 65–89, 1995.
5. Anish Deb, G. Sarkar, M. Bhattacharjee, and S. K. Sen, All-integrator approach to linear SISO control system analysis using block pulse functions (BPF), J. Franklin Inst., vol. **334B**, no. 2, pp. 319–335, 1997.
6. W. L. Chen and S. G. Wu, Analysis of sampled-data systems by block-pulse functions, Int. J. Syst. Sci., vol. **16**, no. 6, pp. 745–752, 1985.

7. Anish Deb, G. Sarkar, M. Bhattacharjee, and S. K. Sen, Analysis of linear discrete SISO control systems via a set of delta functions, Proceedings IEE, Part – D, Control Theory and Applications, vol. **143**, no. 6, pp. 514–518, 1996.

8. P. AM. Dirac, The physical interpretation of the quantum dynamics, Proc. Roy. Soc. A, vol. **113**, no. 765, pp. 621–641, 1927.

9. G. Temple, The theory of generalized functions, Proc. Roy. Soc. A, vol. **228**, no. 1173, pp. 175–190, 1955.

10. C. F. Chen, Y. T. Tsay, and T. T. Wu, Walsh operational matrices for fractional calculus and their application to distributed systems, J. Franklin Inst., vol. **303**, no. 3, pp. 267–284, 1977.

11. G. P. Rao and K. R. Palanisamy, Improved algorithms for parameter identification in lumped continuous systems via Walsh functions, Proceedings IEE, Part – D, Control Theory and Applications, vol. **130**, no. 1, pp. 9–16, 1983.

12. G. P. Rao, *Piecewise constant orthogonal functions and their applications to systems and control*, LNC1S, vol. **55**, Springer-Verlag, Berlin, 1983.

Study Problems

8.1 Determine the output in z-domain for the following systems when a step input is applied through a sampler:

i. $G(s) = \dfrac{1}{(s+4)}$

ii. $G(s) = \dfrac{1}{(s+3)^2}$

iii. $G(s) = \dfrac{1}{(2s^2+5)}$

iv. $G(s) = \dfrac{1}{(s^2+2s+5)}$

8.2 Consider an open loop system having a transfer function $G(s) = (s+4)^{-1}$ and a sampler at the input. Find its output $c(t)$ by the conventional z-transform analysis, for a unit step input $u(t)$ and also using delta operational transfer function (**DOTF**) for a sampling period $h = 0.125$ s. Consider $m = 8$ and $T = 1$ s.

Compare the results obtained using **DOTF** and through conventional z-transform analysis by estimating the percentage errors.

8.3 Consider an open loop system having a transfer function $G(s) = (s+4)^{-1}$ and a sampler placed before $G(s)$. Find its output $c(t)$ by the conventional z-transform analysis for a sampling period $h = 1$ s with a unit step input. Also, determine the output of the system via δ-domain **DOTF** approach. Consider $m = 10$ and $T = 10$ s.

Compare the results obtained by two methods of analysis by estimating the percentage errors.

8.4   Repeat Problem 8.3 considering $G(s) = (s + 2)^{-2}$.

8.5   Consider an open loop system having a transfer function $G(s) = (s^2 + 4)^{-1}$ and a sampler placed before $G(s)$. Find its output $c(t)$ by the conventional z-transform analysis for a sampling period $h = 0.1$ s with a unit step input. Also, determine the output of the system via δ-domain **DOTF** approach. Consider $m = 10$ and $T = 1$ s.

Compare the results obtained by two methods of analysis by estimating the percentage errors.

8.6   Repeat Problem 8.5 considering $G(s) = (s^2 + s + 4)^{-1}$.

9

Non-Optimal Block Pulse Functions (NOBPFs) for System Analysis and Identification

Different piecewise constant basis functions (PCBF) [1–5] have been utilized in different fields of engineering for more than three decades. Of all PCBFs, the block pulse function (BPF) set [6–8] turned out to be the most efficient and fundamental with respect to system analysis and related computational burden. In this Chapter, a different approach for BPF-based analysis as well as identification of control systems is presented. This approach is "different" from the conventional BPF in the sense that the BPF expansion coefficients of a locally square integrable function have been determined in a more "convenient" manner. That is, this approach computes the BPF coefficients from the samples of the function to be expanded.

This not only reduces the computational burden drastically, but also produces far better results than those obtained via traditional BPF analysis for identification of SISO control systems. In case of analysis of such systems, this "non-optimal" BPF approach is marginally better than the traditional approach, as is apparent from several case studies.

In this Chapter, three commonly used linear SISO systems have been studied with three standard inputs to prove the validity of the proposal.

A set of first eight non-optimal block pulse functions (NOBPFs) [9–11] in the semi-open interval $t \in [0, T)$ is shown in Figure 9.1.

9.1 Basic Properties of Non-Optimal Block Pulse Functions

Basic qualitative properties of the NOBPF set are tabulated in Table 9.1 to provide a qualitative appraisal. Of many properties of the NOBPF sets four important properties are presented hereunder.

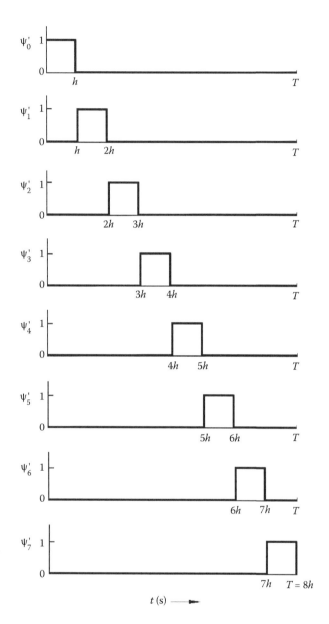

FIGURE 9.1
A set of first eight non-optimal block pulse functions (NOBPFs).

TABLE 9.1

Basic Properties of "Optimal" Block Pulse Function (BPF) and "Non-Optimal" Block Pulse Function (NOBPF)

Properties	BPF	NOBPF
Piecewise constant	Yes	Yes
Orthogonal	Yes	Yes
Finite	Yes	Yes
Disjoint	Yes	Yes
Orthonormal	Can easily be normalized	Can easily be normalized
Implementation	Easily implementable	Easily implementable
Coefficient determination of $f(t)$	Involves integration of $f(t)$ and scaling	Needs only samples of $f(t)$
Accuracy of analysis	Provides staircase solution	Provides staircase solution

9.1.1 Disjointedness

The NOBPFs are disjointed with each other in the interval $t \in [0, T)$. That is

$$\psi_i'(t)\psi_j'(t) = \begin{cases} \psi_i'(t) & \text{for } i = j \\ 0 & \text{for } i \neq j \end{cases} \tag{9.1}$$

where $i, j = 0, 1, 2, \ldots, (m - 1)$.

This property can directly be obtained from the NOBPF components shown in Figure 9.1.

For definition of two block pulses $\psi_i'(t)$ and $\psi_j'(t)$ we have [multiplying the numerator and denominator by $\psi_j'(t)$]

$$\frac{\psi_i'(t)}{\psi_j'(t)} = \frac{\psi_i'(t)\,\psi_j'(t)}{\psi_j'(t)\,\psi_j'(t)}$$

By virtue of Figure 9.1 and equation (9.1)

$$\frac{\psi_i'(t)}{\psi_j'(t)} = \begin{cases} 1 & \text{for } i = j \\ 0 & \text{for } i \neq j \end{cases} \tag{9.2}$$

9.1.2 Orthogonality

The NOBPFs are orthogonal to each other in the interval $t \in [0, T)$. That is

$$\int_0^T \psi_i'(t)\psi_j'(t)\,dt = \begin{cases} h & \text{for } i = j \\ 0 & \text{for } i \neq j \end{cases} \tag{9.3}$$

where $i, j = 0, 1, 2, \ldots, (m - 1)$.

This property can directly be obtained from the disjointedness of NOBPF sets.

9.1.3 Addition

For addition of two time functions $f(t)$ and $g(t)$, the following cases are considered.

Let, $a(t) = f(t) + g(t)$, where, $a(t)$ is the resulting function.

We call the samples of each functions $f(t)$ and $g(t)$ in $t \in [0, T)$ $f_0, f_1, f_2, \ldots, f_i, \ldots, f_m$ and $g_0, g_1, g_2, \ldots, g_i, \ldots g_m$ respectively.

First the continuous functions $f(t)$ and $g(t)$ can be expanded directly into NOBPF domain separately and then added to find the resultant function in NOBPF domain. The NOBPF domain expanded form of $a(t)$ may be called $\bar{a}(t)$. It should be noted that $\bar{a}(t)$ is piecewise constant in nature. Thus

$$a(t) = f(t) + g(t)$$

$$\approx \left[\frac{f_0 + f_1}{2} \quad \frac{f_1 + f_2}{2} \quad \frac{f_2 + f_3}{2} \quad \cdots \quad \frac{f_{(i-1)} + f_i}{2} \quad \cdots \quad \frac{f_{(m-1)} + f_m}{2} \right] \Psi'_{(m)}$$

$$+ \left[\frac{g_0 + g_1}{2} \quad \frac{g_1 + g_2}{2} \quad \frac{g_2 + g_3}{2} \quad \cdots \quad \frac{g_{(i-1)} + g_i}{2} \quad \cdots \quad \frac{g_{(m-1)} + g_m}{2} \right] \Psi'_{(m)}$$

$$= \sum_{i=0}^{m-1} \frac{f_i + f_{(i+1)}}{2} \psi'_i + \sum_{i=0}^{m-1} \frac{g_i + g_{(i+1)}}{2} \psi'_i \tag{9.4}$$

That means

$$\bar{a}(t) = \sum_{i=0}^{m-1} \frac{f_i + g_i}{2} \psi'_i + \sum_{i=0}^{m-1} \frac{f_{(i+1)} + g_{(i+1)}}{2} \psi'_i \tag{9.5}$$

Equation (9.4) shows that the NOBPF coefficients of the sum $a(t)$ are the sums of the NOBPF coefficients of the individual functions $f(t)$ and $g(t)$, in each subinterval.

Now, the continuous functions $f(t)$ and $g(t)$ are first added and then the resulting function $a(t) = f(t) + g(t)$ is expressed in NOBPF domain. In this case, the resulting continuous function expanded in NOBPF domain is given as

$$a(t) \approx \bar{a}(t) = \left[\frac{(f_0 + g_0) + (f_1 + g_1)}{2} \quad \frac{(f_1 + g_1) + (f_2 + g_2)}{2} \quad \cdots \right.$$

$$\left. \cdots \quad \frac{(f_{(m-1)} + g_{(m-1)}) + (f_m + g_m)}{2} \right] \Psi'_{(m)}$$

$$= \left[\frac{a_0 + a_1}{2} \quad \frac{a_1 + a_2}{2} \quad \frac{a_2 + a_3}{2} \quad \cdots \quad \frac{a_{(i-1)} + a_i}{2} \quad \cdots \quad \frac{a_{(m-1)} + a_m}{2} \right] \Psi'_{(m)}$$

$$= \sum_{i=0}^{m-1} \frac{a_i + a_{(i+1)}}{2} \psi'_i \qquad (9.6)$$

where $a_0, a_1, a_2, \cdots, a_i, \cdots, a_m$ are the samples of $a(t)$ at time instants $0, h, 2h, \cdots, ih, \cdots, mh$.

From equations (9.5) and (9.6), it is seen that both the results of addition are identical.

9.1.4 Subtraction

For subtraction of two time functions $f(t)$ and $g(t)$, the following cases are considered.

Let, $s(t) = f(t) - g(t)$, where, $s(t)$ is the resulting function.

As before, the samples of individual functions $f(t)$ and $g(t)$ in time domain are $f_0, f_1, f_2, \cdots, f_i, \cdots, f_m$ and $g_0, g_1, g_2, \cdots, g_i, \cdots, g_m$ respectively.

The functions $f(t)$ and $g(t)$ are expressed in NOBPF domain. Then we subtract the expanded functions to obtain $s(t)$ in NOBPF domain. The NOBPF domain expanded form of $s(t)$ may be called $\bar{s}(t)$.

First the continuous functions $f(t)$ and $g(t)$ can be expanded directly into NOBPF domain separately and then subtracted to find the resultant function in NOBPF domain.

The difference of two time functions $f(t)$ and $g(t)$ can be expressed via NOBPF domain as

$$s(t) = f(t) - g(t)$$

$$\approx \left[\frac{f_0 + f_1}{2} \quad \frac{f_1 + f_2}{2} \quad \frac{f_2 + f_3}{2} \quad \cdots \quad \frac{f_{(i-1)} + f_i}{2} \quad \cdots \quad \frac{f_{(m-1)} + f_m}{2} \right] \Psi'_{(m)}$$

$$- \left[\frac{g_0 + g_1}{2} \quad \frac{g_1 + g_2}{2} \quad \frac{g_2 + g_3}{2} \quad \cdots \quad \frac{g_{(i-1)} + g_i}{2} \quad \cdots \quad \frac{g_{(m-1)} + g_m}{2} \right] \Psi'_{(m)}$$

$$= \sum_{i=0}^{m-1} \frac{f_i + f_{(i+1)}}{2} \psi'_i - \sum_{i=0}^{m-1} \frac{g_i + g_{(i+1)}}{2} \psi'_i \tag{9.7}$$

That means

$$\bar{s}(t) = \sum_{i=0}^{m-1} \frac{f_i - g_i}{2} \psi'_i + \sum_{i=0}^{m-1} \frac{f_{(i+1)} - g_{(i+1)}}{2} \psi'_i \tag{9.8}$$

Equation (9.8) shows that the NOBPF coefficients of the difference of two functions $s(t)$ is the differences of the NOBPF coefficients of the individual functions $f(t)$ and $g(t)$, at each sampling points.

Now, the continuous functions $f(t)$ and $g(t)$ are first subtracted and then the resulting function $s(t) = f(t) - g(t)$ is expressed in NOBPF domain. In this case, the resulting continuous function expanded in NOBPF domain is given as

$$s(t) \approx \bar{s}(t) = \left[\frac{(f_0 - g_0) + (f_1 - g_1)}{2} \quad \frac{(f_1 - g_1) + (f_2 - g_2)}{2} \quad \cdots \right.$$

$$\left. \cdots \quad \frac{(f_{(m-1)} - g_{(m-1)}) + (f_m - g_m)}{2} \right] \Psi'_{(m)}$$

$$= \left[\frac{s_0 + s_1}{2} \quad \frac{s_1 + s_2}{2} \quad \frac{s_2 + s_3}{2} \quad \cdots \quad \frac{s_{(i-1)} + s_i}{2} \quad \cdots \quad \frac{s_{(m-1)} + s_m}{2} \right] \Psi'_{(m)}$$

$$= \sum_{i=0}^{m-1} \frac{s_i + s_{(i+1)}}{2} \psi'_i \tag{9.9}$$

where $s_0, s_1, s_2, \cdots, s_i, \cdots, s_m$ are the samples of $s(t)$ at time instants $0, h, 2h, \cdots, ih, \cdots, mh$.

From equations (9.8) and (9.9), it is seen that both the results of subtraction are identical.

9.1.5 Multiplication

For two functions $f(t)$ and $g(t)$, we determine the NOBPF series of their product using the NOBPF expansion of each function.

That is

$$
f(t)g(t) \approx \left(\sum_{i=0}^{m-1} \frac{f_i + f_{(i+1)}}{2} \psi_i' \right) \left(\sum_{i=0}^{m-1} \frac{g_i + g_{(i+1)}}{2} \psi_i' \right)
$$
$$
= \sum_{i=0}^{m-1} \frac{\left(f_i + f_{(i+1)}\right)}{2} \cdot \frac{\left(g_i + g_{(i+1)}\right)}{2} \psi_i'(t) \tag{9.10}
$$

where either of $f(t)$ or $g(t)$ is not equal to zero.

In deriving equation (9.10), the cross products will be zero, because of the disjointedness property of the NOBPFs.

Now, the product of two functions $f(t)$ or $g(t)$ may be expressed in NOBPF domain as under.

$$
f(t)g(t) \approx \left[\frac{f_0 g_0 + f_1 g_1}{2} \quad \frac{f_1 g_1 + f_2 g_2}{2} \quad \cdots \quad \frac{f_{(m-1)} g_{(m-1)} + f_m g_m}{2} \right] \Psi_{(m)}'
$$
$$
= \sum_{i=0}^{m-1} \frac{\left(f_i g_i + f_{(i+1)} g_{(i+1)}\right)}{2} \psi_i'(t) \tag{9.11}
$$

From equations (9.10) and (9.11), it is clear that the product of two time functions expanded in NOBPF domain is not same as the NOBPF expansion of the product of two time functions.

9.1.6 Division

For functions $f(t)$ and $g(t)$, we have the NOBPF series of their quotient

$$
f(t)/g(t) \approx \left(\sum_{i=0}^{m-1} \frac{f_i + f_{(i+1)}}{2} \psi_i' \right) \bigg/ \left(\sum_{i=0}^{m-1} \frac{g_i + g_{(i+1)}}{2} \psi_i' \right); \quad \text{for } g(t) \neq 0
$$
$$
= \sum_{i=0}^{m-1} \left(f_i + f_{(i+1)}\right) \big/ \left(g_i + g_{(i+1)}\right) \psi_i'(t) \tag{9.12}
$$

Now, the quotient of two time functions $f(t)$ or $g(t)$ may be expressed in NOBPF domain as under.

$$f(t)/g(t) \approx \left[\frac{\frac{f_0}{g_0} + \frac{f_1}{g_1}}{2} \quad \frac{\frac{f_1}{g_1} + \frac{f_2}{g_2}}{2} \quad \cdots \quad \frac{\frac{f_{(m-1)}}{g_{(m-1)}} + \frac{f_m}{g_m}}{2} \right] \Psi'_{(m)}$$

$$= \sum_{i=0}^{m-1} \frac{\frac{f_i}{g_i} + \frac{f_{(i+1)}}{g_{(i+1)}}}{2} \psi'_i(t) \tag{9.13}$$

From equations (9.12) and (9.13), it is clear that the division of two time functions expanded in NOBPF domain is not same as the NOBPF expansion of the division of two time functions.

9.2 From "Optimal" Coefficients to "Non-Optimal" Coefficients [10]

From the standard integration formula equation (1.14) for deriving the coefficient in conventional BPF domain, let us employ the well-known trapezoidal rule [12] for integration to compute the BPF coefficients. Calling these coefficients c'_i s, we get

$$c'_i \approx \frac{1}{h} \frac{\left[f(ih) + f\{(i+1)h\} \right]}{2} h = \frac{\left[f(ih) + f\{(i+1)h\} \right]}{2} \tag{9.14}$$

c'_i s are "non-optimal" coefficients because the integral in equation (1.14) has been computed approximately via equation (9.14). It is observed that c'_i s are, in effect, the average values of consecutive samples of $f(t)$.

Following this method of computation, the function approximated in NOBPF domain may be expressed in the following form

$$f(t) \approx \left[c'_0 \quad c'_1 \quad c'_2 \quad \cdots \quad c'_i \quad \cdots \quad c'_{(m-1)} \right] \Psi'_{(m)}(t) = \mathbf{C}'^{\mathrm{T}} \Psi'_{(m)}(t) \tag{9.15}$$

It is evident that $f(t)$ in equation (9.15) will not be approximated via NOBPF with guaranteed minimum mean integral square error (MISE).

In spite of this shortcoming, there is an obvious advantage using the non-optimal coefficients. Comparing equations (1.14) and (9.14), we find that the hassle of performing exact integration for determination of each coefficient in equation (1.14) is reduced in equation (9.14), because, each coefficient is the average of two consecutive samples of the function taken at the limits of the respective subintervals.

9.3 Function Approximation Using Non-Optimal Block Pulse Functions (NOBF) [9,10]

The "non-optimal" method of BPF coefficient computation has been suggested by Deb et al. [10] which employs trapezoidal [12] integration instead of exact integration. The approach is "new" in the sense that the BPF expansion coefficients of a locally square integrable function have been determined in a more "convenient" manner.

The process of function approximation in NOBPF domain has already been shown in Figure 1.11. For convenience of the reader, the figure is reproduced in Figure 9.2.

A time function $f(t)$ can be approximated in NOBPF domain following equation (9.15).

At first we determine the coefficients of a time function $f(t)$ in NOBPF domain using equation (9.15). We consider the following two examples here.

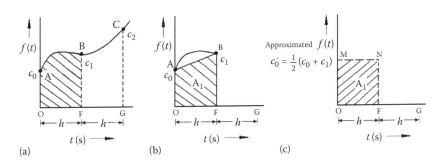

FIGURE 9.2
Function approximation in non-optimal block pulse function (NOBPF) domain where area preserving transformation is employed. (a) The function $f(t)$ to be approximated, (b) its piecewise linear approximation with area under the curve AOFB = A_1 and (c) its NOBPF approximation with area under the curve MOFN = A_1.

9.3.1 Numerical Examples

Example 9.1

Let us consider the function $f(t) = \sin(\pi t)$ for $t \geq 0$. Following equations (1.14) and (9.15), we expand the function $f(t)$ via BPF and NOBPF domain to obtain both optimal as well as non-optimal coefficients. Taking $T = 1$ s, number of component functions $m = 8$ and $h = 0.125$ s, the resulting coefficients are compared along with percentage errors in Table 9.2.

Example 9.2

Let us consider the function $f(t) = t^2$ for $t \geq 0$. Approximation of this function in a similar manner with $T = 1$ s and $m = 8$, yields the results shown in Table 9.3 and are compared along with percentage errors.

The results in the error columns of both the Tables 9.2 and 9.3 are as expected from the theory. The sine curve being the convex nature, c_i is always greater than c'_i, while t^2 is of the concave nature thereby making c_i less than c'_i. This is due to the fact that c_i is computed via exact integration and c'_i is computed via the approximate trapezoidal equivalent of the area under the function in the ith subinterval.

TABLE 9.2

Approximation of the Function $f(t) = \sin(\pi t)$ of Example 9.1 via Block Pulse Function (BPF) to Determine Optimal (c_i) and Non-Optimal (c'_i) Coefficients, for $m = 8$ and $T = 1$ s (Vide Appendix B, Program No. 9.1)

t (sec)	Samples of the Function $f(t) = \sin(\pi t)$	Optimal Expansion Coefficients, c_i	Non-optimal Expansion Coefficients, c'_i	Percentage Error $\in = \dfrac{c_i - c'_i}{c_i} \times 100$
0.000	0.00000000			
		0.19383918	0.19134172	1.28841990
0.125	0.38268343			
		0.55200728	0.54489511	1.28841990
0.250	0.70710678			
		0.82613727	0.81549316	1.28841990
0.375	0.92387953			
		0.97449536	0.96193977	1.28841990
0.500	1.00000000			
		0.97449536	0.96193977	1.28841990
0.625	0.92387953			
		0.82613727	0.81549316	1.28841990
0.750	0.70710678			
		0.55200728	0.54489511	1.28841990
0.875	0.38268343			
		0.19383918	0.19134172	1.28841990
1.000	0.00000000			

TABLE 9.3

Approximation of the Function $f(t) = t^2$ of Example 9.2 via Block Pulse Function (BPF) to Determine Optimal (c_i) and Non-Optimal (c_i') Coefficients, for $m = 8$ and $T = 1$ s

t (sec)	Samples of the Function $f(t) = t^2$	Optimal Expansion Coefficients, c_i	Non-optimal Expansion Coefficients, c_i'	Percentage Error $\in = \dfrac{c_i - c_i'}{c_i} \times 100$
0.000	0.00000000			
		0.00520833	0.00781250	− 50.00000000
0.125	0.01562500			
		0.03645833	0.03906250	− 7.14285714
0.250	0.06250000			
		0.09895833	0.10156250	− 2.63157895
0.375	0.14062500			
		0.19270833	0.19531250	− 1.35135135
0.500	0.25000000			
		0.31770833	0.32031250	− 0.81967213
0.625	0.39062500			
		0.47395833	0.47656250	− 0.54945055
0.750	0.56250000			
		0.66145833	0.66406250	− 0.39370079
0.875	0.76562500			
		0.88020833	0.88281250	− 0.29585799
1.000	1.00000000			

9.4 Operational Matrices for Integration

As was done in case of BPFs, now we present the operational matrix for integration in the NOBPF domain.

The basic functions of the NOBPF set shown in Figure 9.1 are look-alikes of the members of the BPF set. So, the first integrations of the NOBPFs may be expressed as

$$\int \Psi'_{(m)}(t)\,dt \approx N_{(m)}(t) = P1_{n(m)} \Psi'_{(m)}(t) \tag{9.16}$$

where $\Psi'_{(m)}(t)$ is the m-term NOBPF vector and $P1_{n(m)}$ is the operational matrix for integration of order m in the NOBPF domain given by

$$\mathbf{P1}_{n(m)} = h \begin{bmatrix} \frac{1}{2} & 1 & 1 & \cdots & 1 \\ & \frac{1}{2} & 1 & \cdots & 1 \\ & & \frac{1}{2} & \ddots & \vdots \\ \mathbf{0} & & & \ddots & 1 \\ & & & & \frac{1}{2} \end{bmatrix}_{(m \times m)} \tag{9.17}$$

From equation (9.17), it is noted that the first-order integration of the NOBPF vector and that of BPF vector presented in equation (3.10) are identical, because the result of first-order integration of NOBPF set is piecewise linear.

In double integration of the NOBPF set, except the first term all other terms are piecewise linear. That means, integration operational matrix for NOBPF set will remain same as that of BPF domain operational matrix, till the integration(s) of the basis function remain piecewise linear.

9.5 The Process of Convolution and "Deconvolution"

The process of convolution in NOBPF domain is somewhat identical to what we have presented in Section 5.1 for BPF domain convolution. That is, the final equation for convolution in NOBPF domain is of the form

$$\hat{\mathbf{C}}'^{\mathrm{T}} = \frac{h}{2} \mathbf{G}'^{\mathrm{T}} \mathbf{C}'_c \tag{9.18}$$

where $\hat{\mathbf{C}}'$ is the output NOBPF vector obtained through convolution. In fact, $\hat{\mathbf{C}}'$ is given by

$$\hat{\mathbf{C}}' \triangleq \begin{bmatrix} \hat{c}_0 & \hat{c}_1 & \cdots & \hat{c}_i & \cdots & \hat{c}_{(m-1)} \end{bmatrix}^{\mathrm{T}}$$

and $\hat{c}_i$ is the $(i + 1)$th non-optimal BPF coefficients of $\hat{\mathbf{C}}'$; $\mathbf{G}'$ is the NOBPF coefficient vector obtained from the impulse response of the plant $G(s)$; and $\mathbf{C}'_c$ is the convolution matrix.

The difference between $\mathbf{C}_c$ of equation (5.14) and $\mathbf{C}'_c$ of equation (9.18) is, while $\mathbf{C}_c$ has been formed using the *optimal* block pulse coefficients of one of the convolving functions $e(t)$, the matrix $\mathbf{C}'_c$ has been formed by equivalent NOBPF coefficients of the same function.

As the process of convolution is identical for both the cases, Figures 5.2 and 5.3 represent in detail the mechanism of convolution in NOBPF domain as well.

Similarly, the process of "deconvolution" in NOBPF domain is given by

$$\hat{\mathbf{G}}'^{\mathrm{T}} = \frac{2}{h}\mathbf{C}'^{\mathrm{T}}\mathbf{C}_c'^{-1} \tag{9.19}$$

where $\hat{\mathbf{G}}'$ is the NOBPF vector of the impulse response function of the plant obtained through "deconvolution", and $\mathbf{C}'$ is the direct expansion of the output in NOBPF domain.

9.6 Analysis of an Open-Loop System via Convolution

Consider the open-loop system shown in Figure 9.3. Our task is to determine $c(t)$ knowing $r(t)$ and $g(t)$. Using both optimal and non-optimal coefficients of BPF expansion of $r(t)$ and $g(t)$, we can solve for $\mathbf{C}$ and $\mathbf{C}'$ using either of equation (5.14) or equation (9.18).

Now we analyze several open-loop systems with three kinds of inputs for each via convolution using both optimal and non-optimal block pulse coefficients. We choose $T = 1$ s and $m = 8$ for all the cases studied in the following section.

9.6.1 First-Order System

We consider a system having a impulse response of $g_1(t) = \exp(-t)$, $t \geq 0$. Time responses of the system in optimal block pulse function (OBPF) and NOBPF domain, for different inputs, are presented below.

For a unit step input $u(t)$, the response of the system is $c_1(t) = 1-\exp(-t)$, $t \geq 0$. The system is analyzed using the respective convolution matrices both in OBPF and NOBPF domain for $m = 8$ and $T = 1$ s. Also, we have derived the result in these two domains by expanding the output $c_1(t)$ directly for $m = 8$ and $T = 1$ s. The OBPF sets of results are compared with corresponding NOBPF sets of results in Table 9.4. Further, we have computed coefficient-wise percentage error both in OBPF and NOBPF domain for clarity in comparison.

The same system has been analyzed for two more inputs, a unit ramp input and a unit parabolic input. The respective outputs are given by $c_2(t) = t - 1 + \exp(-t)$ and $c_3(t) = t^2 - 2t + 2(1-\exp(-t))$. As was done for Table 9.4, we present

FIGURE 9.3
Plant modeled by impulse response function.

TABLE 9.4

Comparison of Output Coefficients for the First-Order open Loop System Analysis with Unit Step Input Employing Optimal and Non-Optimal Block Pulse Functions (NOBPFs) and Related Percentage Error, for $m = 8$ and $T = 1$ s

t (s)	Output via Convolution, c_{ci}		Output via Direct Expansion, c_{di}		Percentage Error, ε_i	
	OBPF	NOBPF	OBPF	NOBPF	OBPF	NOBPF
0.000						
	0.05875155	0.05882803	0.05997522	0.05875155	2.04029590	– 0.13017444
0.125						
	0.16935116	0.16957161	0.17043104	0.16935116	0.63362090	– 0.13017444
0.250						
	0.26695497	0.26730248	0.26790797	0.26695497	0.35571794	– 0.13017444
0.375						
	0.35309003	0.35354966	0.35393105	0.35309003	0.23762160	– 0.13017444
0.500						
	0.42910396	0.42966254	0.42984615	0.42910396	0.17266517	– 0.13017444
0.625						
	0.49618601	0.49683192	0.49684099	0.49618601	0.13182978	– 0.13017444
0.750						
	0.55538571	0.55610868	0.55596374	0.55538571	0.10396752	– 0.13017444
0.875						
	0.60762927	0.60842025	0.60813937	0.60762927	0.08387919	– 0.13017444
1.000						

Tables 9.5 and 9.6 in the same spirit to bring out the comparison of OBPF and NOBPF analysis for $m = 8$ and $T = 1$ s.

From Table 9.4, it is observed that the percentage error in OBPF domain analysis varies from 0.08387919 to 2.04029590 while for NOBPF domain analysis, it is fixed at –0.13017444. The constancy of error as seen from the last column of the table, is only case specific, and is absent for any input other than step input.

Now, let us define an index called "Average of Mod of Percentage" (AMP) error, which is given by

$$\text{AMP error, } \varepsilon_{\text{av}(m)} \triangleq \frac{\sum\limits_{j=1}^{m} |\varepsilon_j|}{m} \qquad (9.20)$$

where m is the number of subintervals,
ε_j is the percentage error at each subinterval.

TABLE 9.5

Comparison of Output Coefficients for the First-Order Open Loop System Analysis with Unit Ramp Input Employing Optimal and Non-Optimal Block Pulse Functions (NOBPFs) and Related Percentage Error, for $m = 8$ and $T = 1$ s (Vide Appendix B, Program No. 9.2)

t (s)	Output via Convolution, c_{ci}		Output via Direct Expansion, c_{di}		Percentage Error, ε_i	
	OBPF	NOBPF	OBPF	NOBPF	OBPF	NOBPF
0.000						
	0.00367197	0.00367675	0.00252478	0.00374845	− 45.43733626	1.91277741
0.125						
	0.01792839	0.01795173	0.01706896	0.01814884	− 5.03507537	1.08609533
0.250						
	0.04519752	0.04525636	0.04459203	0.04554503	− 1.35784242	0.63381559
0.375						
	0.08395034	0.08405962	0.08356895	0.08440997	− 0.45637004	0.41505889
0.500						
	0.13283746	0.13301038	0.13265385	0.13339604	− 0.13841355	0.28911145
0.625						
	0.19066808	0.19091628	0.19065901	0.19131399	− 0.00476089	0.20788142
0.750						
	0.25639132	0.25672507	0.25653626	0.25711429	0.05650216	0.15137793
0.875						
	0.32907975	0.32950813	0.32936063	0.32987073	0.08527905	0.10992193
1.000						

In Table 9.7, we present the AMP error for optimal as well as non-optimal BPF coefficient based analysis for three different inputs for the open loop first-order system investigated and in the last column of the table the error difference is tabulated.

It is observed that the difference is always positive indicating greater AMP error for optimal BPF coefficient based analysis.

9.6.2 Undamped Second-Order System

We consider a second-order undamped system having a impulse response of $g_2(t) = \sin t, t \geq 0$. Time responses of the system in OBPF and NOBPF domain, for different inputs, are presented below.

For a unit step input $u(t)$, the response of the system is $c_4(t) = 1 - \cos t, t \geq 0$. The system is analyzed using the respective convolution matrices both in OBPF and NOBPF domain for $m = 8$ and $T = 1$ s. Also, we have derived

TABLE 9.6

Comparison of Output Coefficients for the First-Order Open Loop System Analysis with Unit Parabolic Input Employing Optimal and Non-Optimal Block Pulse Functions (NOBPFs) and Related Percentage Error, for $m = 8$ and $T = 1$ s

t (s)	Output via Convolution, c_{ci}		Output via Direct Expansion, c_{di}		Percentage Error, ε_i	
	OBPF	NOBPF	OBPF	NOBPF	OBPF	NOBPF
0.000						
	0.00030600	0.00045959	0.00015877	0.00031560	– 92.72445474	– 45.62665850
0.125						
	0.00271802	0.00316315	0.00232042	0.00276481	– 17.13490917	– 14.40746769
0.250						
	0.01035459	0.01106417	0.00977426	0.01047244	– 5.93723500	– 5.65032705
0.375						
	0.02627376	0.02722866	0.02557043	0.02649256	– 2.75056128	– 2.77851896
0.500						
	0.05317428	0.05436241	0.05240063	0.05352041	– 1.47640542	– 1.57323239
0.625						
	0.09343778	0.09485325	0.09264032	0.09393452	– 0.86081229	– 0.97805008
0.750						
	0.14916604	0.15080841	0.14838580	0.14983393	– 0.52581501	– 0.65037829
0.875						
	0.22221387	0.22408757	0.22148708	0.22307104	– 0.32814311	– 0.45569610
1.000						

TABLE 9.7

Comparison of Average of Mod of Percentage (AMP) Errors for Optimal and Non-Optimal Block Pulse Function (BPF) Coefficients Based Analysis of First-Order System Having Impulse Response of $g_1(t) = \exp(-t)$, for $m = 8$ and $T = 1$ s

Input	AMP Error		Error Difference, $\varepsilon_{av,OBPF} - \varepsilon_{av,NOBPF}$
	OBPF, $\varepsilon_{av,OBPF}$	NOBPF, $\varepsilon_{av,NOBPF}$	
Unit step	0.46994975	0.13017444	0.33977531
Unit ramp	6.57144747	0.60075499	5.97069248
Unit parabolic	15.21729200	9.01504113	6.20225087

the result in these two domains by expanding the output $c_4(t)$ directly for $m = 8$ and $T = 1$ s. The OBPF sets of results are compared with corresponding NOBPF sets of results in Table 9.8. Further, we have computed coefficient-wise percentage error both in OBPF and NOBPF domain for clarity in comparison.

TABLE 9.8

Comparison of Output Coefficients for the Second-Order Undamped Open Loop System Analysis with Unit Step Input Employing Optimal and Non-Optimal Block Pulse Functions and Related Percentage Error, for $m = 8$ and $T = 1$ s

t (s)	Output via Convolution, c_{ci}		Output via Direct Expansion, c_{di}		Percentage Error, ε_i	
	OBPF	**NOBPF**	**OBPF**	**NOBPF**	**OBPF**	**NOBPF**
0.000						
	0.00390117	0.00389609	0.00260213	0.00390117	– 49.92187210	0.13024225
0.125						
	0.01944496	0.01941963	0.01816619	0.01944496	– 7.03924307	0.13024225
0.250						
	0.05028998	0.05022448	0.04905144	0.05028998	– 2.52497543	0.13024225
0.375						
	0.09595491	0.09582993	0.09477592	0.09595491	– 1.24397019	0.13024225
0.500						
	0.15572716	0.15552434	0.15462613	0.15572716	– 0.71206207	0.13024225
0.625						
	0.22867401	0.22837618	0.22766810	0.22867401	– 0.44182846	0.13024225
0.750						
	0.31365714	0.31324862	0.31276206	0.31365714	– 0.28618375	0.13024225
0.875						
	0.40935042	0.40881727	0.40858014	0.40935042	– 0.18852570	0.13024225
1.000						

The same system has been analyzed for two more inputs, a unit ramp input and a unit parabolic input. The respective outputs are given by $c_5(t) = t - \sin (t)$ and $c_6(t) = t^2 - 2(1 - \cos t)$. As was done for Table 9.8, we present Tables 9.9 and 9.10 in the same spirit to bring out the comparison of OBPF and NOBPF analysis for $m = 8$ and $T = 1$ s.

From Table 9.8, it is observed that the percentage error in OBPF domain analysis varies from – 49.92187210 to – 0.18852570 while for NOBPF domain analysis, it is fixed at 0.13024225. The constancy of error as seen from the last column of the table, is only case specific, and is absent for any input other than step input.

As in Table 9.7, we tabulate below the AMP error in Table 9.11, for OBPF as well as NOBPF coefficient based analysis and compute the error difference to judge relative strength. It is noted that NOBPF coefficient based analysis contains less AMP error compared to its OBPF counterpart.

TABLE 9.9

Comparison of Output Coefficients for the Second-Order Undamped Open Loop System Analysis with Unit Ramp Input Employing Optimal and Non-Optimal Block Pulse Functions and Related Percentage Error, for $m = 8$ and $T = 1$ s

	Output via Convolution, c_{ci}		Output via direct Expansion, c_{di}		Percentage Error, ε_i	
t (s)	OBPF	NOBPF	OBPF	NOBPF	OBPF	NOBPF
0.000						
	0.00024382	0.00024351	0.00008134	0.00016263	− 199.76566279	− 49.72661047
0.125						
	0.00170296	0.00170074	0.00121804	0.00146065	− 39.81160883	− 16.43674160
0.250						
	0.00606139	0.00605349	0.00526160	0.00566176	− 15.20045267	− 6.91902952
0.375						
	0.01520169	0.01518190	0.01409952	0.01465097	− 7.81710624	− 3.62385033
0.500						
	0.03093182	0.03089154	0.02954446	0.03023859	− 4.69584660	− 2.15930326
0.625						
	0.05495690	0.05488532	0.05330599	0.05413198	− 3.09702756	− 1.39166466
0.750						
	0.08885259	0.08873687	0.08696391	0.08790887	− 2.17179550	− 0.94188472
0.875						
	0.13404056	0.13386599	0.13194358	0.13299276	− 1.58930293	− 0.65660036
1.000						

Figure 9.4 shows the convolution result with unit step input obtained via NOBPF approach along with the OBPF-based analysis (vide Table 9.10). Also, the actual curve for the output $c_6(t)$ is shown for graphical comparison.

9.6.3 Underdamped Second-Order System

We consider a second-order underdamped system having an impulse response of

$$g_3(t) = \frac{b}{\sqrt{1-a^2}} \left[\exp(-abt) \sin\left\{ \left(b\sqrt{1-a^2} \right) t \right\} \right], t \geq 0.$$

Considering $a = 0.4$ and $b = 5$ rad/sec.

TABLE 9.10

Comparison of Output Coefficients for the Second-Order Undamped Open Loop System Analysis with Unit Parabolic Input Employing Optimal and Non-Optimal Block Pulse Functions and Related Percentage Error, for $m = 8$ and $T = 1$ s

t (s)	Output via Convolution, c_{ci}		Output via direct Expansion, c_{di}		Percentage Error, ε_i	
	OBPF	NOBPF	OBPF	NOBPF	OBPF	NOBPF
0.000						
	0.00002032	0.00003044	0.00000407	0.00001017	− 399.53509595	− 199.37524123
0.125						
	0.00022319	0.00027347	0.00012595	0.00017259	− 77.20694471	− 58.45078404
0.250						
	0.00111340	0.00124275	0.00085545	0.00098254	− 30.15417606	− 26.48268145
0.375						
	0.00365237	0.00389717	0.00315649	0.00340268	− 15.71004811	− 14.53227658
0.500						
	0.00926340	0.00965635	0.00845608	0.00885818	− 9.54722373	− 9.01052761
0.625						
	0.01980953	0.02037846	0.01862213	0.01921449	− 6.37629331	− 6.05776704
0.750						
	0.03756440	0.03833123	0.03593421	0.03674823	− 4.53661089	− 4.30770121
0.875						
	0.06517685	0.06615659	0.06304805	0.06411166	− 3.37646092	− 3.18962846
1.000						

TABLE 9.11

Comparison of Average of Mod of Percentage (AMP) Errors for Optimal and Non-Optimal Block Pulse Function (BPF) Coefficients Based Analysis of Second-Order Undamped System Having Impulse Response of $g_2(t) = \sin t$, for $m = 8$ and $T = 1$ s

Input	AMP Error		Error Difference, $\varepsilon_{av,OBPF} - \varepsilon_{av,NOBPF}$
	OBPF, $\varepsilon_{av,OBPF}$	NOBPF, $\varepsilon_{av,NOBPF}$	
Unit step	7.79483259	0.13024225	7.66459034
Unit ramp	34.26860039	10.23196061	24.03663978
Unit parabolic	68.30535671	40.17582595	28.12953076

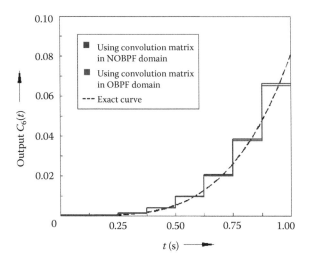

FIGURE 9.4
Analysis of the system having impulse response of $g_2(t) = \sin t$, based on optimal and non-optimal block pulse function (BPF) coefficient with unit parabolic input (vide Table 9.10), for $m = 8$ and $T = 1$ s.

Time responses of the system in OBPF and NOBPF domain, for different inputs, are presented below.

For a unit step input $u(t)$, the response of the system is

$$c_7(t) = 1 - \exp(-abt)\left[\cos\left\{\left(b\sqrt{1-a^2}\right)t\right\} + \frac{a}{\sqrt{1-a^2}}\sin\left\{\left(b\sqrt{1-a^2}\right)t\right\}\right], t \geq 0.$$

The system is analyzed using the respective convolution matrices both in OBPF and NOBPF domain for $m = 8$ and $T = 1$ s. Also, we have derived the result in these two domains by expanding the output $c_7(t)$ directly for $m = 8$ and $T = 1$ s. The OBPF sets of results are compared with corresponding NOBPF sets of results in Table 9.12. Further, we have computed coefficient wise percentage error both in OBPF and NOBPF domain for clarity in comparison.

The same system has been analyzed for two more inputs, a unit ramp input and a unit parabolic input. The respective outputs are given by

$$c_8(t) = t - \frac{2a}{b} + \exp(-abt)\left[\frac{2a}{b}\cos\left\{\left(b\sqrt{1-a^2}\right)t\right\} - \frac{(2a^2-1)}{b\sqrt{1-a^2}}\sin\left\{\left(b\sqrt{1-a^2}\right)t\right\}\right]$$

TABLE 9.12

Comparison of Output Coefficients for the Second-Order Underdamped Open Loop System Analysis with Unit Step Input Employing Optimal and Non-Optimal Block Pulse Functions and Related Percentage Error, for $m = 8$ and $T = 1$ s

t (s)	Output via Convolution, c_{ci}		Output via Direct Expansion, c_{di}		Percentage Error, ε_i	
	OBPF	NOBPF	OBPF	NOBPF	OBPF	NOBPF
0.000						
	0.08064479	0.07196323	0.05664590	0.08064479	−42.36652489	10.76519231
0.125						
	0.33498711	0.31008732	0.32765394	0.33498711	−2.23808451	7.43305993
0.250						
	0.68713579	0.65009972	0.69209296	0.68713579	0.71625803	5.38991979
0.375						
	0.99382112	0.95073589	1.00475770	0.99382112	1.08847914	4.33530999
0.500						
	1.18167377	1.13803141	1.19298274	1.18167377	0.94795724	3.69326676
0.625						
	1.24155304	1.20085039	1.24972275	1.24155304	0.65372241	3.27836511
0.750						
	1.20599475	1.16947804	1.20982942	1.20599475	0.31695983	3.02793303
0.875						
	1.12313144	1.09031095	1.12319573	1.12313144	0.00572337	2.92223097
1.000						

and

$$c_9(t) = 2b^2 \left[\frac{(4a^2 - 1)}{b^4} - \frac{2a}{b^3} t + \frac{1}{2b^2} t^2 + \exp(-abt) \left[\frac{(1 - 5a^2 + 4a^4)}{\left(b^2\sqrt{1-a^2}\right)^2} \cos\left\{\left(b\sqrt{1-a^2}\right)t\right\} \right. \right.$$
$$\left. \left. - \frac{a(4a^2 - 3)}{b^4\sqrt{1-a^2}} \sin\left\{\left(b\sqrt{1-a^2}\right)t\right\} \right] \right]$$

As was done for Table 9.12, we present Tables 9.13 and 9.14 in the same spirit to bring out the comparison of OBPF and NOBPF analysis for $m = 8$ and $T = 1$ s.

TABLE 9.13

Comparison of Output Coefficients for the Second-Order Underdamped Open Loop System Analysis with Unit Ramp Input Employing Optimal and Non-Optimal Block Pulse Functions and Related Percentage Error, for $m = 8$ and $T = 1$ s

t (s)	Output via Convolution, c_{ci}		Output via Direct Expansion, c_{di}		Percentage Error, ε_i	
	OBPF	NOBPF	OBPF	NOBPF	OBPF	NOBPF
0.000						
	0.00504030	0.00449770	0.00182399	0.00354037	– 176.33385675	– 27.04049470
0.125						
	0.03101729	0.02837586	0.02390895	0.02755911	– 29.73086714	– 2.96363912
0.250						
	0.09489997	0.08838755	0.08755639	0.09129329	– 8.38726654	3.18286126
0.375						
	0.19995978	0.18843978	0.19466890	0.19734646	– 2.71788938	4.51321962
0.500						
	0.33592821	0.31898773	0.33346694	0.33470523	– 0. 73808701	4.69592297
0.625						
	0.48737989	0.46516785	0.48737746	0.48737433	– 0. 00049901	4.55635009
0.750						
	0.64035162	0.61331337	0.64185150	0.64109634	0. 23367932	4.33366454
0.875						
	0.78592201	0.75455018	0.78789699	0.78691041	0. 25066504	4.11231376
1.000						

As before, error difference of NOBPF-based analysis with OBPF-based analysis is presented in Table 9.15. Inspection of the last column of the table reveals the superiority of NOBPF-based analysis. From the table, it is noted that AMP error in case of NOBPF-based analysis is much less than that of OBPF-based analysis for all the systems and inputs considered. This indicates that nonoptimal BPF coefficient-based analysis is superior to optimal BPF coefficient-based analysis for the systems studied.

For graphical comparison, Figure 9.5 shows the convolution result with unit step input obtained via NOBPF approach along with the OBPF-based analysis (vide Table 9.12) and the actual curve for the output $c_7(t)$.

TABLE 9.14

Comparison of Output Coefficients for the Second-Order Underdamped Open Loop System Analysis with Unit Parabolic Input Employing Optimal and Non-Optimal Block Pulse Functions and Related Percentage Error, for $m = 8$ and $T = 1$ s

t (s)	Output via Convolution, c_{ci}		Output via Direct Expansion, c_{di}		Percentage Error, ε_i	
	OBPF	NOBPF	OBPF	NOBPF	OBPF	NOBPF
0.000						
	0.00042002	0.00056221	0.00009299	0.00022800	-351.71183784	-146.58598566
0.125						
	0.00426487	0.00467141	0.00259515	0.00344462	-64.34000027	-35.61474570
0.250						
	0.01908748	0.01926683	0.01557285	0.01737778	-22.56893194	-10.87048918
0.375						
	0.05514629	0.05387025	0.05003367	0.05265594	-10.21835512	-2.30611302
0.500						
	0.12164309	0.11729869	0.11556029	0.11867292	-5.26374032	1.15800198
0.625						
	0.22440067	0.21531814	0.21801973	0.22127847	-2.92677107	2.69359078
0.750						
	0.36545970	0.35012829	0.35927950	0.36243209	-1.72016616	3.39478867
0.875						
	0.54395970	0.52111123	0.53822564	0.54115065	-1.06536388	3.70311308
1.000						

TABLE 9.15

Comparison of Average of Mod of Percentage (AMP) Errors for Optimal and Non-Optimal Block Pulse Function (BPF) Coefficients Based Analysis of Second-Order Underdamped System Having the Impulse Response of $g_3(t)$, for $m = 8$ and $T = 1$ s

Input	AMP Error		Error Difference, $\varepsilon_{av,OBPF} - \varepsilon_{av,NOBPF}$
	OBPF, $\varepsilon_{av,OBPF}$	NOBPF, $\varepsilon_{av,NOBPF}$	
Unit step	6.04171368	5.10565974	0.93605394
Unit ramp	27.29910127	6.92480826	20.37429301
Unit parabolic	57.47689582	25.79085351	31.68604231

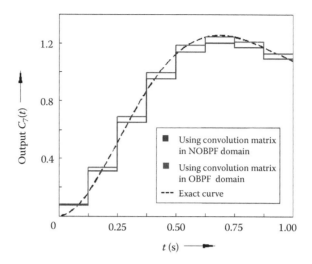

FIGURE 9.5

Analysis of a underdamped system having the impulse response $g_3(t)$, based on optimal and non-optimal block pulse function (BPF) coefficient with unit step input (vide Table 9.12), for $m = 8$ and $T = 1$ s.

9.7 Identification of an Open-Loop System via "Deconvolution"

Consider the open loop-system shown in Figure 9.3. Our task is to determine $g(t)$ using $r(t)$ and $c(t)$ employing optimal and non-optimal coefficients of BPF expansions of $r(t)$ and $c(t)$. We can solve for **G** (OPBF-based approach) or **G'** (NOPBF-based approach), using either of equations (5.17) and (9.19).

Thus, we identify the following first- and second-order systems with three kinds of inputs for each via "deconvolution" using both optimal and non-optimal BPF-based approach. We choose $m = 8$ and $T = 1$ s for all the cases studied.

9.7.1 First-Order System

For the first-order plant to be identified, the output is $c_1(t) = 1-\exp(-t)$, $t \geq 0$, for a step input $u(t)$. The system is identified in OBPF and NOBPF domain, using the respective "deconvolution" operation for $m = 8$ and $T = 1$ s.

The NOBPF sets of "deconvolution" results are compared with corresponding OBPF sets of results in Table 9.16. Also, we have derived the result in these two domains by expanding the impulse response $g_1(t)$ directly for $m = 8$ and $T = 1$ s. Further, we have computed coefficient-wise percentage error both in OBPF and NOBPF domain for clarity in comparison.

The same system has been identified using two more inputs, a unit ramp input and a unit parabolic input. The respective outputs are given

TABLE 9.16

Identification of the First-Order Open Loop System with Unit Step Input and Comparison of Optimal and Non-Optimal Block Pulse Functions (NOBPFs) Coefficients and Related Percentage Error of the Impulse Response, for $m = 8$ and $T = 1$ s

t (s)	System via "Deconvolution", g_{ci}		Direct Expansion of Impulse Response, g_{di}		Percentage Error, ε_i	
	OBPF	NOBPF	OBPF	NOBPF	OBPF	NOBPF
0.000						
	0.95960353	0.94002478	0.94002478	0.94124845	– 2.08279100	0.13000520
0.125						
	0.80768964	0.82956896	0.82956896	0.83064884	2.63743180	0.13000520
0.250						
	0.75194111	0.73209203	0.73209203	0.73304503	– 2.71128136	0.13000520
0.375						
	0.62442820	0.64606895	0.64606895	0.64690997	3.34960468	0.13000520
0.500						
	0.59021345	0.57015385	0.57015385	0.57089604	– 3.51827897	0.13000520
0.625						
	0.48170404	0.50315901	0.50315901	0.50381399	4.26405277	0.13000520
0.750						
	0.46425983	0.44403626	0.44403626	0.44461429	– 4.55448440	0.13000520
0.875						
	0.37055036	0.39186063	0.39186063	0.39237073	5.43822735	0.13000520
1.000						

by $c_2(t) = t - 1 + \exp(-t)$ and $c_3(t) = t^2 - 2t + 2(1 - \exp(-t))$. As was done for Table 9.16, we present Tables 9.17 and 9.18 in the same spirit to bring out the comparison of OBPF and NOBPF-based "deconvolution" results for $m = 8$ and $T = 1$ s.

From the percentage error column in Table 9.16, and last two columns of Tables 9.17 and 9.18, the oscillation in system identification is noted.

The error difference of NOBPF-based analysis with OBPF-based analysis is presented in Table 9.19. Inspection of the last column of the table reveals the superiority of NOBPF-based analysis for all the inputs considered in first-order system.

Figure 9.6 shows the graphical comparison of the "deconvolution" results, with unit step input on first-order system, obtained via NOBPF coefficient-based analysis along with the OBPF-based analysis. Also, the exact curve of the impulse response of the plant $g_1(t)$ is shown for clarity.

9.7.2 Undamped Second-Order System

Now an undamped second-order system is to be identified with a unit step input, where the output of the system is $c_4(t) = 1 - \cos t$, $t \geq 0$. The system is

TABLE 9.17

Identification of the First-Order Open Loop System with Unit Ramp Input and Comparison of Optimal and Non-Optimal Block Pulse Functions (NOBPFs) Coefficients and Related Percentage Error of the Impulse Response, for $m = 8$ and $T = 1$ s

t (s)	System via "Deconvolution", g_{ci}		Direct Expansion of Impulse Response, g_{di}		Percentage Error, ε_i	
	OBPF	NOBPF	OBPF	NOBPF	OBPF	NOBPF
0.000						
	0.64634351	0.95960353	0.94002478	0.94124845	31.24186501	− 1.95007806
0.125						
	1.78427874	0.80768964	0.82956896	0.83064884	− 115.08504180	2.76400821
0.250						
	− 0.89230223	0.75194111	0.73209203	0.73304503	221.88388754	− 2.57775135
0.375						
	2.93250883	0.62442820	0.64606895	0.64690997	− 353.90028701	3.47525523
0.500						
	− 2.38507292	0.59021345	0.57015385	0.57089604	518.32093647	− 3.38369982
0.625						
	4.12122351	0.48170404	0.50315901	0.50381399	− 719.06980911	4.38851448
0.750						
	− 3.84211609	0.46425983	0.44403626	0.44461429	965.27078936	− 4.41855813
0.875						
	5.34146762	0.37055036	0.39186063	0.39237073	− 1263.10393013	5.56116257
1.000						

identified, using the respective "deconvolution" operation both in OBPF and NOBPF domain for $m = 8$ and $T = 1$ s.

The NOBPF sets of "deconvolution" results are compared with corresponding OBPF sets of results in Table 9.20. Also, in these two domains we have compared the results by expanding the impulse response $g_2(t)$ directly for $m = 8$ and $T = 1$ s. Further, we have computed coefficient-wise percentage error both in OBPF and NOBPF domain for clarity in comparison.

The same system has been identified using two more inputs, a unit ramp input and a unit parabolic input. The respective outputs are given by $c_5(t) = t -$ sin t and $c_6(t) = t^2 - 2(1 - \cos t)$. As was done for Table 9.20, we present Tables 9.21 and 9.22 in the same spirit to bring out the comparison of OBPF- and NOBPF-based "deconvolution" results for $m = 8$ and $T = 1$ s.

From the percentage error column of OBPF approach in Table 9.20, and last two columns of Tables 9.21 and 9.22, the oscillation in system identification is noted.

The error difference of NOBPF-based analysis with OBPF-based analysis is presented in Table 9.23. Inspection of the last column of the table reveals the superiority of NOBPF-based analysis for all the inputs considered in undamped second-order system.

TABLE 9.18

Identification of the First-Order Open Loop System with Unit Parabolic Input and Comparison of Optimal and Non-Optimal Block Pulse Functions (NOBPFs) Coefficients and Related Percentage Error of the Impulse Response, for $m = 8$ and $T = 1$ s

t (s)	System via "Deconvolution", g_{ci}		Direct Expansion of Impulse response, g_{di}		Percentage Error, ε_i	
	OBPF	NOBPF	OBPF	NOBPF	OBPF	NOBPF
0.000						
	0. 48775584	0.64634351	0.94002478	0.94124845	0.0481 e03	0.0313 e03
0.125						
	3.22628698	1.78427874	0.82956896	0.83064884	− 0.2889 e03	− 0.1148 e03
0.250						
	− 0.0084 e03	− 0.89230223	0.73209203	0.73304503	1.2563 e03	0.2217 e03
0.375						
	0.0350 e03	2.93250883	0.64606895	0.64690997	− 5.3294 e03	− 0.3533 e03
0.500						
	− 0.1280 e03	− 2.38507292	0.57015385	0.57089604	2.2553 e04	0.5178 e03
0.625						
	0.4805 e03	4.12122351	0.50315901	0.50381399	− 9.5390 e04	− 0.7180 e03
0.750						
	− 1.7909 e03	− 3.84211609	0.44403626	0.44461429	4.0342 e05	0.9641 e03
0.875						
	6.6858 e03	5.34146762	0.39186063	0.39237073	− 1.7061 e06	− 1.2613 e03
1.000						

TABLE 9.19

Comparison of Average of Mod of Percentage (AMP) Errors for Optimal and Non-Optimal Block Pulse Function (BPF) Coefficients Based Identification of First-Order System Using "Deconvolution" Operation having Impulse Response of $g_1(t) = \exp(-t)$, for $m = 8$ and $T = 1$ s

Input	AMP Error		Error Difference, $\varepsilon_{av,OBPF} - \varepsilon_{av,NOBPF}$
	OBPF, $\varepsilon_{av,OBPF}$	NOBPF, $\varepsilon_{av,NOBPF}$	
Unit step	3.56951904	0.13000520	3.43951384
Unit ramp	523.48456830	3.56487848	519.91968982
Unit parabolic	279295.09650576	522.80401112	278772.29249464

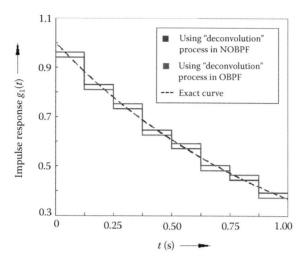

FIGURE 9.6

Identification of a first-order system having the impulse response $g_1(t) = \exp(-t)$, using "deconvolution" process based on optimal and non-optimal block pulse function (BPF) coefficient with unit step input (vide Table 9.16), for $m = 8$ and $T = 1$ s.

TABLE 9.20

Identification of the Second-Order Undamped System with Unit Step Input and Comparison of Optimal and Non-Optimal Block Pulse Functions (NOBPFs) Coefficients and Related Percentage Error of the Impulse Response, for $m = 8$ and $T = 1$ s (Vide Appendix B, Program No. 9.3)

t (s)	System via "Deconvolution", g_{ci}		Direct Expansion of Impulse Response, g_{di}		Percentage Error, ε_i	
	OBPF	**NOBPF**	**OBPF**	**NOBPF**	**OBPF**	**NOBPF**
0.000						
	0.04163413	0.06241866	0.06241866	0.06233737	33.29859172	− 0.13041211
0.125						
	0.20739084	0.18628196	0.18628196	0.18603935	− 11.33167738	− 0.13041211
0.250						
	0.28677314	0.30723840	0.30723840	0.30683824	6.66103619	− 0.13041211
0.375						
	0.44481858	0.42340048	0.42340048	0.42284903	− 5.05859191	− 0.13041211
0.500						
	0.51278464	0.53295554	0.53295554	0.53226141	3.78472442	− 0.13041211
0.625						
	0.65588701	0.63419401	0.63419401	0.63336802	− 3.42056256	− 0.13041211
0.750						
	0.70561634	0.72553609	0.72553609	0.72459113	2.74552162	− 0.13041211
0.875						
	0.82747290	0.80555642	0.80555642	0.80450724	− 2.72066358	− 0.13041211
1.000						

TABLE 9.21

Identification of the Second-Order Undamped System with Unit Ramp Input and Comparison of Optimal and Non-Optimal Block Pulse Functions (NOBPFs) Coefficients and Related Percentage Error of the Impulse Response, for $m = 8$ and $T = 1$ s

t (s)	System via "Deconvolution", g_{ci}		Direct Expansion of Impulse Response, g_{di}		Percentage Error, ε_i	
	OBPF	NOBPF	OBPF	NOBPF	OBPF	NOBPF
0.000						
	0.02082249	0.04163413	0.06241866	0.06233737	66.64060884	33.21160501
0.125						
	0.22852724	0.20739084	0.18628196	0.18603935	– 22.67813264	– 11.47686737
0.250						
	0.26628119	0.28677314	0.30723840	0.30683824	13.33075917	6.53931088
0.375						
	0.46626462	0.44481858	0.42340048	0.42284903	– 10.12378084	– 5.19560104
0.500						
	0.49258744	0.51278464	0.53295554	0.53226141	7.57438457	3.65924805
0.625						
	0.67760830	0.65588701	0.63419401	0.63336802	– 6.84558596	– 3.55543550
0.750						
	0.68567061	0.70561634	0.72553609	0.72459113	5.49462373	2.61869000
0.875						
	0.84941796	0.82747290	0.80555642	0.80450724	– 5.44487523	– 2.85462376
1.000						

Figure 9.7 shows the graphical comparison of the "deconvolution" results with unit step input on undamped second-order system, obtained via NOBPF coefficient-based analysis along with the OBPF-based analysis. Also, the exact curve of the impulse response of the plant $g_2(t)$ is shown for clarity.

9.7.3 Underdamped Second-Order System

In this section, an undamped second-order system is to be identified with a unit step input, where the output of the system is

$$c_7(t) = 1 - \exp(-abt)\cos\left\{\left(b\sqrt{1-a^2}\right)t\right\} - \frac{a}{\sqrt{1-a^2}}\exp(-abt)\sin\left\{\left(b\sqrt{1-a^2}\right)t\right\}, t \geq 0.$$

The system is identified in OBPF and NOBPF domain, using the respective "deconvolution" operation for $m = 8$ and $T = 1$ s.

The NOBPF sets of "deconvolution" results are compared with corresponding OBPF sets of results in Table 9.24. Also, in these two domains we have compared the results by expanding the impulse response $g_3(t)$ directly for $m = 8$

TABLE 9.22

Identification of the Second-Order Undamped System with Unit Parabolic Input and Comparison of Optimal and Non-Optimal Block Pulse Functions (NOBPFs) Coefficients and Related Percentage Error of the Impulse Response, for $m = 8$ and $T = 1$ s

t (s)	System via "Deconvolution", g_{ci} OBPF	NOBPF	Direct Expansion of Impulse Response, g_{di} OBPF	NOBPF	Percentage Error, ε_i OBPF	NOBPF
0.000						
	0.01249535	0.02082249	0.06241866	0.06233737	0.0080 e04	66.59710416
0.125						
	0.28694712	0.22852724	0.18628196	0.18603935	– 0.0054 e04	– 22.83811978
0.250						
	0.00748827	0.26628119	0.30723840	0.30683824	0.0098 e04	13.21773199
0.375						
	0.0148 e02	0.46626462	0.42340048	0.42284903	– 0.0249 e04	– 10.26739558
0.500						
	– 0.0332 e02	0.49258744	0.53295554	0.53226141	0.0723 e04	7.45385038
0.625						
	0.1496 e02	0.67760830	0.63419401	0.63336802	– 0.2259 e04	– 6.98492553
0.750						
	– 0.5267 e02	0.68567061	0.72553609	0.72459113	0.7360 e04	5.37137727
0.875						
	2.0002 e02	0.84941796	0.80555642	0.80450724	– 2.4731 e04	– 5.58238811
1.000						

TABLE 9.23

Comparison of Average of Mod of Percentage (AMP) Errors for Optimal and Non-Optimal Block Pulse Function (BPF) Coefficients Based Identification of Second-Order Undamped System Using "Deconvolution" Operation Having Impulse Response of $g_2(t) = \sin t$, for $m = 8$ and $T = 1$ s

Input	AMP Error OBPF, $\varepsilon_{av,OBPF}$	NOBPF, $\varepsilon_{av,NOBPF}$	Error Difference, $\varepsilon_{av,OBPF} - \varepsilon_{av,NOBPF}$
Unit step	8.62767117	0.13041211	8.49725906
Unit ramp	17.26659387	8.63892270	8.62767117
Unit parabolic	4444.22269262	17.28911160	4426.93358102

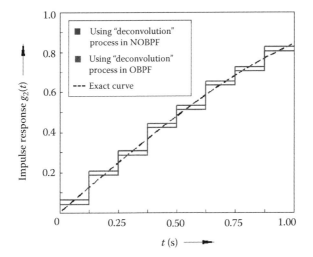

FIGURE 9.7
Identification of a second-order undamped system having the impulse response $g_2(t) = \sin t$, using "deconvolution" process-based on optimal and non-optimal block pulse function (BPF) coefficient with unit step input (vide Table 9.20), for $m = 8$ and $T = 1$ s.

and $T = 1$ s. Similar to the previous cases, we have computed coefficient-wise percentage error both in OBPF and NOBPF domain for clarity in comparison.

The same system has been identified using two more inputs, a unit ramp input and a unit parabolic input. The respective outputs are given by

$$c_8(t) = t - \frac{2a}{b} + \exp(-abt)\left[\frac{2a}{b}\cos\left\{\left(b\sqrt{1-a^2}\right)t\right\} - \frac{(2a^2-1)}{b\sqrt{1-a^2}}\sin\left\{\left(b\sqrt{1-a^2}\right)t\right\}\right]$$

and

$$c_9(t) = 2b^2\left[\frac{(4a^2-1)}{b^4} - \frac{2a}{b^3}t + \frac{1}{2b^2}t^2 + \exp(-abt)\left[\frac{(1-5a^2+4a^4)}{\left(b^2\sqrt{1-a^2}\right)^2}\cos\left\{\left(b\sqrt{1-a^2}\right)t\right\}\right.\right.$$
$$\left.\left. - \frac{a(4a^2-3)}{b^4\sqrt{1-a^2}}\sin\left\{\left(b\sqrt{1-a^2}\right)t\right\}\right]\right]$$

As was done for Table 9.24, we present Tables 9.25 and 9.26 in the same spirit to bring out the comparison of OBPF and NOBPF-based "deconvolution" results for $m = 8$ and $T = 1$ s.

TABLE 9.24

Identification of the Second-Order Underdamped System with Unit Step Input and Comparison of Optimal and Non-Optimal Block Pulse Functions (NOBPFs) Coefficients and Related Percentage Error of the Impulse Response, for $m = 8$ and $T = 1$ s

t (s)	System via "Deconvolution", g_{ci}		Direct Expansion of Impulse Response, g_{di}		Percentage Error, ε_i	
	OBPF	NOBPF	OBPF	NOBPF	OBPF	NOBPF
0.000						
	0.90633433	1.29031669	1.29031669	1.15141161	0.0029 e04	– 12.063893660
0.125						
	3.42979436	2.77916039	2.77916039	2.65857385	– 0.0023 e04	– 4.53576029
0.250						
	2.40123000	2.85521847	2.85521847	2.78162461	0.0016 e04	– 2.64571526
0.375						
	2.60140578	2.05174680	2.05174680	2.02855413	– 0.0027 e04	– 1.14331021
0.500						
	0.41019488	0.95389564	0.95389564	0.96817409	0.0057 e04	1.47478152
0.625						
	0.49764537	0.00417259	0.00417259	0.03692970	– 1.1827 e04	88.70126956
0.750						
	– 1.13593866	– 0.57310516	– 0.57310516	– 0.53888743	– 0.0098 e04	– 6.34969992
0.875						
	– 0.25020048	– 0.75270776	– 0.75270776	– 0.72778600	0.0067 e04	– 3.42432558
1.000						

From the percentage error column of OBPF approach in Table 9.24, and last two columns of Tables 9.25 and 9.26, the oscillation in system identification is noted.

The error difference of NOBPF-based analysis with OBPF-based analysis is presented in Table 9.27. Inspection of the last column of the table reveals the superiority of NOBPF-based analysis for all the inputs considered in underdamped second-order system.

It is observed that the AMP error in case of NOBPF-based identification is much less than that of the OBPF-based identification of all the systems and inputs considered. In Table 9.28, the results of system identification via both the approaches are compared qualitatively for all the inputs considered.

From the Table 9.28 it is noted that the optimal coefficient-based approach has failed to identify the systems completely in all the three cases while in case of non-optimal coefficient based approach all the three systems could be identified successfully.

TABLE 9.25

Identification of the Second-Order Underdamped System with Unit Ramp Input and Comparison of Optimal and Non-Optimal Block Pulse Functions (NOBPFs) Coefficients and Related Percentage Error of the Impulse Response, for $m = 8$ and $T = 1$ s

t (s)	System via "Deconvolution", g_{ci}		Direct Expansion of Impulse Response, g_{di}		Percentage Error, ε_i	
	OBPF	NOBPF	OBPF	NOBPF	OBPF	NOBPF
0.000						
	0.1450 e02	20.64506702	1.29031669	1.15141162	– 0.0010 e06	– 0.1693 e04
0.125						
	0.2587 e02	3.17643223	2.77916039	2.65857385	– 0.0008 e06	– 0.0019 e04
0.250						
	– 0.4233 e02	– 1.95950291	2.85521847	2.78162461	0.0016 e06	0.0170 e04
0.375						
	0.4553 e02	– 10.89604386	2.05174680	2.02855413	– 0.0021 e06	0.0637 e04
0.500						
	– 0.8059 e02	– 6.66957471	0.95389564	0.96817409	0.0085 e06	0.0789 e04
0.625						
	0.8199 e02	– 8.52599414	0.00417259	0.03692970	– 1.9649 e06	2.3187 e04
0.750						
	– 1.0813 e02	– 0.71044988	– 0.57310516	– 0.53888743	– 0.0188 e06	– 0.0032 e04
0.875						
	1.2230 e02	– 2.16319164	– 0.75270776	– 0.72778600	0.0163 e06	– 0.0197 e04
1.000						

Figure 9.8 shows the graphical comparison of the "deconvolution" results with unit step input on underdamped second-order system, obtained via NOBPF coefficient based analysis along with the OBPF-based analysis. Also, the exact curve of the impulse response of the plant $g_3(t)$ is shown for clarity.

OBPF-based approach shows oscillation in three cases while the results obtained via NOBPF approach oscillate in one case only. But the degree of oscillation is severe in case of OBPF, as it is evident from different Tables.

From Table 9.28, it is noted that the OBPF coefficient based approach has not been successful in identifying the systems, while its contender has succeeded in three cases. Also, percentage error and AMP error are more in case of optimal coefficient-based approach as evident from above Tables.

For the three cases with ramp input, NOBPF coefficient based approach identifies the first-order system and the undamped system but fails to identify the underdamped systems.

TABLE 9.26

Identification of the Second-Order Underdamped System with Unit Parabolic Input and Comparison of Optimal and Non-Optimal Block Pulse Functions (NOBPFs) Coefficients and Related Percentage Error of the Impulse Response, for $m = 8$ and $T = 1$ s

t (s)	System via "Deconvolution", g_{ci}		Direct Expansion of Impulse Response, g_{di}		Percentage Error, ε_i	
	OBPF	NOBPF	OBPF	NOBPF	OBPF	NOBPF
0.000						
	0.0017 e05	1.6516 e02	1.29031669	1.15141162	– 0.0001 e08	– 0.0142 e06
0.125						
	– 0.0039 e05	– 3.0491 e02	2.77916039	2.65857385	0.0001 e08	0.0115 e06
0.250						
	0.0069 e05	2.6382 e02	2.85521847	2.78162461	– 0.0002 e08	– 0.0094 e06
0.375						
	– 0.0212 e05	– 3.3531 e02	2.05174680	2.02855413	0.0010 e08	0.0166 e06
0.500						
	0.0735 e05	3.6913 e02	0.95389564	0.96817409	– 0.0077 e08	– 0.0380 e06
0.625						
	– 0.2684 e05	– 3.8398 e02	0.00417259	0.03692970	6.4316 e08	1.0399 e06
0.750						
	0.9967 e05	4.4650 e02	– 0.57310516	– 0.53888743	0.1739 e08	0.0830 e06
0.875						
	– 3.7134 e05	– 4.5812 e02	– 0.75270776	– 0.72778600	– 0.4933 e08	– 0.0628 e06
1.000						

TABLE 9.27

Comparison of Average of Mod of Percentage (AMP) Errors for Optimal and Non-Optimal Block Pulse Function (BPF) Coefficients Based Identification of Second-Order Underdamped System Using "Deconvolution" Operation Having Impulse Response of $g_3(t)$, for $m = 8$ and $T = 1$ s

Input	AMP Error		Error Difference, $\varepsilon_{av,OBPF} - \varepsilon_{av,NOBPF}$
	OBPF, $\varepsilon_{av,OBPF}$	NOBPF, $\varepsilon_{av,NOBPF}$	
Unit step	1518.04577595	15.04234450	1503.00343145
Unit ramp	251768.39571530	3340.63956272	248427.75615258
Unit parabolic	88851164.70989263	159438.72213053	88691725.98776209

TABLE 9.28

Qualitative Comparison Between Optimal Coefficient Based Approach
and Non-Optimal Coefficient Based Approach for System Identification
with Different Inputs

| | | Remarks | |
| | | OBPF-Based | NOBPF-Based |
Plant	Nature of Input	Approach	Approach
First order	Unit step	Shows oscillation	Successfully identified
	Unit ramp	Fails to identify	Shows oscillation
	Unit Parabolic	Fails to identify	Fails to identify
Second order	Unit step	Shows oscillation	Successfully identified
(undamped)	Unit ramp	Shows oscillation	Shows oscillation
	Unit Parabolic	Fails to identify	Shows oscillation
Second order	Unit step	Fails to identify	Successfully identified
(underdamped)	Unit ramp	Fails to identify	Fails to identify
	Unit Parabolic	Fails to identify	Fails to identify

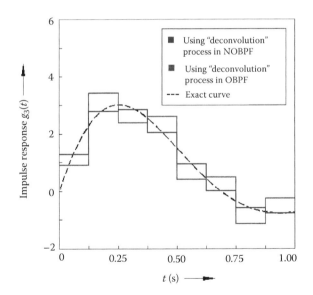

FIGURE 9.8
Identification of a second-order underdamped system having the impulse response $g_3(t)$, using
"deconvolution" process-based on optimal and non-optimal block pulse function (BPF) coefficient with unit step input (vide Table 9.24), for $m = 8$ and $T = 1$ s.

9.8 Identification of a Closed-Loop System via "Deconvolution"

Following the method proposed by Kwong and Chen [13], we identify the plant modeled by the impulse response function $g(t)$ as shown in Figure 9.9.

9.8.1 Using "Optimal" BPF Coefficients

We employ optimal BPF coefficients to determine the solution for $g(t)$. For the system shown in Figure 9.9, let

$$r(t) \approx \sum_{i=0}^{m-1} r_i \psi_i(t) = \begin{bmatrix} r_0 & r_1 & r_2 & \cdots & r_i & \cdots & r_{(m-1)} \end{bmatrix} \boldsymbol{\Psi}_{(m)}(t) \triangleq \mathbf{R}^{\mathrm{T}} \boldsymbol{\Psi}_{(m)}(t)$$

$$g(t) \approx \sum_{i=0}^{m-1} g_i \psi_i(t) = \begin{bmatrix} g_0 & g_1 & g_2 & \cdots & g_i & \cdots & g_{(m-1)} \end{bmatrix} \boldsymbol{\Psi}_{(m)}(t) \triangleq \mathbf{G}^{\mathrm{T}} \boldsymbol{\Psi}_{(m)}(t)$$

$$c(t) \approx \sum_{i=0}^{m-1} c_i \psi_i(t) = \begin{bmatrix} c_0 & c_1 & c_2 & \cdots & c_i & \cdots & c_{(m-1)} \end{bmatrix} \boldsymbol{\Psi}_{(m)}(t) \triangleq \mathbf{C}^{\mathrm{T}} \boldsymbol{\Psi}_{(m)}(t)$$

$$h(t) \approx \sum_{i=0}^{m-1} h_i \psi_i(t) = \begin{bmatrix} h_0 & h_1 & h_2 & \cdots & h_i & \cdots & h_{(m-1)} \end{bmatrix} \boldsymbol{\Psi}_{(m)}(t) \triangleq \mathbf{H}^{\mathrm{T}} \boldsymbol{\Psi}_{(m)}(t)$$

$$z(t) \approx \sum_{i=0}^{m-1} z_i \psi_i(t) = \begin{bmatrix} z_0 & z_1 & z_2 & \cdots & z_i & \cdots & z_{(m-1)} \end{bmatrix} \boldsymbol{\Psi}_{(m)}(t) \triangleq \mathbf{Z}^{\mathrm{T}} \boldsymbol{\Psi}_{(m)}(t)$$

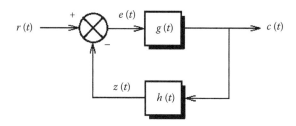

FIGURE 9.9
Closed-loop system modeled by impulse response function.

Knowing $\mathbf{R}$, $\mathbf{C}$, $\mathbf{H}$, and $\mathbf{Z}$, we are to determine $\mathbf{G}$ where the time functions have been expanded in BPF domain following equations (1.13) and (1.14).

Since $z(t)$ is the function obtained after convolution of $h(t)$ and $c(t)$, we write

$$\mathbf{Z}^{\mathrm{T}} = \frac{h}{2} \mathbf{H}^{\mathrm{T}} \mathbf{C}_{\mathrm{c}} = \frac{h}{2} \mathbf{C}^{\mathrm{T}} \mathbf{C}_{\mathrm{h}}$$

where $\mathbf{C}_{\mathrm{c}}$ and $\mathbf{C}_{\mathrm{h}}$ are convolution matrices formed by the elements of the matrices $\mathbf{C}$ and $\mathbf{H}$ respectively, in BPF domain with optimal expansion coefficients.

Again $e_i = r_i - z_i$, for $i = 0, 1, 2, \cdots, (m-1)$.

Therefore

$$e(t) \approx \sum_{i=0}^{m-1} e_i \psi_i(t) = \begin{bmatrix} e_0 & e_1 & e_2 & \cdots & e_i & \cdots & e_{(m-1)} \end{bmatrix} \mathbf{\Psi}_{(m)}(t)$$

$$= [\mathbf{R} - \mathbf{Z}]^{\mathrm{T}} \mathbf{\Psi}_{(m)}(t) \triangleq \mathbf{E}^{\mathrm{T}} \mathbf{\Psi}_{(m)}(t)$$

Then $\mathbf{G}$ is given by

$$\mathbf{G}^{\mathrm{T}} = \frac{2}{h} \mathbf{E}^{\mathrm{T}} \mathbf{C}_{\mathrm{c}}^{-1} = \frac{2}{h} \mathbf{C}^{\mathrm{T}} \mathbf{C}_{\mathrm{e}}^{-1} \tag{9.21}$$

where $\mathbf{C}_{\mathrm{c}}^{-1}$ and $\mathbf{C}_{\mathrm{e}}^{-1}$ are "deconvolution" matrices formed by optimal expansion coefficients in BPF domain.

9.8.2 Using "Non-Optimal" BPF Coefficients

As before, for the closed-loop system shown in Figure 9.9, we define

$$r(t) \approx \sum_{i=0}^{m-1} r_i' \psi_i'(t) = \begin{bmatrix} r_0' & r_1' & r_2' & \cdots & r_i' & \cdots & r_{(m-1)}' \end{bmatrix} \mathbf{\Psi}_{(m)}'(t) \triangleq \mathbf{R}'^{\mathrm{T}} \mathbf{\Psi}_{(m)}'(t)$$

$$g(t) \approx \sum_{i=0}^{m-1} g_i' \psi_i'(t) = \begin{bmatrix} g_0' & g_1' & g_2' & \cdots & g_i' & \cdots & g_{(m-1)}' \end{bmatrix} \mathbf{\Psi}_{(m)}'(t) \triangleq \mathbf{G}'^{\mathrm{T}} \mathbf{\Psi}_{(m)}'(t)$$

$$c(t) \approx \sum_{i=0}^{m-1} c_i' \psi_i'(t) = \begin{bmatrix} c_0' & c_1' & c_2' & \cdots & c_i' & \cdots & c_{(m-1)}' \end{bmatrix} \mathbf{\Psi}_{(m)}'(t) \triangleq \mathbf{C}'^{\mathrm{T}} \mathbf{\Psi}_{(m)}'(t)$$

$$h(t) \approx \sum_{i=0}^{m-1} h'_i \psi'_i(t) = \begin{bmatrix} h'_0 & h'_1 & h'_2 & \cdots & h'_i & \cdots & h'_{(m-1)} \end{bmatrix} \Psi'_{(m)}(t) \triangleq \mathbf{H}'^{\mathrm{T}} \Psi'_{(m)}(t)$$

$$z(t) \approx \sum_{i=0}^{m-1} z'_i \psi'_i(t) = \begin{bmatrix} z'_0 & z'_1 & z'_2 & \cdots & z'_i & \cdots & z'_{(m-1)} \end{bmatrix} \Psi'_{(m)}(t) \triangleq \mathbf{Z}'^{\mathrm{T}} \Psi'_{(m)}(t)$$

Knowing $\mathbf{R}'$, $\mathbf{C}'$, $\mathbf{H}'$ and $\mathbf{Z}'$, we are to determine $\mathbf{G}'$ where the time functions have been expanded with non-optimal BPF coefficients following equations (9.14) and (9.15).

Then $\mathbf{G}'$ is given by

$$\mathbf{G}'^{\mathrm{T}} = \frac{2}{h} \mathbf{E}'^{\mathrm{T}} \mathbf{C}'^{-1}_{\mathrm{c}} = \frac{2}{h} \mathbf{C}'^{\mathrm{T}} \mathbf{C}'^{-1}_{\mathrm{e}} \tag{9.22}$$

where $\mathbf{C}'^{-1}_{\mathrm{c}}$ and $\mathbf{C}'^{-1}_{\mathrm{e}}$ are "deconvolution" matrices formed by non-optimal expansion coefficients in NOBPF domain and $\mathbf{E}'^{\mathrm{T}} = [\mathbf{R}' - \mathbf{Z}']^{\mathrm{T}}$.

Consider a closed-loop system [13] with unit step input, $r(t) = u(t)$, feedback $h(t) = 4u(t)$, output $c(t) = \exp(-2t)\sin(2t)$, and the impulse response of the system is $g(t) = 2\exp(-4t)$.

The BPF and NOBPF domain results using optimal and non-optimal coefficients based "deconvolution" for $m = 16$ and $T = 1$ s are presented in Table 9.29.

From Table 9.29, comparing g_c and g_d, obtained using OBPF approach, we find that not only the result contains large error, but it has also gone negative in two occasions. For a first-order plant, such negative coefficients are simply impossible and this proves the OBPF coefficient based method to be unreliable for identification even for a simple first-order plant.

However, comparing g_c and g_d, obtained using NOBPF approach, shows surprisingly good result which has not only much less error than the optimal coefficient based method, but also contains no oscillation as observed in OBPF based result.

As a numerical estimate Table 9.30 compares the AMP error for OBPF and NOBPF based approaches, for identification of the closed-loop system using "deconvolution" process.

Figure 9.10 compares the percentage error between OBPF coefficient and NOBPF coefficient (vide Table 9.29) based identification.

TABLE 9.29

Identification of the Closed-Loop System with Unit Step Input and Comparison of Optimal and Non-Optimal Block Pulse Functions (NOBPFs) Coefficients and Related Percentage Error of the Impulse Response, for $m = 16$ and $T = 1$ s

t (s)	System via "Deconvolution", g_{ci}		Direct Expansion of Impulse Response, g_{di}		Percentage Error, ε_i	
	OBPF	NOBPF	OBPF	NOBPF	OBPF	NOBPF
0.0000						
	1.85183592	1.77259039	1.76959374	1.77880078	– 0.0465 e02	0.34913356
0.0625						
	1.29217298	1.37765173	1.37816099	1.38533144	0.0624 e02	0.55435948
0.1250						
	1.15317233	1.07057851	1.07331286	1.07889721	– 0.0744 e02	0.77103769
0.1875						
	0.74876684	0.83182259	0.83589689	0.84024599	0.1042 e02	1.00249290
0.2500						
	0.73052223	0.64618485	0.65099715	0.65438424	– 0.1222 e02	1.25299267
0.3125						
	0.42003832	0.50184766	0.50699709	0.50963496	0.1715 e02	1.52801471
0.3750						
	0.47484396	0.38962254	0.39484973	0.39690410	– 0.2026 e02	1.83459015
0.4375						
	0.22117703	0.30236525	0.30750928	0.30910923	0.2807 e02	2.18174406
0.5000						
	0.32017375	0.23452100	0.23948847	0.24073451	– 0.3369 e02	2.58106098
0.5625						
	0.10087772	0.18177081	0.18651381	0.18748422	0.4591 e02	3.04741082
0.6250						
	0.22660743	0.14075657	0.14525710	0.14601286	– 0.5600 e02	3.59988055
0.6875						
	0.02810377	0.10886730	0.11312634	0.11371493	0.7516 e02	4.26297027
0.7500						
	0.17000535	0.08407288	0.08810288	0.08856128	– 0.9296 e02	5.06812838
0.8125						
	– 0.01592016	0.06479486	0.06861460	0.06897159	1.2320 e02	6.05572217
0.8750						
	0.13576445	0.04980597	0.05343710	0.05371513	– 1.5406 e02	7.27756731
0.9375						
	– 0.04255203	0.03815197	0.04161686	0.04183338	2.0225 e02	8.80017471
1.0000						

TABLE 9.30

Comparison of Average of Mod of Percentage (AMP) Errors for Optimal and Non-Optimal Block Pulse Function (BPF) Coefficients Based Identification of Closed-Loop System Using "Deconvolution" Operation, for $m = 8$ and $T = 1$ s

| Input | AMP Error | | Error Difference, $\varepsilon_{av,OBPF} - \varepsilon_{av,NOBPF}$ |
	OBPF, $\varepsilon_{av,OBPF}$	NOBPF, $\varepsilon_{av,NOBPF}$	
Unit step	55.60591185	3.13545502	52.47045683

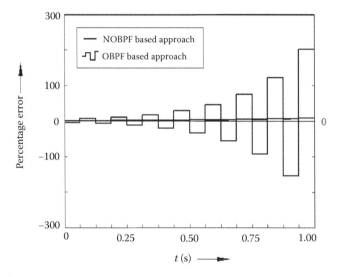

FIGURE 9.10

Comparison of percentage error between optimal block pulse function (OBPF) coefficient and non-optimal block pulse function (NOBPF) coefficient based identification.

9.9 Error Analysis

This section deduces the maximum possible error estimate [14–16] for NOBPF coefficient based representation of a locally square integrable function $f(t)$ to have a quantitative judgment about the efficiency of approximation.

Let us consider m sample points of a function $f(t)$, having sampling period h, denoted by $f(ih), i = 0, 1, 2, \ldots, m$. Then non-optimal BPF representation of the function $f(t)$ in any subinterval is obtained by a horizontal line representing the average height of two consecutive sample points of the function itself.

We know that the MISE over the interval $[0, T)$ is given by

$$\text{MISE} = [\epsilon^2] = \frac{1}{T}\int_0^T \left[f(t) - \sum_{i=0}^{m-1} c_i \psi_i(t) \right]^2 dt \tag{9.23}$$

Following equation (9.14), the NOBPF coefficient in the ith interval is given by

$$c_i' = \frac{[f(ih) + f\{(i+1)h\}]}{2} \tag{9.24}$$

and the NOBPF representation of $f(t)$ in the ith interval is given by

$$f(t) \approx \hat{f}(t) \Big|_{ih}^{(i+1)h} = c_i' \psi_i \tag{9.25}$$

Then integral square error (ISE) in the ith interval is

$$\left[\epsilon_i^2 \right] = \int_{ih}^{(i+1)h} \left[f(t) - c_i' \psi_i(t) \right]^2 dt \tag{9.26}$$

Let the function $f(t)$ be expanded via Taylor series in the ith interval around a point μ_i [14,15]. Considering first-order approximation, we have

$$f(t) \approx f(\mu_i) + \dot{f}(\mu_i)(t - \mu_i) \quad \text{where } \mu_i \in \left[ih, (i+1)h \right] \tag{9.27}$$

Equation (9.26) may now be written as

$$\left[\epsilon_i^2 \right] = \int_{ih}^{(i+1)h} \left[\{ f(\mu_i) + \dot{f}(\mu_i)(t - \mu_i) \} - \frac{[f(ih) + f\{(i+1)h\}]}{2} \right]^2 dt \tag{9.28}$$

Let

$$A = f(\mu_i) - \frac{[f(ih) + f\{(i+1)h\}]}{2}$$

and $B = \dot{f}(\mu_i)$

Then

$$\left[\varepsilon_i^2\right] = \int_{ih}^{(i+1)h} [A + B(t - \mu_i)]^2 dt \tag{9.29}$$

Following equation (9.29), the ISE over m subintervals is given by

$$\sum_{i=0}^{m-1}\left[\varepsilon_i^2\right] = A^2 mh + B^2\left[\frac{m^3 h^3}{3} - m^2 h^2 \mu_i + mh\mu_i^2\right] + 2AB\left[\frac{m^2 h^2}{2} - mh\mu_i\right] \tag{9.30}$$

Putting

$$\left.\begin{aligned}
f_{max} &\triangleq \max[f(ih), f\{(i+1)h\}, f(\mu_i)] \\
\dot{f}_{max} &\triangleq \max\left[\dot{f}(\mu_i)\right] \\
\mu_{max} &\triangleq \max[\mu_i]
\end{aligned}\right\} \tag{9.31}$$

Then, we find that A = 0 and B = $\dot{f}_{max}$ the upper bound of ISE over m subintervals is

$$\sum_{i=0}^{m-1}\left[\varepsilon_i^2\right] = \dot{f}_{max}^2\left[\frac{m^3 h^3}{3} - m^2 h^2 \mu_{max} + mh\mu_{max}^2\right] \tag{9.32}$$

9.10 Conclusion

This Chapter presented an approach using BPFs with non-optimal coefficients approximated square integrable functions of Lebesgue measure reasonable well but does not guarantee minimum MISE. However, it has the advantage of simplicity with regards to computation of the expansion coefficients by exact integration.

It has seen from Table 9.2 that in case of approximation of the function $f(t) = \sin(\pi t)$ via non-optimal coefficients, the error is only less than 1.28841990% compared to equivalent optimal coefficient-based expansion. However, for the expansion of t^2, the magnitude of percentage error increases for the first four coefficients, as observed from Table 9.3. Furthermore, the error for the first coefficient goes up to 50%. The reason for such increased error is, the more the function deviates from linear nature, the more error creeps into

NOBPF coefficient-based approximation. The percentage error reduces drastically, as we move along the time axis.

In case of analysis of SISO systems via convolution, the NOBPF coefficient-based method gives superior results compared to its OBPF counterpart. For a first-order system, with step input, the OBPF coefficient-based results show a maximum error of 2.04029590% while for NOBPF approach it is only – 0.13017444%. These are apparent from Table 9.4.

Inspection of Tables 9.7, 9.11, and 9.15 reveal that the average of mod of percentage (AMP) error for OBPF coefficient-based analysis is more than the analysis based upon NOBPF coefficients.

Superiority of NOBPF coefficient-based analysis compared to OBPF analysis is strongly focused during solution of identification problems for a SISO open-loop system via "deconvolution".

The OBPF approach produces oscillatory as well as impossible results while NOBPF coefficient-based approach solves the problem decently with reliability. These are summarized in Table 9.28.

This "weakness" of OBPF coefficient-based analysis becomes more clear when we try to identify the plant of a closed-loop system. While OBPF-based identification produces totally unacceptable results, the NOBPF coefficient-based approach comes up with correct solution. Table 9.29 shows the results of OBPF and NOBPF-based identification.

As a qualitative assessment, percentage errors from Table 9.29 for OBPF and NOBPF-based identification are plotted in Figure 9.10. The highly oscillatory behavior of OBPF coefficient-based results is quite apparent. Also, from Table 9.29, the difference of AMP error of OBPF and NOBPF coefficient based identification is obtained as 52.47045683 and is presented in Table 9.30.

Finally, a formal error analysis has been presented for NOBPF coefficient-based approximation of any square integrable function and the expression for the upper bound of integral squared error is derived.

It can be concluded that in all the applications presented, NOBPF-based approach proves to be a strong contender of the OBPF-based approach.

References

1. Alfred Haar, Zur theorie der orthogonalen funktionen systeme, Math. Annalen, vol. **69**, pp. 331–371, 1910.
2. H. F. Harmuth, *Transmission of information by orthogonal functions* (2nd Ed.), Springer-Verlag, Berlin, 1972.
3. K. G. Beauchamp, *Walsh functions and their applications*, Academic Press, vol. **3**, London, 1975.
4. K. G. Beauchamp, *Applications of Walsh and related functions: With an introduction to sequency theory*, Academic Press, London, 1984.

5. Anish Deb and Suchismita Ghosh, *Power electronic systems: Walsh analysis with MATLAB®*, CRC Press, Boca Raton, 2014.
6. G. P. Rao, *Piecewise constant orthogonal functions and their applications to systems and control*, LNCIS, vol. **55**, Springer-Verlag, Berlin, 1983.
7. Z. H. Jiang and W. Schaufelberger, *Block pulse functions and their applications in control systems*, LNCIS, vol. **179**, Springer-Verlag, Berlin, 1992.
8. Anish Deb, Gautam Sarkar, and Sunit K. Sen, Block pulse functions, the most fundamental of all piecewise constant basis functions, Int. J. Syst. Sci., vol. **25**, no. 2, pp. 351–363, 1994.
9. Anish Deb, Gautam Sarkar, Priyaranjan Mandal, Amitava Biswas, and Anindita Sengupta, Optimal block pulse function (OBPF) vs. Non-optimal block Pulse function (NOBPF), Proceedings of Int. Conf. of IEE (PEITSICON) 2005, pp. 195–199, Kolkata, 28–29th Jan., 2005.
10. Anish Deb, Gautam Sarkar, Priyaranjan Mandal, Amitava Biswas, and Anindita Sengupta, Non-optimal block pulse function (NOBPF) based analysis and identification of SISO control systems, J. Autom. & Syst. Engg., pp. 24–58, 2013.
11. Anish Deb, Srimanti Roychoudhury, and Gautam Sarkar, *Analysis and identification of time-invariant Systems, time-varying systems and multi-delay systems using orthogonal hybrid functions: Theory and algorithms with MATLAB®*, vol. **46**, Springer, Switzerland, 2016.
12. S. C. Chapra and R. P. Canale, *Numerical methods for engineers* (6th Ed.), McGraw-Hill Education (India) Pvt. Ltd., New Delhi, 2012.
13. C. P. Kwong and C. F. Chen, Linear feedback system identification via block-pulse functions, Int. J. Syst. Sci., vol. **12**, no. 5, pp. 635–642, 1981.
14. G. P. Rao and T. Srinivasan, Analysis and synthesis of dynamic systems containing time delays via block-pulse functions, IEE Proc., vol. **125**, no. 10, pp. 1064–1068, 1978.
15. Anish Deb, Gautam Sarkar, and Sunit K. Sen, Linearly pulse-width modulated block pulse functions and their application to linear SISO feedback control system identification, Proc. IEE, Part D, Control Theory and Appl., vol. **142**, no. 1, pp. 44–50, 1995.
16. Anish Deb, G. Sarkar, M. Bhattacharjee, and S. K. Sen, A new set of piecewise constant orthogonal functions for the analysis of linear SISO systems with sample-and-hold, J. Franklin Inst., vol. **335B**, no. 2, pp. 333–358, 1998.

Study Problems

9.1 Consider an open-loop system having a transfer function $G(s) = (s + 2)^{-1}$. Find its output $c(t)$ in block pulse function (BPF) domain as well as in non-optimal block pulse function (NOBPF) domain, for a step input $u(t)$ using the convolution matrix. Consider $m = 4$ and $T = 1$ s. Determine the percentage errors of different coefficients. Finally, compare the two results of the output and discuss.

9.2 Consider the first-order system of Problem 9.1. Identify the system using "deconvolution" process for the same value of m, T and similar input in NOBPF and optimal block pulse function (OBPF) approaches. Compared to direct expansions in respective domains, compute percentage error for each segment and make your observation with respect to oscillation in the result.

9.3 Consider an undamped system having a transfer function $G(s) = (s^2 + 2)^{-1}$. Find its output $c(t)$ in BPF domain as well as in NOBPF domain, for a ramp input $r(t)$ using the convolution matrix. Consider $m = 4$ and $T = 1$ s. Determine the percentage errors of different coefficients. Finally, compare the outputs and discuss.

9.4 Consider the second-order undamped system of Problem 9.3. Identify the system using "deconvolution" process for the same value of m, T and similar input in NOBPF and OBPF approaches. Compared to direct expansions in respective domains, compute percentage error for each segment and make your observation with respect to oscillation in the result.

9.5 Consider an underdamped system having a transfer function $G(s) = (s^2 + 2s + 2)^{-1}$. Find its output $c(t)$ in BPF domain as well as in NOBPF domain, for a step input $u(t)$ using the convolution matrix. Consider $m = 4$ and $T = 1$ s. Determine the percentage errors of different coefficients. Finally, compare the two results of the output and discuss.

9.6 Consider the second-order underdamped system of Problem 9.5. Identify the system using "deconvolution" process for the same value of m, T and similar input in NOBPF and OBPF approaches. Compared to direct expansions in respective domains, compute percentage error for each segment and make your observation with respect to oscillation in the result.

9.7 Consider an output of a closed-loop system as $c(t) = \exp(-t)\sin(3t)$, having the feedback element $h(t) = 2u(t)$. Identify the system $g(t)$ in non-optimal and optimal BPF domain for a step input $u(t)$ using the "deconvolution" matrix. Consider $m = 8$ and $T = 1$ s. Finally, compare the result graphically with the direct BPF expansion of $g(t)$ in respective domains. Is there any oscillation in the result?

10

System Analysis and Identification Using Linearly Pulse-Width Modulated Generalized Block Pulse Functions (LPWM-GBPF)

Among all the piecewise constant basis functions (PCBFs) available, like Haar functions [1], Rademacher functions [2], Walsh functions [3–5], etc., the block pulse functions (BPF) [6,7] seem to be the simplest and most advantageous for system analysis and applications. The BPF set has been found to be more effective and elegant than other PCBFs of the same class for a score of applications [8–10].

The conventional BPF set was introduced into the literature by Chen et al. in 1977 [11]. A generalized version of the BPF set was presented by Wang et al. in 1987 [12]. But the advantages of a GBPF set over a conventional BPF set were not clearly unraveled by researchers and hence, the conventional BPF set continued to enjoy its popularity.

In this chapter, a new set of linearly pulse-width modulated generalized block pulse functions (LPWM-GBPF) [13] has been presented. It has been shown that characterization of a particular member of this LPWM-GBPF set is possible in a unique manner which proves advantageous for analytical purposes. In 1981, Kwong and Chen [14] developed a convolution matrix with conventional BPFs and subsequently used it to solve linear feedback system identification problems. But their technique suffered from some defect which led to failure by way of oscillatory results. This chapter uses a set of LPWM-GBPF to form a generalized convolution matrix (**GCVM**) that is used to solve linear feedback system identification problems successfully.

The representational error with LPWM-GBPF has also been estimated and found to be less than that for conventional BPF. This has been established theoretically as well as supported by relevant computations.

10.1 Conversion of a GBPF Set to a LPWM-GBPF Set [13]

To approximate a function $f(t)$ by a set of GBPF is not simple and also the selection of widths h_i of different block pulses is difficult because it needs *a priori* knowledge of the nature of variation of $f(t)$ so that the widths of component BPFs may be selected accordingly.

To simplify matters, we present a new kind of BPF set having members of monotonically linearly increasing or decreasing width, where magnitude of increase or decrease, δ (say), is constant. For instance, if the first member of a monotonically increasing set has a width h_0, the next members of the set will be of widths $(h_0 + \delta)$, $(h_0 + 2\delta)$, $(h_0 + 3\delta)$, ..., $[h_0 + (m-1)\,\delta]$ for an m-set LPWM-GBPF. In general, the width of the $(i+1)$th block pulse may be represented as $h_i = h_0 + i\delta$, where $i = 0, 1, 2, \cdots, (m-1)$.

Considering this set of GBPF to be a *positive pulse width modulated set*, let us designate this set to be a linearly positive pulse-width modulated (LPPWM) GBPF set. Hence, for a linearly negative pulse-width modulated (LNPWM) set of m members, the widths of the functions will be h_0, $(h_0 - \delta)$, $(h_0 - 2\delta)$, ..., $[h_0 - (m-1)\,\delta]$ and in general, $h_i = h_0 - i\delta$.

Let $f(t)$ be a square integrable function of Lebesgue measure. It is to be approximated by a set of LPWM BPFs, and either increasing or decreasing pulse-width modulation has to be chosen for such approximation. This may be done simply be investigating the nature of $f(t)$ through computing the derivatives $\dot{f}(0)$ and $\dot{f}(\infty)$, and comparing their magnitudes. It is assumed that $\dot{f}(0)$ and $\dot{f}(\infty)$ do exist. If $\left|\dot{f}(0)\right| > \left|\dot{f}(\infty)\right|$, the LPPWM-GBPF set is chosen, and if $\left|\dot{f}(0)\right| < \left|\dot{f}(\infty)\right|$ the LNPWM-GBPF set is chosen. For such cases, $\left|\dot{f}(t)\right| \neq 0$.

When $\left|\dot{f}(0)\right| = \left|\dot{f}(\infty)\right|$, one obviously sets $\delta = 0$ and works with a conventional BPF set. Like the GBPF analysis, the LPWM-GBPF set also needs *a priori* knowledge to analyze any system. But in this case, the amount of *a priori* knowledge is commensurate with the compatibility of the set.

Total time period of the LPWM-GBPF set is

$$T = mh_0 + \frac{1}{2}\Big[m(m-1)\Big]\delta \tag{10.1}$$

where δ will assume positive value for LPPWM-GBPF set and negative value for LNPWM-GBPF set. However, for the sake of simplicity, in the following analysis, a restriction is put on the choice of δ, given by

$$(m-1)\left|\delta\right| < h_0 \tag{10.2}$$

10.2 Representation of Time Functions via LPWM-GBPF Set

It is well known [6] that an absolutely integrable real-valued function $f(t)$ of Lebesgue measure defined in the semi-open interval $[0,T)$ can be expanded in an m-term conventional BPF series. If $f(t)$ is expanded in terms of a LPWM-GBPF set $\Psi_{g(m)}(t)$ having m members,

$$f(t) \underset{\cong}{\Delta} \mathbf{F}^T \Psi_{g(m)}(t) \tag{10.3}$$

where $\mathbf{F} \triangleq \begin{bmatrix} f_0 & f_1 & \cdots & f_{(m-1)} \end{bmatrix}^T$ and $[\cdots]^T$ denotes transpose.

The coefficients f_i, $i = 0,1,2,\ldots,(m-1)$, associated with respective LPWM-GBPF members, are given by

$$f_i = \frac{1}{(h_0 + i\delta)} \int_{ih_0 + i(i-1)\delta/2}^{(i+1)h_0 + i(i+1)\delta/2} f(t) \, \mathrm{d}t \tag{10.4}$$

When $\delta = 0$, one has a conventional BPF set and equation (10.4) gives the respective coefficients for BPF expansion of $f(t)$.

10.3 Convolution Process in LPWM-GBPF Domain

Now the LPWM-GBPFs are applied to determine the convolution of two real-valued functions $f_1(t)$ and $f_2(t)$. If the functions $f_1(\tau)$ and $f_1(\tau)$ are defined and continuous for all τ, the convolution is defined by the integral [15]

$$CV_{12}(t) = \int_{-\infty}^{+\infty} f_1(\tau) f_2(t-\tau) \, \mathrm{d}\tau \tag{10.5}$$

Obviously, it is assumed that the integral given by equation (10.5) does exist.

For convolution of two functions, we assume one function, ($f_1(\tau)$ say), to be the *static function* (STF) and the other function, ($f_2(\tau)$ say) to be the *scanning function* (SCF).

Let $f_1(t)$ and $f_2(t)$ be two time-varying BPFs, as shown in Figure 10.1(a). To evaluate equation (10.5), functions $f_1(\tau)$ and $f_2(t-\tau)$ are required. The functions $f_1(\tau)$ and $f_2(\tau)$ are obtained by changing the variable t to the pseudo variable τ and the function $f_2(\tau)$ is *folded* to obtain $f_2(-\tau)$. When the function $f_2(-\tau)$ is shifted by t, we get $f_2(t-\tau)$. The result of convolution of $f_1(t)$ and $f_2(t)$, as per equation (10.5), is the triangular function shown in Figure 10.1(b).

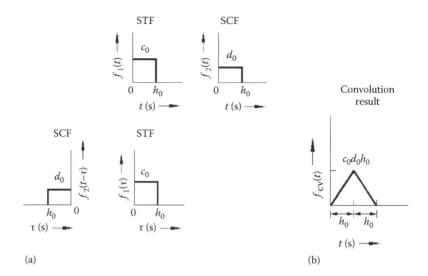

FIGURE 10.1
(a) Two block pulses, each of width h_0, and having amplitudes c_0 and d_0, and (b) the result of their convolution.

To illustrate the *process* of convolution graphically, we consider two more BPFs of unequal amplitudes as well as unequal widths (h_0 and ($h_0 + \delta$)), shown in Figure 10.2(a), and plot their convolution result in Figure 10.2(b). It is noted that the result is a trapezium.

Also, we consider another case of convolution where one convolving function $f_2(t)$ is shifted to the right by h_0 and it convolves with the function $f_1(t)$. These functions are shown in Figure 10.3(a). Their convolution result is shown in Figure 10.3(b).

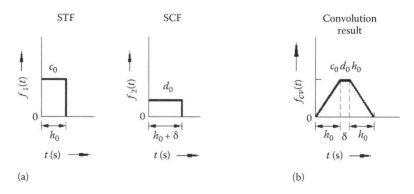

FIGURE 10.2
(a) Two block pulses, having amplitudes c_0 and d_0, and widths h_0 and ($h_0 + \delta$) respectively, and (b) the result of their convolution.

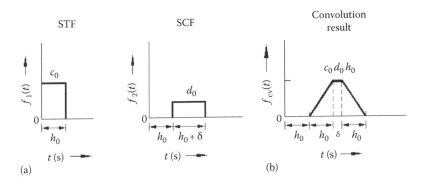

FIGURE 10.3
(a) Two block pulses, one delayed by time h_0, having amplitudes c_0 and d_0, and widths h_0 and $(h_0 + \delta)$ respectively, and (b) the result of their convolution.

Now consider two BPFs $f_1(t)$ and $f_2(t)$ delayed by d_1 and d_2 from the origin, and having amplitudes c and d, respectively. Let $f_1(t)$ has a width h_0, while $f_2(\tau)$ is of duration $(h_0 + \delta)$. For these two functions, $CV_{12}(t)$ as per equation (10.5) is

$$CV_{12}(t) = \begin{cases} 0 & \text{for } 0 \le t < (d_1 + d_2) \\ cd\left[t - (d_1 + d_2)\right] & \text{for } (d_1 + d_2) \le t < (d_1 + d_2 + h_0) \\ cdh_0 & \text{for } (d_1 + d_2 + h_0) \le t < (d_1 + d_2 + h_0 + \delta) \\ cd[2h_0 + \delta - t + d_1 + d_2] & \text{for } (d_1 + d_2 + h_0 + \delta) \le t \le (d_1 + d_2 + 2h_0 + \delta) \\ 0 & \text{for } t > (d_1 + d_2 + 2h_0 + \delta) \end{cases}$$

(10.6)

Let any two real-valued functions $f_1(t)$ and $f_2(t)$ be expanded in an m-term LPWM-GBPF series as per equation (10.3). Thus,

$$f_1(t) \approx \begin{bmatrix} c_0 & c_1 & c_2 & c_3 & \cdots & c_{(m-1)} \end{bmatrix} \mathbf{\Psi}_{g(m)}(\tau) \triangleq \mathbf{C}^{\mathrm{T}} \mathbf{\Psi}_{g(m)}(\tau) \qquad (10.7)$$

and

$$f_2(t) \approx \begin{bmatrix} d_0 & d_1 & d_2 & d_3 & \cdots & d_{(m-1)} \end{bmatrix} \mathbf{\Psi}_{g(m)}(\tau) \triangleq \mathbf{D}^{\mathrm{T}} \mathbf{\Psi}_{g(m)}(\tau) \qquad (10.8)$$

In equations (10.7) and (10.8), the coefficients $c_0, c_1, c_2, \ldots, c_{(m-1)}$ and $d_0, d_1 d_2, \ldots, d_{(m-1)}$ are determined using equation (10.4). To obtain the convolution of the two functions $f_1(t)$ and $f_2(t)$ in the LPWM-GBPF domain, equations (10.7) and (10.8) may be used.

Let us first consider two elements from each of $\mathbf{C}$ and $\mathbf{D}$ vectors as shown in Figure 10.4(a). Figure 10.4(b) shows the four components of their convolution

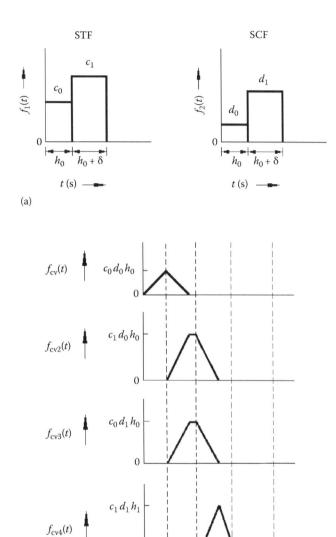

FIGURE 10.4
(a) Two time functions $f_1(t)$ and $f_2(t)$ for convolution over two segments h_0 and $(h_0 + \delta)$ and (b) the four components of their convolution spectrum.

spectrum and Figure 10.5 shows the result of convolution of these vectors graphically.

Convolutions between corresponding block pulses of the two LPWM-GBPF pulse trains, having magnitude vectors **C** and **D** would result in *trapezoidal* and *triangular* functions, having definite patterns and recursive characteristics.

Such two trains are considered, as shown in Figure 10.6 and their convolution process and subsequent result comprised of several components are presented in detail in Figure 10.7.

Superposition of these component convolution results yields the desired function $CV_{12}(t)$ as per equation (10.5). But, to carry out the whole analysis in the LPWM-GBPF domain, one finds the average value of the overall convolution result in each of the m subintervals. These average values will represent the LPWM-GBPF coefficients of the convolution function $CV_{12}(t)$.

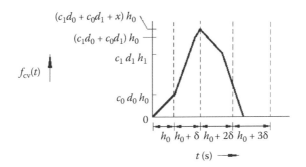

FIGURE 10.5
Convolution result of two time functions $f_1(t)$ and $f_2(t)$, obtained via superposition of component results of Figure 10.4.

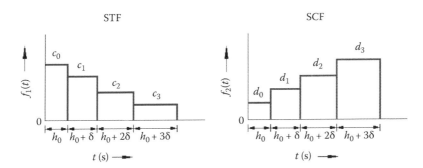

FIGURE 10.6
Two LPWM-GBPF trains $f_1(t)$ and $f_2(t)$ for convolution.

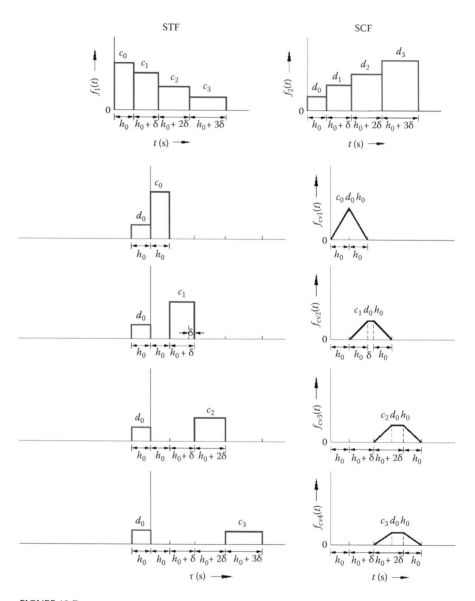

FIGURE 10.7
Process of convolution of two LPWM-GBPF trains shown in Figure 10.6.   (*Continued*)

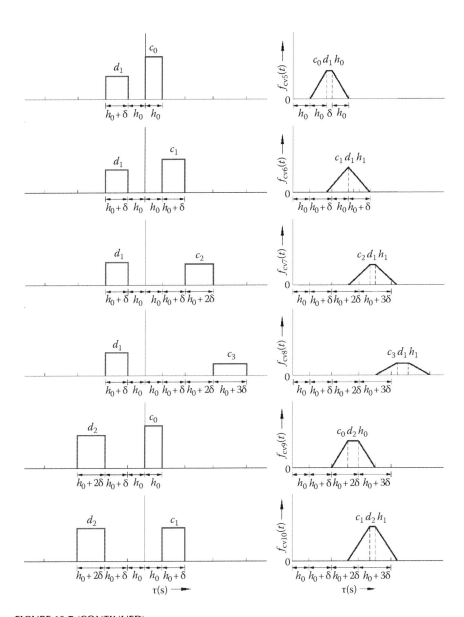

FIGURE 10.7 (CONTINUED)
Process of convolution of two LPWM-GBPF trains shown in Figure 10.6. *(Continued)*

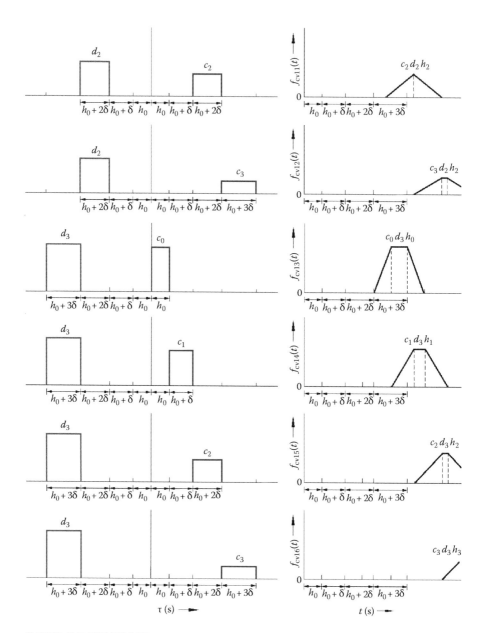

FIGURE 10.7 (CONTINUED)
Process of convolution of two LPWM-GBPF trains shown in Figure 10.6.

Suppose,

$$CV_{12}(t) \approx \begin{bmatrix} r_0 & r_1 & r_2 & \cdots & r_{(m-1)} \end{bmatrix} \mathbf{\Psi}_{(m)}(t) \triangleq \mathbf{R}^{\mathrm{T}} \mathbf{\Psi}_{g(m)}(t) \tag{10.9}$$

where, the desired coefficients $r_0, r_1, r_2, \ldots, r_{(m-1)}$ may be determined by employing equation (10.4).

It can be shown that

$$\mathbf{R}^{\mathrm{T}} \approx \mathbf{D}^{\mathrm{T}} \mathbf{GCVM}_{(m)} \tag{10.10}$$

where, $\mathbf{GCVM}_{(m)}$ is the generalized convolution matrix of order m formed by the elements of the vector $\mathbf{C}$, and given by

$$\mathbf{GCVM}_{(m)} = \begin{bmatrix} a_{11} & a_{12} & \cdots & a_{1j} & \cdots & a_{1m} \\ & a_{22} & \cdots & a_{2j} & \cdots & a_{2m} \\ & & \ddots & \vdots & & \vdots \\ & & & a_{ij} & & \vdots \\ & \mathbf{0} & & & \ddots & \vdots \\ & & & & & a_{mm} \end{bmatrix} \tag{10.11}$$

where

$$\begin{aligned}
a_{ij} = \frac{1}{h_0 + (j-1)\delta} &\left[\frac{1}{2}\left\{ h_0 + (i-1)(i-j+1)\delta \right\}^2 c_{j-i-1} + \left[\frac{1}{2}\left\{ h_0 + (i-1)(i-j+1)\delta \right\} \right.\right. \\
&\times \left\{ h_0 - (i-1)(i-j-1)\delta \right\} - (2i-j-1)\delta \left\{ h_0 + (i-1)\delta - \frac{1}{2}(i-1) \right\} \\
&\times (i-j-1)\delta \left\{ 2h_0 + (i-1)(i-j+1)\delta \right\} \right] c_{j-1} + \frac{1}{2}\left\{ (i-1)(i-j-1)\delta \right\}^2 c_{j-i+1} \Bigg]
\end{aligned} \tag{10.12}$$

where $c_n = 0$ for $n < 0$, and $a_{ij} = 0$ for $i > j$.

The constraint of equation (10.2) had been adhered to while forming $\mathbf{GCVM}_{(m)}$. Hence, knowing the LPWM-GBPF vectors of equations (10.7) and (10.8) one can solve for the LPWM-GBPF vector $\mathbf{R}$ for convolution of $f_1(t)$ and $f_2(t)$. Also, if $\mathbf{R}$ is known and we know the static function $\mathbf{C}^{\mathrm{T}} \mathbf{\Psi}_{g(m)}(t)$, the scanning function $\mathbf{D}^{\mathrm{T}} \mathbf{\Psi}_{g(m)}(t)$ can be solved by using equation (10.10). That is

$$\mathbf{D}^{\mathrm{T}} = \mathbf{R}^{\mathrm{T}} \mathbf{GCVM}_{(m)}^{-1} \tag{10.13}$$

which is equivalent to the "deconvolution" operation. Equation (10.13) will be computable if $c_0 \neq 0$. The matrix $\mathbf{GCVM}_{(m)}$ for conventional BPF analysis may be obtained by putting $\delta = 0$ in equation (10.11), which yields

$$\overline{\mathbf{GCVM}}_{(m)} \triangleq \frac{h_0}{2} \underbrace{\left[\begin{array}{ccccc} \overline{c}_0 & \overline{c}_1 + \overline{c}_2 & \cdots & \overline{c}_{(m-2)} + \overline{c}_{(m-1)} & \overline{c}_{(m-1)} + \overline{c}_m \end{array} \right.}_{m \text{ terms}} \Big]_{(m \times m)} \tag{10.14}$$

where

$$\begin{bmatrix} a & b & c \end{bmatrix} = \begin{bmatrix} a & b & c \\ 0 & a & b \\ 0 & 0 & a \end{bmatrix}$$

A bar symbol above the c_i's indicate that they are related to the conventional BPF domain.

Thus, equation (10.10) is converted to the following special case with $\delta = 0$,

$$\overline{\mathbf{R}}^\mathsf{T} = \overline{\mathbf{D}}^\mathsf{T} \, \overline{\mathbf{GCVM}}_{(m)} \tag{10.15}$$

where all the vectors are related to the conventional BPF domain. $\overline{\mathbf{GCVM}}_{(m)}$ is the conventional matrix developed by Kwong and Chen [14].

10.3.1 Numerical Example

Example 10.1 [13]

Consider the functions $f_1(t)$ and $f_2(t)$ as shown in Figure 10.8. That is, for $0 \le t \le 1$, $f_1(t) = u(t)$ and $f_2(t) = \exp(-t)$. The result of their convolution is given by

$$CV_{12}(t) = \begin{cases} 1 - \exp(-t) & \text{for } 0 \le t < 1 \\ \exp\big(-(t-1)\big) - \exp(-1) & \text{for } 1 \le t \le 2 \end{cases} \tag{10.16}$$

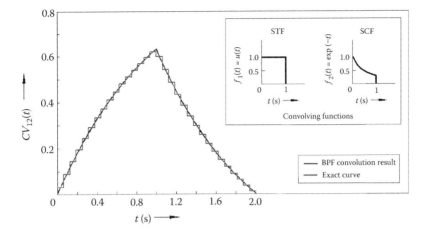

FIGURE 10.8
Comparison of BPF domain convolution result for $\delta = 0$ and $m = 32$, with the exact convolution of the functions $f_1(t)$ and $f_2(t)$ of Example 10.1 (vide Appendix B, Program no. 10.1).

Direct expansion of $CV_{12}(t)$ in BPF domain, for $T = 2$ s and $\delta = 0$, is

$$CV_{12}(t) \approx \begin{bmatrix} 0.03060901 & 0.08934144 & 0.14451545 & 0.19634664 \\ 0.24503753 & 0.29077840 & 0.33374796 & 0.37411413 \\ 0.41203464 & 0.44765766 & 0.48112239 & 0.51255960 \\ 0.54209212 & 0.56983535 & 0.59589771 & 0.62038103 \\ 0.60151155 & 0.54277912 & 0.48760511 & 0.43577392 \\ 0.38708302 & 0.34134216 & 0.29837260 & 0.25800643 \\ 0.22008592 & 0.18446290 & 0.15099817 & 0.11956096 \\ 0.09002844 & 0.06228521 & 0.03622285 & 0.01173953 \end{bmatrix} \Psi_{(32)}$$

(10.17)

It is noted from Figure 10.8 that the convolution result $CV_{12}(t)$ spans a region from 0 to 2 seconds. Hence, using an m-term BPF expansion of $f_1(t)$ and $f_2(t)$ for the analysis with $T = 1$ second will result in a truncated solution of $CV_{12}(t)$ terminated at $t = 1$ s. To take care of this situation, consider the functions $f_1(t)$ and $f_2(t)$ up to 2 seconds, where, from $t = 1$ s to $t = 2$ s, their BPF coefficients are taken to be zero. Now, deriving the vectors **C** and **D** of equations (10.7) and (10.8) from $f_1(t)$ and $f_2(t)$ of Figure 10.8, respectively, the convolution result $CV_{12}(t)$ is

$$CV_{12}(t) \approx \left[\begin{array}{cccc} 0.03029347 & 0.08904502 & 0.14423699 & 0.19608505 \\ 0.24479179 & 0.29054755 & 0.33353110 & 0.37391041 \\ 0.41184326 & 0.44747787 & 0.48095350 & 0.51240093 \\ 0.54194307 & 0.56969534 & 0.59576618 & 0.62025747 \\ 0.60182709 & 0.54307554 & 0.48788357 & 0.43603551 \\ 0.38732876 & 0.34157301 & 0.29858946 & 0.25821015 \\ 0.22027730 & 0.18464269 & 0.15116706 & 0.11971962 \\ 0.09017749 & 0.06242522 & 0.03635438 & 0.01186309 \end{array} \right] \Psi_{(32)}$$

$$\triangleq Y^T \Psi_{(32)}(t) \tag{10.18}$$

The result **Y** of equation (10.18) is compared with the actual solution in Figure 10.8. It is evident that the BPF solution closely matches the exact solution. Also, direct expansion of $CV_{12}(t)$ or equation (10.17) gives coefficients which are close to the elements of **Y** in equation (10.18).

Next, consider the same example with nonzero value of δ. For conventional BPF expansion of $f_2(t) = \exp(-t)$ over a period of 1 second and $m = 16$, the MISE is found to be $1.40678237 \times 10^{-4}$. And for LPWM-GBPF analysis involving this function, it is obvious that one must choose $\delta > 0$. For a value of $\delta = 0.00245833$, the MISE for the expansion of $f_2(t)$ assumes the minimum value of $1.26941836 \times 10^{-4}$. Considering the exponential nature of $CV_{12}(t)$ in equation (10.16) within the interval $0 \le t \le 1$ second, it is apparent that this value of δ may possibly minimize the MISE for the convolution result. With this value of δ, the output vector **Y** is

$$Y^T = \left[\begin{array}{cccc} 0.02155294 & 0.06485385 & 0.10843161 & 0.15206444 \\ 0.19554074 & 0.23866065 & 0.28123732 & 0.32309804 \\ 0.36408504 & 0.40405617 & 0.44288528 & 0.48046243 \\ 0.51669388 & 0.55150186 & 0.58482422 & 0.61661387 \end{array} \right] \Psi_{g(16)}$$

$$\tag{10.19}$$

Using the exact solution given by equation (10.16), the LPWM-GBPF expansion coefficients (with $\delta = 0.00245833$) are found to be in close agreement with the elements of **Y**. For nonzero δ, the convolution spectrum has been determined for m blocks only, as the end of the convolution spectrum could not be located exactly in terms of LPWM-GBPF cells. It is found that for positive δ, the spectrum will cover less than $2m$ cells, for negative δ more than $2m$ cells and for $\delta = 0$ the spectrum will cover exactly $2m$ cells. However, following the constraint of equation (10.2), the convolution spectrum always covers more than m cells for any δ. Thus, the operational method will always give correct result for m cells.

10.4 Linear Feedback System Identification Using Generalized Convolution Matrix (GCVM)

In control theory, the problem of system identification [16] is common and there are several approaches for solving such problems. The classical approach uses a nonparametric model involving the impulse response function to design a dynamic system. The basic system identification problem is to determine the unknown plant for the input–output data available.

Consider the linear single input, single output (SISO) feedback control system shown in Figure 10.9. For this system, the output is

$$c(t) = \int_0^t \left[r(\tau) - \int_0^\tau c(\sigma)h(\tau - \sigma) \ d\sigma \right] g(t - \tau) \ d\tau \tag{10.20}$$

It is required to solve $g(t)$, the unknown plant, from a knowledge of $r(t)$, $c(t)$ and $h(t)$. Using equation (10.4), let the time functions $r(t)$, $c(t)$ and $h(t)$ be represented by their respective LPWM-GBPF **R**, **C** and **H**, each having m components.

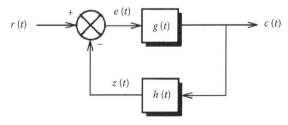

FIGURE 10.9
Linear single input, single output (SISO) feedback control system.

Following a procedure similar to Reference [14] and using the results derived in Section 10.3, it is possible to show that the LPWM-GBPF vector representing $g(t)$ is given by

$$\mathbf{G}^T = \begin{bmatrix} g_0 & g_1 & g_2 & \cdots & g_{(m-1)} \end{bmatrix} \triangleq \mathbf{C}^T \mathbf{GCVME}_{(m)}^{-1} \tag{10.21}$$

where $\mathbf{GCVME}_{(m)}$ is the generalized convolution matrix formed by the elements of the LPWM-GBPF vector $\mathbf{E}$, determined from the following relation

$$\mathbf{E}^T = \mathbf{R}^T - \mathbf{Z}^T \tag{10.22}$$

In equation (10.22), $\mathbf{Z}$ is given by

$$\mathbf{Z}^T = \mathbf{H}^T \mathbf{GCVMC}_{(m)} \tag{10.23}$$

where $\mathbf{GCVMC}_{(m)}$ is formed by the elements of $\mathbf{C}$.

10.4.1 Numerical Example

Example 10.2 [13,14]

Consider the example treated by Kwong and Chen [14]. Referring to Figure 10.9 in Laplace domain, $R(s) = \dfrac{1}{s}$, $G(s) = \dfrac{2}{(s+4)}$ and $H(s) = \dfrac{4}{s}$.

It is apparent that the output is $c(t) = \exp(-2t) \sin 2t$. We consider $m = 16$ and $T = 1$ s, as was considered by Kwong and Chen [14].

We have considered this problem in Chapter 5 for identification of the system in conventional BPF domain. It had been noted there that the results obtained in conventional BPF domain and the direct expansion of the system $g(t)$ were not very close. Also the elements of BPF domain identification results showed noticeable deviations from the desired solution.

So now we apply the LPWM-GBPF approach, using equations (10.11)–(10.13) for identification of the same system using a nonzero value of δ.

The direct BPF expansion of $g(t)$ gives

$$g(t) \approx \begin{bmatrix} 1.769594 & 1.378161 & 1.073313 & 0.835897 \\ & 0.650997 & 0.506997 & 0.394850 & 0.307509 \\ & 0.239488 & 0.186514 & 0.145257 & 0.113126 \\ & & 0.088103 & 0.068615 & 0.053437 & 0.041617 \end{bmatrix} \Psi_{(16)}$$

$$\tag{10.24}$$

Now consider a nonzero value of δ and launch upon the LPWM-GBPF analysis for $m = 16$ and $T = 1$ s. It was found that the MISE ($= 9.265842 \times 10^{-5}$) for the LPWM-GBPF expansion of $c(t)$ was minimized for $\delta = 0.004917$ and the LPWM-GBPF coefficients of $c(t)$ are

$$\mathbf{C}^{\mathrm{T}} = \begin{bmatrix} 0.024761 & 0.074998 & 0.126665 & 0.176903 \\ 0.222930 & 0.262213 & 0.292646 & 0.312694 \\ 0.321504 & 0.318973 & 0.305744 & 0.283148 \\ 0.253079 & 0.217823 & 0.179844 & 0.141577 \end{bmatrix}$$

(10.25)

With this $\mathbf{C}$, we solve for the vector $\mathbf{G}$ to obtain

$$\mathbf{G}^{\mathrm{T}} = \begin{bmatrix} 1.935021 & 1.681760 & 1.514493 & 1.266814 \\ 1.097820 & 0.880190 & 0.737898 & 0.565906 \\ 0.463758 & 0.340746 & 0.278829 & 0.197768 \\ 0.167156 & 0.115554 & 0.104078 & 0.069315 \end{bmatrix}$$

(10.26)

For $\delta = 0.004917$, $m = 16$ and $T = 1$ s, the direct LPWM-GBPF expansion of the time function $g(t)$ gives the vector $\hat{\mathbf{G}}$ as

$$\hat{\mathbf{G}}^{\mathrm{T}} = \begin{bmatrix} 1.900914 & 1.699248 & 1.489443 & 1.280159 \\ 1.078889 & 0.891584 & 0.722471 & 0.574053 \\ 0.447256 & 0.341691 & 0.255966 & 0.188021 \\ 0.135426 & 0.095646 & 0.066238 & 0.044980 \end{bmatrix}$$

(10.27)

Table 10.1 shows that the vectors $\mathbf{G}$ and $\hat{\mathbf{G}}$ are very close and there is no appreciable deviations. From the error column of the table it is observed that there is slight oscillation but the amplitudes of oscillation have been much reduced compared to the earlier solution, vide Table 5.5. Thus, the LPWM-GBPF analysis proves to be superior to the conventional BPF analysis. Incidentally, for $m = 16$ and $T = 1$ s, the function $g(t)$ has a minimum MISE of 1.134125×10^{-3} for $\delta = 0.005875$. And for $\delta = 0.004917$, $m = 16$ and $T = 1$ s, the MISE is 1.173436×10^{-3}.

TABLE 10.1

Identification of the Forward Path Element of Example 10.2 for $m = 16$ and $T = 1$ s in LPWM-GBPF Domain Using the "Deconvolution" Operation and the Result is Compared with the Direct LPWM-GBPF Expansion of the Equivalent Impulse Response

t (s)	Direct Expansion of $g(t)$ in LPWM-GBPF Domain	"Deconvolution" Result in LPWM-GBPF Domain using Equation (10.21)	Percentage Error
0.0000			
	1.900914	1.935021	– 1.794242
0.0256			
	1.699248	1.681760	1.029161
0.0562			
	1.489443	1.514493	– 1.681837
0.0916			
	1.280158	1.266814	1.042371
0.1320			
	1.078888	1.097820	– 1.754770
0.1773			
	0.891584	0.880190	1.277950
0.2275			
	0.722471	0.737898	– 2.168325
0.2826			
	0.574053	0.565906	1.419207
0.3427			
	0.447256	0.463758	– 3.689610
0.4076			
	0.341691	0.340746	0.276566
0.4775			
	0.255966	0.278829	– 8.932046
0.5523			
	0.188020	0.197768	– 5.184555
0.6320			
	0.135426	0.167156	– 23.429770
0.7166			
	0.095646	0.115554	– 20.814253
0.8062			
	0.066238	0.104078	– 57.127329
0.9006			
	0.044980	0.069315	– 54.101824
1.0000			

10.5 Error Analysis

The representational error for conventional BPF expansion of any square integrable function $f(t)$ of Lebesgue measure was formally investigated by Rao and Srinivasan [17]. For m blocks of equal width h over an interval $[0,T)$ such that $T = mh$, the mean integral square error (MISE) in the ith interval is

$$\left[\varepsilon_{ie}\right]^2 = \int_{(i-1)h}^{ih} \left[f(t)\right]^2 dt - \frac{1}{h}\left[\int_{(i-1)h}^{ih} f(t)\ dt\right]^2 \tag{10.28}$$

Here, we have considered $i = 1,2,\ldots,m$, otherwise the error contribution of the first sub-interval will become zero.

Expanding $f(t)$ in the ith interval around any point μ_i, by Taylor's series, Rao and Srinivasan neglected second and higher-order derivatives of $f(t)$ to arrive at the following simple result from equation (10.28) for the upper bound of MISE:

$$\sum_{i=1}^{m}\left[\varepsilon_{ie}\right]^2 \triangleq E_e^2 = \frac{mh^3}{12}\left(\dot{f}_{max}\right)^2 \tag{10.29}$$

where $\dot{f}_{max}$ is the largest among all $\dot{f}(\mu_i)$. Equation (10.29) takes into consideration all functions of the ramp type in each sub-interval in $[0,T)$.

To present an error analysis in the LPWM-GBPF domain, a similar track as in Reference [17] has been followed, but with required modifications.

Now, assuming that an m-set BPF has been converted to an m-set LPWM-GBPF over the same interval $[0,T)$ for expanding $f(t)$, then following equation (10.28), the total MISE is

$$\sum_{i=1}^{m}\left[\varepsilon_{ie}\right]^2 \triangleq E_u^2 = \sum_{i=1}^{m}\left[\frac{\left\{h_0 + (i-1)\delta\right\}^3}{12}\right]\left[\dot{f}(\mu_i)\right]^2$$

Considering $\dot{f}_{max}$ to be the largest of all $\dot{f}(\mu_i)$, the upper bound of MISE, using equation (10.1) and equating T with mh, is

$$E_u^2 = \left[\frac{mh^3}{12} + \frac{mh\delta^2(m^2-1)}{48}\right]\left(\dot{f}_{max}\right)^2 \tag{10.30}$$

Comparing equations (10.29) and (10.30), one finds that

$$\sigma \triangleq \left[E_e^2 - E_u^2 \right] = -\frac{mh\delta^2(m^2-1)}{48}\left(\dot{f}_{max} \right)^2 < 0 \tag{10.31}$$

Thus, the upper limit of the error bound has increased with LPWM-GBPF showing no apparent improvement, as stated earlier. This is due to the fact that in this case $f(t)$, approximated by first-order Taylor expansion, is at the most a combination of step and ramp functions, and hence, LPWM-GBPF unnecessarily introduces more error than the conventional BPF.

Also, the additional error is dependent on the value of δ and vanishes when $\delta = 0$.

To bring out the advantages of LPWM-GBPF over the conventional BPF, consider at least up to the second-order derivative in the Taylor series expansion of $f(t)$. For the LPWM-GBPF, the MISE is given by

$$\left[\varepsilon_{iu} \right]^2 = \int_{(i-1)h_0 + (i-1)(i-2)\delta/2}^{ih_0 + i(i-1)\delta/2} \left[f(t) \right]^2 dt - \frac{1}{h_0 + (i-1)\delta} \left[\int_{(i-1)h_0 + (i-1)(i-2)\delta/2}^{ih_0 + i(i-1)\delta/2} f(t) dt \right]^2 \tag{10.32}$$

Replacing $f(t)$ by its Taylor series expansion, it is possible to show that the upper bound of the total MISE over m cells is

$$\sum_{i=1}^{m} \left[\varepsilon_{iu} \right]^2 \triangleq E_u^2 = \frac{\left[\dot{f}_{max} \right]^2}{12} \left[mh_0^3 + 3h_0^2\delta \frac{m(m-1)}{2} \right] + \frac{\left[\ddot{f}_{max} \right]^2}{180} 4h_0^5 m + 5h_0^5 m(m^2-1)$$

$$+ \frac{5}{2}\delta h_0^4 m(m-1)(3m-1)(2m+1) - 15h_0^4\mu_a m^2 + 15h_0^3\mu_a^2 m$$

$$- 5h_0^3\mu_a\delta m(m-1)(7m+1) + \frac{45}{2}h_0^2\mu_a^2\delta m(m-1)$$

$$+ \frac{\dot{f}_{max}\ddot{f}_{max}}{12}\left[h_0^4 m^2 + \frac{1}{3}h_0^3\delta m(m-1)(7m+1) \right.$$

$$\left. - 2h_0^3\mu_a m - 3h_0^2\mu_a\delta m(m-1) \right] \tag{10.33}$$

where μ_a is the average value of μ_i's, $\dot{f}_{max}$ and $\ddot{f}_{max}$ are the largest among all $\dot{f}(\mu_i)$ and $\ddot{f}(\mu_i)$, respectively, and terms containing δ^2 and higher powers of δ have been neglected.

For the conventional BPF set, $\delta = 0$ and $h_0 = h$. Substituting these in equation (10.33), we obtain the total MISE as

$$\sum_{i=1}^{m}\left[\varepsilon_{ie}\right]^2 \triangleq E_e^2 = \frac{1}{12}\left[\dot{f}_{max}\right]^2 mh^3 + \frac{1}{180}\left[\ddot{f}_{max}\right]\left[4h^5m + \frac{5}{3}h^5m(m^2-1)\right.$$
$$\left.-15h^4\mu_a m^2 + 15h^3\mu_a^2 m\right] + \frac{\dot{f}_{max}\ddot{f}_{max}}{12}\left[h^4m^2 - 2h^3\mu_a m\right] \tag{10.34}$$

Now comparing equations (10.33) and (10.34) as before and simplifying,

$$\sigma = \frac{1}{36}\left(\ddot{f}_{max}\right)^2 m(m^2-1)\delta h_0^3\left(\mu_a - \frac{mh_0}{2}\right) - \frac{1}{36}\dot{f}_{max}\ddot{f}_{max}m(m^2-1)\delta h_0^3 \tag{10.35}$$

In deducing equation (10.35), use is made of equation (10.1) and the approximate relation $\left(h^p - h_0^p\right) \simeq \frac{p}{2}(m-1)\delta h_0^{p-1}$ where $p = 3, 4, 5$. Consider that all μ_i's lie at a fixed distance α from the lower limit of t in the ith interval. Then

$$\mu_\alpha \triangleq \frac{1}{m}\sum_{i=1}^{m}\mu_i \simeq \alpha + \frac{1}{2}(m-1)h_0 + \frac{1}{6}(m-1)(m-2)\delta \tag{10.36}$$

Substituting equation (10.36) in equation (10.35) and simplifying,

$$\sigma = \frac{1}{36}\left(\ddot{f}_{max}\right)^2 m(m^2-1)\delta h_0^3\left[\left(\alpha - \frac{h_0}{2}\right) + \frac{1}{6}(m-1)(m-2)\delta\right]$$
$$-\frac{1}{36}\dot{f}_{max}\ddot{f}_{max}m(m^2-1)\delta h_0^3 \tag{10.37}$$

Equation (10.37) is of main interest and if the expression on the RHS is greater than zero, it can be concluded that LPWM-GBPF analysis introduces less MISE than the conventional BPF.

Case 1: $\delta > 0$

For this case note that excepting the $(h_0/2)$ term within the square bracket, all other terms connected with $\left(\ddot{f}_{max}\right)^2$ are positive if m is greater than 2. On close scrutiny, one finds that $(\alpha - h_0/2) \approx 0$, h_0 being the thinnest pulse of the set.

Also, for monotonically rising or decaying functions, the product $\dot{f}_{max}\ddot{f}_{max}$ is negative. Other terms associated with it being positive, we conclude that

$$-\frac{1}{36}\dot{f}_{max}\ddot{f}_{max}m(m^2-1)\delta h_0^3 > 0$$

Thus, for this case, when the LPPWM-GBPF set is used, we have $\sigma > 0$.

Case 2: $\delta < 0$

For such a case, h_0 is the thickest pulse and δ is negative. So $(\alpha - h_0/2) < 0$, keeping in mind that α is taken to be the same for the thinnest pulse also. Thus,

$$\frac{1}{36}\left(\dot{f}_{max}\right)^2 m(m^2-1)\delta h_0^3 \left[\left(\alpha - h_0/2\right) + \frac{(m-1)(m-2)\delta}{6}\right] > 0$$

For functions having $|\dot{f}(0)| < |\dot{f}(\infty)|$, note that $\dot{f}_{max}$ and $\ddot{f}_{max}$ are of the same sign. Also, δ being negative, the second term with the negative sign on the RHS of equation (10.37) turns out to be greater than zero. Hence, for LNPWM-GBPF analysis again $\sigma > 0$. This result has also been supported by the numerical examples treated in this chapter.

10.6 Conclusion

From a set of GBPFs, a set of LPPWM/LNPWM-GBPF has been defined. This LPWM-GBPF set has been used to form the generalized convolution operational matrix **GCVM**. Using this matrix, convolution of time-varying functions has been determined in the LPWM-GBPF domain. It has been shown that the conventional operational technique proposed by Kwong and Chen [14] is only a special case of the LPWM-GBPF technique presented here.

In support of the theory, examples have been treated successfully to determine the convolution spectrums of convolving functions. Further, the method has been applied to solve the linear SISO feedback control system identification problem. An error analysis has also been carried out to show that the LPWM-GBPF domain analysis introduces less error than the conventional BPF method.

The presented method is better suited to deal with monotonic functions and nonoscillatory systems. Also, the method is computer compatible and considering the broad application areas of convolution integral, it is a worthwhile technique suitable for its possible applications in the fields of control theory and design.

References

1. Alfred Haar, Zur theorie der orthogonalen funktionen systeme, Math. Annalen, vol. **69**, pp. 331–371, 1910.
2. H. Rademacher, Einige sätze von allegemeinen orthogonal funktionen, Math. Annalen, vol. **87**, pp. 122–138, 1922.
3. K. G. Beauchamp, *Walsh functions and their applications*, Academic Press, London, 1975.
4. K. G. Beauchamp, *Application of Walsh and related functions*, Academic Press, London, 1984.
5. Anish Deb and Suchismita Ghosh, *Power electronic systems: Walsh analysis with MATLAB®*, CRC Press, Boca Raton, 2014.
6. Z. H. Jiang and W. Schaufelberger, *Block pulse functions and their applications in control systems*, LNCIS, vol. **179**, Springer-Verlag, Berlin, 1992.
7. Anish Deb, Gautam Sarkar and Sunit K. Sen Block pulse functions, the most fundamental of all piecewise constant basis functions, Int. J. Syst. Sci., vol. **25**, no. 2, pp. 351–363, 1994.
8. Anish Deb, Gautam Sarkar, Sunit K. Sen and Asit K. Dutta, Cross-/autocorrelation of time-varying signals using block pulse functions and determination of power spectral densities, Int. J. Syst. Sci., vol. **25**, no. 1, pp. 113–128, 1994.
9. G. P. Rao and S. G. Tzafestas, A decade of piecewise constant orthogonal functions in systems and control, Math. and Computers in Simulation, vol. **27**, no. 5 & 6, pp. 389–407, 1985.
10. S. G. Tzafestas (Ed.), Walsh functions in signal and systems analysis and design, Van Nostrand Reinhold Co., New York, 1985.
11. C. F. Chen, Y. T. Tsay and T. T. Wu, Walsh operational matrices for fractional calculus and their application to distributed systems, J. Franklin Inst., vol. **303**, no. 3, pp. 267–284, 1977.
12. M. L. Wang, S. Y. Yang and R. Y. Chang, Analysis of systems with multiple time-varying delays via generalized block pulse functions, Int. J. Syst. Sci., vol. **18**, no. 3, pp. 543–552, 1987.
13. Anish Deb, Gautam Sarkar and Sunit K. Sen, Linearly pulse-width modulated block pulse functions and their application to linear SISO feedback control system identification, Proc. IEE, Part D, Control Theory and Appl., vol. **142**, no. 1, pp. 44–50, 1995.
14. C. P. Kwong and C. F. Chen, Linear feedback system identification via block pulse functions, Int. J. Syst. Sci., vol. **12**, no. 5, pp. 635–642, 1981.

15. E. O. Brigham, *The fast Fourier transform and its applications*, Prentice-Hall, New York, 1988.
16. H. Unbehauen and G. P. Rao, *Identification of continuous systems*, North-Holland, Amsterdam, 1987.
17. G. P. Rao and T. Srinivasan, Analysis and synthesis of dynamic systems containing time delays via block pulse functions, IEE Proc., vol. **125**, no. 9, pp. 1064–1068, 1978.

Study Problems

10.1 Consider an open loop system having a transfer function $G(s) = (s + 1)^{-1}$. Find its output $c(t)$ in LPWM-GBPF domain for a step input $u(t)$ using the convolution matrix. Consider $m = 8$ and $T = 1$ s, and choose the values of h_0 and δ to make MISE a minimum.

Compare the convolution results with direct expansion of the output and determine the percentage errors of different coefficients.

10.2 Consider the system of Problem 10.1. For the same values of m and T, find its output $c(t)$ in conventional BPF domain using the convolution matrix. Finally, compare the convolution result with the result obtained above in LPWM-GBPF domain with respect to MISE.

10.3 Consider a closed loop system having the forward path transfer function $G(s) = (s + 1)^{-1}$ and the feedback path transfer function $H(s) = \dfrac{4}{s}$. Find its output $c(t)$ in LPWM-GBPF domain for a step input $u(t)$ using the convolution matrix. Consider $m = 4$ and $T = 1$ s choosing h_0 and δ.

Finally, compare the result graphically with the direct expansion of $c(t)$.

10.4 Consider the feedback system of Problem 10.3. Double the value of m for the same interval T and find its output $c(t)$ in BPF domain using the convolution matrix. Finally, comment on the convolution result.

10.5 Consider the open loop system of Problem 10.1. Knowing the input and output, identify the system both in BPF domain and in LPWM-GBPF domain using the "deconvolution" matrix, for $m = 8$, $T = 1$ s and suitable choice of h_0 and δ. Comment on the results obtained using these two approaches. Is there any oscillation in the result?

10.6 Consider the closed loop system of Problem 10.3. Knowing the input output and feedback path transfer function, identify the system both in BPF domain and in LPWM-GBPF domain using the convolution matrix, for $m = 8$, $T = 1$ s, and your choice of h_0 and δ.

Comment on the results obtained using LPWM-GBPF and BPF approaches.

10.7 Consider the closed loop system of Problem 10.3. Identify the system in LPWM-GBPF domain for double the value of m keeping the time interval T same and new choices for h_0 and δ. Compute percentage error for each LPWM-GBPF coefficient with reference to respective direct expansion coefficient.

Make your observation with respect to oscillation in the result.

Appendix A: Introduction to Linear Algebra

A matrix is a rectangular array of variables, mathematical expressions, or simply numbers. Commonly a matrix is written as

$$\mathbf{A} = \begin{bmatrix} a_{11} & a_{12} & \cdots & a_{1n} \\ a_{21} & a_{22} & \cdots & a_{2n} \\ \vdots & \vdots & \ddots & \vdots \\ a_{m1} & a_{m2} & \cdots & a_{mn} \end{bmatrix}. \tag{A.1}$$

The size of the matrix, with m rows and n columns, is called an m-by-n (or, $m \times n$) matrix, where, m and n are called its dimensions.

A matrix with one row [a $(1 \times n)$ matrix] is called a row vector, and a matrix with one column [an $(m \times 1)$ matrix] is called a column vector. Any isolated row or column of a matrix is a row or column vector, obtained by removing all other rows or columns respectively from the matrix.

Square Matrices

A square matrix is a matrix with $m = n$, i.e., the same number of rows and columns. An n-by-n matrix is known as a square matrix of order n. Any two square matrices of the same order can be added, subtracted, or multiplied.

For example, each of the following matrices is a square matrix of order 4, with four rows and four columns:

$$\mathbf{A} = \begin{bmatrix} 1 & 2 & 7 & 0 \\ 3 & 4 & 3 & 1 \\ 8 & 3 & 1 & 1 \\ 2 & 4 & 0 & 3 \end{bmatrix}, \quad \mathbf{B} = \begin{bmatrix} 3 & 1 & 0 & -1 \\ 2 & -3 & 4 & 2 \\ 1 & 0 & 9 & 1 \\ -3 & 2 & -1 & 0 \end{bmatrix}$$

Then, $\mathbf{A} + \mathbf{B} = \begin{bmatrix} 4 & 3 & 7 & -1 \\ 5 & 1 & 7 & 3 \\ 9 & 3 & 10 & 2 \\ -1 & 6 & -1 & 3 \end{bmatrix}$ and $\mathbf{A} - \mathbf{B} = \begin{bmatrix} -2 & 1 & 7 & 1 \\ 1 & 7 & -1 & -1 \\ 7 & 3 & -8 & 0 \\ 5 & 2 & 1 & 3 \end{bmatrix}$

Determinant

The determinant [written as det($\mathbf{A}$) or $|\mathbf{A}|$] of a square matrix $\mathbf{A}$ is a number encoding certain properties of the matrix. $\mathbf{A}$ matrix is invertible, if and only if, its determinant is nonzero.

The determinant of a 2-by-2 matrix is given by

$$\det \begin{pmatrix} a & b \\ c & d \end{pmatrix} = ad - bc \tag{A.2}$$

Properties

The determinant of a product of square matrices equals the product of their determinants: $\det(\mathbf{AB}) = \det(\mathbf{A}) \cdot \det(\mathbf{B})$.

Adding a multiple of any row to another row, or a multiple of any column to another column, does not change the determinant.

Interchanging two rows or two columns makes the determinant to be multiplied by –1.

Using these operations, any matrix can be transformed to a lower (or, upper) triangular matrix, and for such matrices the determinant equals the product of the entries on the main diagonal.

Orthogonal Matrix

An orthogonal matrix is a square matrix with real entries whose columns and rows are orthogonal vectors, i.e., orthonormal vectors.

Equivalently, a matrix $\mathbf{A}$ is orthogonal if its transpose is equal to its inverse. That is

$$\mathbf{A}^\mathrm{T} = \mathbf{A}^{-1}$$

which implies $\mathbf{A}^\mathrm{T} \mathbf{A} = \mathbf{A} \mathbf{A}^\mathrm{T} = \mathbf{I}$, where $\mathbf{I}$ is the identity matrix.

An orthogonal matrix $\mathbf{A}$ is necessarily invertible with inverse $\mathbf{A}^{-1} = \mathbf{A}^\mathrm{T}$. The determinant of any orthogonal matrix is either $+\mathbf{I}$ or $-\mathbf{I}$.

Trace of a Matrix

In equation (A.1), the entries $a_{i,i}$ form the main diagonal of the matrix $\mathbf{A}$. The trace, $\mathrm{tr}(\mathbf{A})$, of the square matrix $\mathbf{A}$ is the sum of its diagonal entries. The trace of the product of two matrices is independent of the order of the factors $\mathbf{A}$ and $\mathbf{B}$. That is

$$\mathrm{tr}(\mathbf{AB}) = \mathrm{tr}(\mathbf{BA}).$$

Also, the trace of a matrix is equal to that of its transpose, i.e., $\mathrm{tr}(\mathbf{A}) = \mathrm{tr}(\mathbf{A}^{\mathrm{T}})$.

Diagonal, Lower Triangular and Upper Triangular Matrices

If all entries of a matrix except those of the main diagonal are zero, the matrix is called a *diagonal matrix*. If only all entries above (or, below) the main diagonal are zero, it is called a *lower triangular matrix* (or, *upper triangular matrix*).
For example, a diagonal matrix of 3rd order is

$$\begin{bmatrix} d_{11} & 0 & 0 \\ 0 & d_{22} & 0 \\ 0 & 0 & d_{33} \end{bmatrix}$$

A lower triangular matrix of 3rd order is

$$\begin{bmatrix} l_{11} & 0 & 0 \\ l_{21} & l_{22} & 0 \\ l_{31} & l_{32} & l_{33} \end{bmatrix}$$

and an upper triangular matrix of similar order is

$$\begin{bmatrix} u_{11} & u_{12} & u_{13} \\ 0 & u_{22} & u_{23} \\ 0 & 0 & u_{33} \end{bmatrix}$$

Symmetric Matrix

A square matrix $\mathbf{A}$ that is equal to its transpose, i.e., $\mathbf{A} = \mathbf{A}^T$, is a symmetric matrix. If instead, $\mathbf{A}$ was equal to the negative of its transpose, i.e., $\mathbf{A} = -\mathbf{A}^T$, then $\mathbf{A}$ is a skew-symmetric matrix.

Singular Matrix

If the determinant of a square matrix $\mathbf{A}$ is equal to zero, it is called a *singular matrix* and its inverse does not exist. Examples of two singular matrices are

$$
\begin{bmatrix} 4 & 2 \\ 6 & 3 \end{bmatrix} \text{ and } \begin{bmatrix} 2 & -1 & \dfrac{2}{7} \\ -7 & 0 & 3 \\ -2 & -1 & 2 \end{bmatrix}
$$

Identity Matrix or Unit Matrix

If $\mathbf{A}$ is a square matrix, then

$$\mathbf{AI} = \mathbf{IA} = \mathbf{A}$$

where $\mathbf{I}$ is the identity matrix of the same order.

The identity matrix $\mathbf{I}_{(n)}$ of size n is the n-by-n matrix in which all the elements on the main diagonal are equal to 1 and all other elements are equal to 0. An identity matrix of order 3 is

$$
\mathbf{I}_{(3)} = \begin{bmatrix} 1 & 0 & 0 \\ 0 & 1 & 0 \\ 0 & 0 & 1 \end{bmatrix}.
$$

It is called identity matrix because multiplication with it leaves a matrix unchanged. If $\mathbf{A}$ is an $(m \times n)$ matrix then

$$\mathbf{A}_{(m \times n)} \mathbf{I}_{(n)} = \mathbf{I}_{(m)} \mathbf{A}_{(m \times n)}$$

Transpose of a Matrix

The transpose $\mathbf{A}^{\mathrm{T}}$ of a square matrix $\mathbf{A}$ can be obtained by reflecting the elements along its main diagonal. Repeating the process on the transposed matrix returns the elements to their original position.

The transpose of a matrix may be obtained by any one of the following equivalent actions:

   i. Reflect $\mathbf{A}$ over its main diagonal to obtain $\mathbf{A}^{\mathrm{T}}$

   ii. Write the rows of $\mathbf{A}$ as the columns of $\mathbf{A}^{\mathrm{T}}$

   iii. Write the columns of $\mathbf{A}$ as the rows of $\mathbf{A}^{\mathrm{T}}$

Formally, the ith row, jth column element of $\mathbf{A}^{\mathrm{T}}$ is the jth row, ith column element of $\mathbf{A}$. That is

$$[\mathbf{A}^{\mathrm{T}}]_{ij} = \mathbf{A}_{ji}$$

If $\mathbf{A}$ is an $m \times n$ matrix then $\mathbf{A}^{\mathrm{T}}$ is an $(n \times m)$ matrix.

Properties

For matrices $\mathbf{A}$, $\mathbf{B}$, and scalar c we have the following properties of transpose:

   i. $(\mathbf{A}^{\mathrm{T}})^{\mathrm{T}} = \mathbf{A}$

   ii. $(\mathbf{A} + \mathbf{B})^{\mathrm{T}} = \mathbf{A}^{\mathrm{T}} + \mathbf{B}^{\mathrm{T}}$

   iii. $(\mathbf{AB})^{\mathrm{T}} = \mathbf{B}^{\mathrm{T}}\mathbf{A}^{\mathrm{T}}$

Note that the order of the factors above reverses. From this one can deduce that a square matrix $\mathbf{A}$ is invertible if and only if $\mathbf{A}^{\mathrm{T}}$ is invertible, and in this case, we have $(\mathbf{A}^{-1})^{\mathrm{T}} = (\mathbf{A}^{\mathrm{T}})^{-1}$. By induction this result extends to the general case of multiple matrices, where we find that

$$(\mathbf{A}_1\mathbf{A}_2\ldots\mathbf{A}_{k-1}\mathbf{A}_k)^{\mathrm{T}} = \mathbf{A}_k^{\mathrm{T}}\mathbf{A}_{k-1}^{\mathrm{T}}\ldots\mathbf{A}_2^{\mathrm{T}}\mathbf{A}_1^{\mathrm{T}}.$$

   iv. $(c\mathbf{A})^{\mathrm{T}} = c\mathbf{A}^{\mathrm{T}}$

      The transpose of a scalar is the same scalar.

   v. $\det(\mathbf{A}^{\mathrm{T}}) = \det(\mathbf{A})$

   vi. $\det(\mathbf{A}^{-1}) = \dfrac{1}{\det(\mathbf{A})}$

Matrix Multiplication

Matrix multiplication is a binary operation that takes a pair of matrices, and produces another matrix. This term normally refers to the matrix product.

Multiplication of two matrices is defined only if the number of columns of the left matrix is the same as the number of rows of the right matrix. If $\mathbf{A}$ is an m-by-n matrix and $\mathbf{B}$ is an n-by-p matrix, then their matrix product $\mathbf{AB}$ is the m-by-p matrix whose entries are given by dot product of the corresponding row of $\mathbf{A}$ and the corresponding column of $\mathbf{B}$. That is

$$[\mathbf{AB}]_{i,j} = A_{i,1}B_{1,j} + A_{i,2}B_{2,j} + \cdots + A_{i,n}B_{n,j} = \sum_{r=1}^{n} A_{i,r}B_{r,j}$$

where $1 \le i \le m$ and $1 \le j \le p$.

Matrix multiplication satisfies the rules

  i. $(\mathbf{AB})\mathbf{C} = \mathbf{A}(\mathbf{BC})$ (associativity),

  ii. $(\mathbf{A} + \mathbf{B})\mathbf{C} = \mathbf{AC} + \mathbf{BC}$

  iii. $\mathbf{C}(\mathbf{A} + \mathbf{B}) = \mathbf{CA} + \mathbf{CB}$ (left and right distributivity),

whenever the size of the matrices is such that the various products are defined.

The product $\mathbf{AB}$ may be defined without $\mathbf{BA}$ being defined, namely, if $\mathbf{A}$ and $\mathbf{B}$ are m-by-n and n-by-k matrices, respectively, and $m \ne k$.

Even if both products are defined, they need not be equal, i.e., generally one has

$$\mathbf{AB} \ne \mathbf{BA},$$

i.e., matrix multiplication is not commutative, in marked contrast to (rational, real, or complex) numbers whose product is independent of the order of the factors. An example of two matrices not commuting with each other is:

$$\begin{bmatrix} 5 & 2 \\ 3 & 3 \end{bmatrix} \begin{bmatrix} 1 & 0 \\ 0 & 0 \end{bmatrix} = \begin{bmatrix} 5 & 0 \\ 3 & 0 \end{bmatrix}$$

whereas

$$\begin{bmatrix} 1 & 0 \\ 0 & 0 \end{bmatrix} \begin{bmatrix} 5 & 2 \\ 3 & 3 \end{bmatrix} = \begin{bmatrix} 5 & 2 \\ 0 & 0 \end{bmatrix}$$

Since det(**A**) and det(**B**) are just numbers and so commute, det(**AB**) = det(**A**) det(**B**) = det(**B**)det(**A**) = det(**BA**), even when **AB** ≠ **BA**.

A few properties of matrix multiplication

i. **Associative**

$$A(BC) = (AB)C$$

ii. **Distributive over matrix addition**

$$A(B + C) = AB + AC, (A + B)C = AC + BC$$

iii. **Scalar multiplication is compatible with matrix multiplication**

$$\lambda(AB) = (\lambda A)B \text{ and } (AB)\lambda = A(B\lambda)$$

where λ is a scalar. If the entries of the matrices are real or complex numbers, then all four quantities are equal.

Inverse of a Matrix

If **A** is a square matrix, there may be an inverse matrix $A^{-1} = B$ such that

$$AB = BA = I$$

If this property holds, then **A** is an invertible matrix. If not, **A** is a *singular* or *degenerate* matrix.

Analytic Solution of the Inverse

Inverse of a square non-singular matrix **A** may be computed from the transpose of a matrix **C** formed by the *cofactors* of **A**. Thus, C^T is known as the *adjoint matrix* of **A**. The matrix C^T is divided by the determinant of **A** to compute A^{-1}. That is

$$A^{-1} = \frac{C^T}{\det(A)} = \frac{1}{\det(A)} \begin{bmatrix} c_{11} & c_{21} & \cdots & c_{n1} \\ c_{12} & c_{22} & \cdots & c_{n2} \\ \vdots & \vdots & \ddots & \vdots \\ c_{1n} & c_{2n} & \cdots & c_{nn} \end{bmatrix} \qquad (A.3)$$

where **C** is the matrix of cofactors, and C^T denotes the transpose of **C**.

Cofactors

Let a (3 × 3) matrix be given by

$$\mathbf{A} = \begin{bmatrix} a & b & c \\ d & e & f \\ g & h & k \end{bmatrix}$$

Then a (3 × 3) matrix $\mathbf{P}$ formed by the cofactors of $\mathbf{A}$ is

$$\mathbf{P} = \begin{bmatrix} A & B & C \\ D & E & F \\ G & H & K \end{bmatrix}$$

where

$$A = (ek - fh), \quad B = (fg - dk), \qquad C = (dh - eg),$$
$$D = (ch - bk), \quad E = (ak - cg), \qquad F = (gb - ah),$$
$$G = (bf - ce), \quad H = (cd - af) \quad \text{and} \quad K = (ae - bd).$$

Inversion of a 2 × 2 Matrix

The cofactor equation listed above yields the following result for a 2 × 2 matrix. Let the matrix to be inverted be

$$\mathbf{A} = \begin{bmatrix} a & b \\ c & d \end{bmatrix}.$$

Then

$$\mathbf{A}^{-1} = \begin{bmatrix} a & b \\ c & d \end{bmatrix}^{-1} = \frac{1}{\det(\mathbf{A})} \begin{bmatrix} d & -b \\ -c & a \end{bmatrix} = \frac{1}{(ad - bc)} \begin{bmatrix} d & -b \\ -c & a \end{bmatrix}.$$

using equations (A.2) and (A.3).

Inversion of a 3 × 3 Matrix

Let the matrix to be inverted be

$$\mathbf{A} = \begin{bmatrix} a & b & c \\ d & e & f \\ g & h & k \end{bmatrix}.$$

Then, its inverse is given by

$$\mathbf{A}^{-1} = \begin{bmatrix} a & b & c \\ d & e & f \\ g & h & k \end{bmatrix}^{-1} = \frac{1}{\det(\mathbf{A})} \begin{bmatrix} A & B & C \\ D & E & F \\ G & H & K \end{bmatrix}^{\mathrm{T}} = \frac{1}{\det(\mathbf{A})} \begin{bmatrix} A & D & G \\ B & E & H \\ C & F & K \end{bmatrix} \qquad (A.4)$$

where, $A, B, C, D, E, F, G, H, K$ are the cofactors of the matrix $\mathbf{A}$.
The determinant of $\mathbf{A}$ can be computed as follows:

$$\det(\mathbf{A}) = a(ek - fh) - b(dk - fg) + c(dh - eg).$$

If the determinant is non-zero, the matrix is invertible and determination of the cofactors subsequently lead to the computation of the inverse of $\mathbf{A}$.

Similarity Transformation

Two n-by-n matrices $\mathbf{A}$ and $\mathbf{B}$ are called similar if

$$\mathbf{B} = \mathbf{P}^{-1} \mathbf{A} \, \mathbf{P} \qquad (A.5)$$

for some invertible n-by-n matrix $\mathbf{P}$.
Similar matrices represent the same linear transformation under two different bases, with $\mathbf{P}$ being the change of basis matrix.
The determinant of the similarity transformation of a matrix is equal to the determinant of the original matrix $\mathbf{A}$.

$$\det(\mathbf{B}) = \det(\mathbf{P}^{-1} \mathbf{A} \, \mathbf{P}) = \det(\mathbf{P}^{-1}) \det(\mathbf{A}) \det(\mathbf{P}) = \frac{\det(\mathbf{A})}{\det(\mathbf{P})} \det(\mathbf{P}) = \det(\mathbf{A}) \qquad (A.6)$$

Also, the eigenvalues of the matrices **A** and **B** are also same. That is

$$
\begin{aligned}
\det(\mathbf{B} - \lambda \mathbf{I}) &= \det(\mathbf{P}^{-1}\mathbf{A}\mathbf{P} - \lambda \mathbf{I}) \\
&= \det(\mathbf{P}^{-1}\mathbf{A}\mathbf{P} - \mathbf{P}^{-1}\lambda \mathbf{I}\mathbf{P}) \\
&= \det\left(\mathbf{P}^{-1}(\mathbf{A} - \lambda \mathbf{I})\mathbf{P}\right) \\
&= \det(\mathbf{P}^{-1})\det(\mathbf{A} - \lambda \mathbf{I})\det(\mathbf{P}) \\
&= \det(\mathbf{A} - \lambda \mathbf{I})
\end{aligned}
\tag{A.7}
$$

where λ is a scalar.

The eigenvalues of an $n \times n$ matrix **A** are the roots of the characteristic equation

$$
\left| \lambda \mathbf{I} - \mathbf{A} \right| = 0
$$

Hence, eigenvalues are also called the *characteristic roots*. Also, the eigenvalues are invariant under any linear transformation.

Appendix B: Selected MATLAB Programs

2.1 Approximation of the function $f(t) = \sin(\pi t)$ in BPF domain and estimation of the MISE.
(Chapter 2, Figure 2.2, pages 41 and 42)

```
clc
clear all
format long

%%---- Number of Sub-intervals and Total Time ----%%

m = input('Enter the number of sub-intervals chosen:\n')
T = input('Enter the total time period:\n')
h = T/m;
t=0:h:T;

%%---- Function to be approximated ----%%

syms x
f = sin(pi*x);

j = 0:0.01:T;
plot(j,sin(pi*j),'--k','LineWidth',2)    % exact plot
hold on

%%---- BPF Based Representation ----%%

C = zeros(1,m);
for i=1:m
    C(i)=(m/T)*int(f,t(i),t(i+1)); % BPF Coefficients
end
Coeff = [C C(m)]       % BPF Coefficients for plotting

%%---- Function Plotting ----%%

stairs(t,Coeff,'-k','LineWidth',2)
ylim([0 1])

%%---- Calculation of MISE ----%%

for i=1:m
    Fd = (f-Coeff(i))^2;
    ise(i)=double(int(Fd,t(i),t(i+1)));
end
```

```
ise

ISE=sum(ise)    % ISE over m sub-intervals

MISE=ISE/T      % Mean Integral Square Error (MISE)
```

2.2 Approximation of the function *f(t)* = sin(π*t*) in GBPF domain and comparison of the MISE's for BPF and GBPF approximations. (Chapter 2, Figure 2.3, Table 2.2, pages 42 and 43)

```
clc
clear all
format long

%%---- Number of Sub-intervals and Total Time ----%%

m = input('Enter the number of sub-intervals chosen:\n')
T = input('Enter the total time period:\n')
h=T/m;
t=0:h:T;

%%---- For GBPF based approximation ----%%

h1=[0.1 0.1 0.12 0.18 0.18 0.12 0.1 0.1];

T1=sum(h1,2);

%%---- Exact Plot of Function ----%%

syms x
f=sin(pi*x);
j=0:0.001:T;
plot(j,sin(pi*j),'--k','LineWidth',2) % exact plot
hold on

%%---- Conventional BPF Based Representation ----%%

C_BPF=zeros(1,m);
for i=1:m
    C_BPF(i)=(m/T)*int(f,t(i),t(i+1)); % Coefficients
end
Coeff_BPF=[C_BPF C_BPF(m)];

%%---- Function Plotting ----%%

stairs(t,Coeff_BPF,'-m','LineWidth',2)
ylim([0 1])
```

```
%%---- Calculation of MISE ----%%

for i=1:m
      Fd_BPF=(f-Coeff_BPF(i))^2;
      ise_BPF(i)=double(int(Fd_BPF,t(i),t(i+1)));
end

ise_BPF

ISE_BPF=sum(ise_BPF)    % Integral Square Error over m
                                            sub-intervals
MISE_BPF=ISE_BPF/T      % MISE

%%---- GBPF Based Representation ----%%

C_GBPF=zeros(1,m);

t1(1)=0;
for i=1:m
     t1(i+1)=t1(i)+h1(i);
end
t1

for i=1:m
    t_Lower(i)=t1(i);
    t_Upper(i)=t1(i+1);
    T1(i)=h1(i);
    C_GBPF(i)=(1/h1(i))*int(f,x,t_Lower(i),t_Upper(i));

GBPF Coefficients %
end
Coeff_GBPF=[C_GBPF C_GBPF(m)];

%%---- Function Plotting ----%%

hold on
stairs(t1,Coeff_GBPF,'-b','LineWidth',2)
ylim([0 1.001])

%%-- Calculation of MISE for GBPF based Approximation --%%

for i=1:m
    Fd_GBPF=(f-Coeff_GBPF(i))^2;
ise_GBPF(i)=double(int(Fd_GBPF,x,t_Lower(i),t_Upper(i)));
end

ise_GBPF
```

```
ISE_GBPF=sum(ise_GBPF) % Integral Square Error (ISE)

MISE_GBPF=ISE_GBPF/T % Mean Integral Square Error (MISE)
```

2.3 Approximation of the function $f(t) = 1-\exp(-t)$ in PWM-GBPF domain and comparison of the MISEs for BPF and PWM-GBPF approximations.
(Chapter 2, Figure 2.4, Table 2.3, pages 44 to 45)

```
clc
clear all
format long

%%--- Number of Sub-intervals and Total Time ---%%

m = input('Enter the number of sub-intervals chosen:\n')
T = input('Enter the total time period:\n')

%%--- For Conventional BPF based Approximation ---%%

h=T/m;
t=0:h:T;

%%--- For PWMGBPF based Approximation ---%%

delta=0.1;
h0=(T-((m*(m-1)*delta)/2))/m;

for i=1:m
    h1(i)=h0+(i-1)*delta;
end
h1;

T1=sum(h1);

%%------------- Exact Plot of Function -------------%%

syms x
f=exp(-x);

j=0:0.001:T;
plot(j,exp(-j),'--k','LineWidth',2)  % exact plot
hold on

%%--- Conventional BPF Based Representation ---%%

C_BPF=zeros(1,m);
for i=1:m
```

```
   C_BPF(i)=(m/T)*int(f,t(i),t(i+1)); % Coefficients
end
Coeff_BPF=[C_BPF C_BPF(m)];

%%---- Function Plotting ----%%

stairs(t,Coeff_BPF,'-m','LineWidth',2)
ylim([0 1])

%%---------- Calculation of MISE ---------%%

for i=1:m
    Fd_BPF=(f-Coeff_BPF(i))^2;
    ise_BPF(i)=double(int(Fd_BPF,t(i),t(i+1)));
end

ise_BPF;

ISE_BPF=sum(ise_BPF);    % ISE over m sub-intervals

MISE_BPF=ISE_BPF/T; % Mean Integral Square Error (MISE)

%%---- PWMGBPF Based Approximation ----%%

C_PWMGBPF=zeros(1,m);

t1(1)=0;
for i=1:m
    t1(i+1)=t1(i)+h1(i);
end
t1;

for i=1:m
    t_Lower(i)=((i-1)*h0)+((i-1)*(i-2)*delta/2);
    t_Upper(i)=(i*h0)+((i-1)*i*delta/2);
    T1(i)=h0+(i-1)*delta;
    C_PWMGBPF(i)=(1/T1(i))*int(f,x,t_Lower(i),t_Upper(i));
end
Coeff_PWMGBPF=[C_PWMGBPF C_PWMGBPF(m)];

%%--------------- Function Plotting ---------------%%

hold on
stairs(t1,Coeff_PWMGBPF,'-b','LineWidth',2)
ylim([0 1])

%%---- Calculation of MISE ----%%

for i=1:m
Fd_PWMGBPF=(f-Coeff_PWMGBPF(i))^2;
```

```
    ise_PWMGBPF(i)=double(int(Fd_PWMGBPF,x,t_
    Lower(i),t_Upper(i)));
end

ise_PWMGBPF;
ISE_PWMGBPF=sum(ise_PWMGBPF);   % ISE

MISE_PWMGBPF=ISE_PWMGBPF/T       % MISE
```

2.4 Approximation of the function $f(t)=\sin(\pi t)$ in NOBPF domain and comparison of the MISEs for BPF and NOBPF approximations. (Chapter 2, Figure 2.5, Table 2.4, pages 45 and 46)

```
clc
clear all
format long

%%-- Number of Sub-intervals and Total Time --%%

m = input('Enter the number of sub-intervals chosen:\n')
T = input('Enter the total time period:\n')

%%-- For Conventional BPF based Approximation --%%

h=T/m;
t=0:h:T;

%%--- Exact Plot of Function ---%%

syms x
f=sin(pi*x);

j=0:0.001:T;
plot(j,sin(pi*j),'--k','LineWidth',2)
hold on

%%--- Conventional BPF Based Representation ---%%

C_BPF=zeros(1,m);
for i=1:m
    C_BPF(i)=(m/T)*int(f,t(i),t(i+1));   % coefficients
end
Coeff_BPF=[C_BPF C_BPF(m)];

%%---- Function Plotting ----%%

stairs(t,Coeff_BPF,'-m','LineWidth',2)
ylim([0 1])
```

```
%%---- Calculation of MISE ----%%

for i=1:m
        Fd_BPF=(f-Coeff_BPF(i))^2;
        ise_BPF(i)=double(int(Fd_BPF,t(i),t(i+1)));
end

ise_BPF;

ISE_BPF=sum(ise_BPF);   % ISE over m sub-intervals

MISE_BPF=ISE_BPF/T   % Mean Integral Square Error (MISE)

%%---- NOBPF Based Representation ----%%

C_NOBPF=subs(f,t);

for i=1:m
    Coeff_NOBPF(i)=(C_NOBPF(i)+C_NOBPF(i+1))/2;
end

Coeff_NOBPF=[Coeff_NOBPF Coeff_NOBPF(m)]

%%---- Function Plotting ----%%

hold on
stairs(t,Coeff_NOBPF,'-b','LineWidth',2)
ylim([0 1])

%%--- MISE for NOBPF based Approximation ---%%

for i=1:m
        Fd_NOBPF=(f-Coeff_NOBPF(i))^2;

        ise_NOBPF(i)=double(int(Fd_NOBPF,x,t(i),t(i+1)));
end

ise_NOBPF;

ISE_NOBPF=sum(ise_NOBPF); % ISE over m sub-intervals

MISE_NOBPF=ISE_NOBPF/T % Mean Integral Square Error (MISE)
```

2.5 Approximation of the function $f(t) = \sin(\pi t)$ in DUSF domain and comparison of the MISEs for BPF and DUSF approximations.

(Chapter 2, Example 2.6, pages 47)

```
clc
clear all
format long
```

```
%%-- Number of Sub-intervals and Total Time --%%

m = input('Enter the number of sub-intervals chosen:\n')
T = input('Enter the total time period:\n')

%%-- For Conventional BPF based Approximation --%%

h=T/m;
t=0:h:T;

%%---- Exact Plot of Function ----%%

syms x
f=sin(pi*x);

j=0:0.001:T;
plot(j,sin(pi*j),'--k','LineWidth',2)  % exact plot
hold on

%%-- Conventional BPF Based Representation --%%

C_BPF=zeros(1,m);
for i=1:m
    C_BPF(i)=(m/T)*int(f,t(i),t(i+1));  % Coefficients
end
Coeff_BPF=[C_BPF C_BPF(m)]

%%---- Function Plotting ----%%

stairs(t,Coeff_BPF,'-m','LineWidth',4)
ylim([0 1])

%%---- Calculation of MISE ----%%

for i=1:m
        Fd_BPF=(f-Coeff_BPF(i))^2;
        ise_BPF(i)=double(int(Fd_BPF,t(i),t(i+1)));
end

ise_BPF;

ISE_BPF=sum(ise_BPF);  % ISE over m sub-intervals

MISE_BPF=ISE_BPF/T  % Mean Integral Square Error (MISE)

%%---- DUSF Based Representation ----%%

fi=C_BPF;

gi(1)=fi(1);
```

```
for i=2:m
    gi(i)=fi(i)-fi(i-1);
end
gi;

G=zeros(m);
for i=1:m
    G(i,(i:m))=gi(i);
end
G;

C_DUSF=sum(G)        % DUSF coefficients

%%---- Function Plotting ----%%

C_DUSF_plot=[C_DUSF C_DUSF(m)];
hold on
stairs(t,C_DUSF_plot,'-b','LineWidth',2)
ylim([0 1])
```

2.6 Approximation of the function $f(t) = \sin(\pi t)$ in SHF domain and comparison of the MISEs for BPF and SHF approximations. (Chapter 2, Figure 2.6, Table 2.5, page 48)

```
clc
clear all
format long

%%-- Number of Sub-intervals and Total Time --%%

m = input('Enter the number of sub-intervals chosen:\n')
T = input('Enter the total time period:\n')

%%-- For Conventional BPF based Approximation --%%

h=T/m;
t=0:h:T;

%%---- Exact Plot of Function ----%%

syms x
f=1-exp(-x);

j=0:0.001:T;
plot(j,(1-exp(-j)),'--k','LineWidth',2)   % exact plot
hold on
```

```
%%--- Conventional BPF Based Representation ---%%

C_BPF=zeros(1,m);
for i=1:m
    C_BPF(i)=(m/T)*int(f,t(i),t(i+1));
end
Coeff_BPF=[C_BPF C_BPF(m)];

%%--- Function Plotting ---%%

stairs(t,Coeff_BPF,'-m','LineWidth',2)
ylim([0 0.8])

%%---------- Calculation of MISE ---------%%

for i=1:m
        Fd_BPF=(f-Coeff_BPF(i))^2;
        ise_BPF(i)=double(int(Fd_BPF,t(i),t(i+1)));
end

ise_BPF;

ISE_BPF=sum(ise_BPF);  % ISE over m sub-intervals

MISE_BPF=ISE_BPF/T      % Mean Integral Square Error

%%--- SHF Based Representation ---%%

Coeff_SHF=subs(f,t);

%%----- Function Plotting -----%%

hold on
stairs(t,Coeff_SHF,'-b','LineWidth',2)

%%-- Calculation of MISE for SHF based Approx. --%%

for i=1:m
        Fd_SHF=(f-Coeff_SHF(i))^2;
        ise_SHF(i)=double(int(Fd_SHF,x,t(i),t(i+1)));
end

ise_SHF;

ISE_SHF=sum(ise_SHF);  % ISE over m sub-intervals

MISE_SHF=ISE_SHF/T    % Mean Integral Square Error (MISE)
```

3.1 Integration of the function $f(t) = t^2/2$ in BPF domain using operational matrix for integration.
(Chapter 3, Figure 3.5(a), Table 3.1, page 63)

```
clc
clear all
format long

%%-- Number of Sub-intervals and Total Time --%%

m = input('Enter the number of sub-intervals chosen:\n')
T = input('Enter the total time period:\n')
h=T/m;
tt=0:h:T;

%%--- Integration of a time-function ---%%

syms t
f=t^2/2;        % function to be integrated

fi=int(f,t);    % integrated function

t1=0:0.01:T;
Fi=subs(fi,t1);
plot(t1,Fi,'--k','LineWidth',2)   % exact plot
hold on

%%--- BPF Coefficients of Direct Expansion ---%%

Cd=zeros(1,m);
for i=1:m
    Cd(i)=(m/T)*int(fi,tt(i),tt(i+1));   % Calculating BPF
Coefficients
end
Coeff_direct=[Cd Cd(m)];

%%---- Function Plotting ----%%

stairs(tt,Coeff_direct,'-k','LineWidth',2)

%%-- Operational matrix in BPF domain P --%%

P=zeros(m,m);
for i=1:m
    P(i,i)=h/2;
    for j=(i+1):m
        P(i,j)=h;
    end
end
P
```

```
%%-- Integration using Operational matrix P --%%

C_bpf=zeros(1,m);
for i=1:m
    C_bpf(i)=(m/T)*int(f,tt(i),tt(i+1));
end
C_bpf;

C_int_BPF=C_bpf*P    % Coefficients of the integrated function

  %%-- Calculation of Percentage Error --%%

C_direct=Coeff_direct(1:m);

Percentage_Error=(C_direct-C_int_BPF)./C_direct*100
```

3.2 Integration of the function $f(t) = t^2/2$ in GBPF domain using operational matrix for integration.
(Chapter 3, Figure 3.5(b), Table 3.2, pages 66 and 67)

```
clc
clear all
format long

%%-- Number of Sub-intervals and Total Time --%%

m = input('Enter the number of sub-intervals chosen:\n')
T = input('Enter the total time period:\n')
h=T/m;
tt=0:h:T;

h1=[0.2 0.17 0.15 0.12 0.1 0.08 0.07 0.05 0.04 0.02];

T1=sum(h1,2)

%%--- Integration of a time-function ---%%

syms t
f=t^2/2;       % function to be integrated

fi=int(f,t);   % integrated function
t1=0:0.01:T;
Fi=subs(fi,t1);
plot(t1,Fi,'--k','LineWidth',2)
hold on
```

```
%%--- GBPF coefficients of the function ---%%

C_GBPF=zeros(1,m);

t2(1)=0;
for i=1:m
    t2(i+1)=t2(i)+h1(i);
end
t2

for i=1:m
    t_Lower(i)=t2(i);
    t_Upper(i)=t2(i+1);
    T1(i)=h1(i);
    C_GBPF(i)=(1/h1(i))*int(f,t,t_Lower(i),t_Upper(i));
end
C_GBPF

%%-- GBPF Coefficients of Direct Expansion --%%

Cd_GBPF=zeros(1,m);
for i=1:m
    Cd_GBPF(i)=(1/h1(i))*int(fi,t2(i),t2(i+1));
end
Coeff_direct_GBPF=[Cd_GBPF Cd_GBPF(m)];

%%--- Operational matrix in GBPF domain ---%%

PG=zeros(m,m);
for i=1:m
    PG(i,i)=h1(i)/2;
    for j=(i+1):m
        PG(i,j)=h1(i);
    end
end
PG

%%-- Integration using Operational matrix PG --%%

C_int_GBPF=C_GBPF*PG;

C_int_GBPF_plot=[C_int_GBPF  C_int_GBPF(m)];

plot(t1,Fi,'--k','LineWidth',2)   % exact plot
hold on
stairs(t2,C_int_GBPF_plot,'-m','LineWidth',2)

Coeff_direct_GBPF

C_int_GBPF
```

```
%%--- Calculation of MISE for GBPF ---%%

for i=1:m
    Fdi_GBPF=(fi-C_int_GBPF(i))^2;
    ise_GBPF(i)=double(int(Fdi_GBPF,t2(i),t2(i+1)));
end
ise_GBPF;

ISE_GBPF=sum(ise_GBPF);  % ISE over m sub-intervals

  MISE_GBPF=ISE_GBPF/T % Mean Integral Square Error (MISE)
```

3.3 Integration of the function $f(t)$ = exp(2t) in BPF domain using two improved operational matrices P1 and P2, and also the operational matrix P with comparison of the results. (Chapter 3, Table 3.3, pages 74)

```
clc
clear all
format long

%%-- Number of Sub-intervals and Total Time --%%

m = input('Enter the number of sub-intervals chosen:\n')
T = input('Enter the total time period:\n')
h=T/m;
tt=0:h:T;

%%---- Integration of a time-function ----%%

syms t
f=exp(2*t);      % function to be integrated
fi=int(f,0,t);   % integrated function

t1=0:0.01:T;
Fi=subs(fi,t1);
plot(t1,Fi,'--k','LineWidth',2)   % exact plot
hold on

%%-- BPF Coefficients of Direct Expansion --%%

Cd=zeros(1,m);
for i=1:m
    Cd(i)=(m/T)*int(fi,tt(i),tt(i+1));  % Coefficients
end
Coeff_direct=[Cd Cd(m)];

%%---- Function Plotting ----%%

stairs(tt,Coeff_direct,'-k','LineWidth',2)
```

```
%%--- Operational matrix in BPF domain P ---%%

P=zeros(m,m);
for i=1:m
    P(i,i)=h/2;
    for j=(i+1):m
        P(i,j)=h;
    end
end
P;

%%--- Operational matrix in BPF domain P1 ---%%

P1=zeros(m,m);
for i=1:m
    P1(i,i)=h*(5/12);
    if (i+1)<=m
        P1(i,(i+1))=h*(13/12);
    end
    for j=(i+2):m
        P1(i,j)=h;
    end
end
P1(1,1)=h/2;
P1;

%%-- Operational matrix in BPF domain P2 --%%

P2=P1;
for i=1:m
    if i>=3
        P2(i,i)=h*(9/24);
    end
    if i>=2 & (i+1)<=m
        P2(i,(i+1))=h*(28/24);
    end
    if (i+2)<=m
        P2(i,(i+2))=h*(23/24);
    end
end
P2

%%-- Function Integration using Operational matrices --%%

C_bpf=zeros(1,m);
for i=1:m
    C_bpf(i)=(m/T)*int(f,tt(i),tt(i+1));
end
C_bpf

C_direct=Coeff_direct(1:m)
```

```
C_int_BPF_P=C_bpf*P

C_int_BPF_P1=C_bpf*P1

C_int_BPF_P2=C_bpf*P2

%%-- Calculation of Percentage Error --%%

Percentage_Error_P=(C_direct-C_int_BPF_P)./C_direct*100

Percentage_Error_P1=(C_direct-C_int_BPF_P1)./C_direct*100

Percentage_Error_P2=(C_direct-C_int_BPF_P2)./C_direct*100
```

3.4 **To determine the second-order integration of the function** $f(t) =$
$1-\exp(-t)$ **in BPF domain using (a) first-order operational matrix**
twice and (b) one-shot operational matrix for double integration
once and comparison of the results.
(Chapter 3, Figure 3.6, Table 3.5, pages 77 and 78)

```
format long
clc
clear all

%%-- Number of Sub-intervals and Total Time --%%

m = input('Enter the number of sub-intervals chosen:\n')
T = input('Enter the total time period:\n')
h=T/m;
tt=0:h:T;

%%--- Approximation of a time-function ---%%

syms t
x1=1-exp(-t);

%%-- BPF Coefficients of Direct Expansion --%%

C_d=zeros(1,m);
for i=1:m
    C_d(i)=(m/T)*int(x1,tt(i),tt(i+1));
end
Coeff_d=[C_d C_d(m)];

%%-- Double Integration and Expansion in BPF --%%

xi_int2=int(int(x1,0,t),0,t);
```

```
t1=0:0.01:T;
X1=subs(xi_int2,t1);
plot(t1,X1,'--k','LineWidth',2)
hold on

C_direct=zeros(1,m);
for i=1:m
    C_direct(i)=(m/T)*int(xi_int2,tt(i),tt(i+1));
end
Coeff_direct=[C_direct C_direct(m)];

%%---- Function Plotting ----%%

stairs(tt,Coeff_direct,'-k','LineWidth',2)
%%--- Operational matrix in BPF domain P ---%%

P=zeros(m,m);
for i=1:m
    P(i,i)=h/2;
    for j=(i+1):m
        P(i,j)=h;
    end
end
P;

%%--- Solution of the Function using P ---%%

Coeff_P_FirstOrder=C_d*P^2;

Coeff_P_FirstOrder_plot=[Coeff_P_FirstOrder Coeff_P_First
Order(m)];

hold on
stairs(tt,Coeff_P_FirstOrder_plot,'-m','LineWidth',2)

%%-- Formation of One Shot Operational Matrix --%%

n=2;
pn=zeros(m,m);

for i=1:m
    for j=1:m
        if j-i==0
            pn(i,j)=1;
        elseif j-i>0
            pn(i,j)=(j-i+1)^(n+1)-2*(j-i)^(n+1)+
                                        (j-i-1)^(n+1);
        end
    end
end
end
```

```
Pn=((h^n)/factorial(n+1))*pn;

%%-- Solution of the Function using One Shot Pn --%%

Coeff_Pn_OneShot=C_d*Pn;

Coeff_Pn_OneShot_plot=[Coeff_Pn_OneShot
Coeff_Pn_OneShot(m)];

hold on
stairs(tt,Coeff_Pn_OneShot_plot,'-b','LineWidth',2)

Coeff_P_FirstOrder

C_direct

Coeff_Pn_OneShot
```

3.5 Differentiation of the function $f(t) = t{\wedge}2/2$ in BPF domain using operational matrix for differentiation.
(Chapter 3, Figure 3.7, Table 3.6, pages 79 and 80)

```
clc
clear all
format long

%%--- Number of Sub-intervals and Total Time ---%%

m = input('Enter the number of sub-intervals chosen:\n')
T = input('Enter the total time period:\n')
h=T/m;
tt=0:h:T;

%%--- Differentiation of a time-function ---%%

syms t
f=t^2/2;        % function to be differentiated

fd=diff(f,t);   % differentiated function
t1=0:0.01:T;
Fd=subs(fd,t1);
plot(t1,Fd,'--k','LineWidth',2)
hold on
```

```
%%--- BPF Coefficients of Direct Expansion ---%%

Cd=zeros(1,m);
for i=1:m
    Cd(i)=(m/T)*int(fd,tt(i),tt(i+1));
end
Coeff_direct=[Cd Cd(m)]

%%---- Function Plotting ----%%

stairs(tt,Coeff_direct,'-k','LineWidth',2)

%%-- Operational matrix D in BPF domain --%%

d=zeros(m,m);

for i=1:m
    for j=1:m
        if j-i==0
            d(i,j)=2;
        elseif j-i>0
            d(i,j)=4*(-1)^(j-i);
        else
            d(i,j)=0;
        end
    end
end
D=(1/h)*d

%%-- Differentiation using Operational matrix D --%%

C_bpf=zeros(1,m);
for i=1:m
    C_bpf(i)=(m/T)*int(f,tt(i),tt(i+1));
end
C_bpf;

C_diff_BPF=C_bpf*D
C_diff_BPF_plot=[C_diff_BPF C_diff_BPF(m)];

hold on
stairs(tt,C_diff_BPF_plot,'-m','LineWidth',2)

%%-------- Calculation of Percentage Error ----------%%

C_direct=Coeff_direct(1:m);

Percentage_Error=(C_direct-C_diff_BPF)./C_direct*100
```

3.6 Differentiation of a function $f(t) = t^4/24$ in GBPF domain using operational matrix for differentiation.
(Chapter 3, Figure 3.8, Table 3.7, pages 81 and 82)

```
format long
clc
clear all

%%-- Number of Sub-intervals and Total Time --%%
m = input('Enter the number of sub-intervals chosen:\n')
T = input('Enter the total time period:\n')
h=T/m;
tt=0:h:T;


h1=[0.14 0.13 0.12 0.11 0.1 0.1 0.09 0.08 0.07 0.06];

%%-- Differentiation of a time-function --%%

syms t
f=t^4/24;        % function to be differentiated

fd=diff(f,t);  % differentiated function

t1=0:0.01:T;
Fd=subs(fd,t1);
plot(t1,Fd,'--k','LineWidth',2)
hold on

%%-- BPF Coefficients of Direct Expansion --%%

Cd=zeros(1,m);
for i=1:m
    Cd(i)=(m/T)*int(fd,tt(i),tt(i+1));   % Coefficients
end
Coeff_direct=[Cd Cd(m)]

%%---- Function Plotting ----%%

stairs(tt,Coeff_direct,'-k','LineWidth',2)

%%-- Operational matrix D in BPF domain --%%

d=zeros(m,m);

for i=1:m
    for j=1:m
        if j-i==0
            d(i,j)=2;
        elseif j-i>0
```

```
                    d(i,j)=4*(-1)^(j-i);
             else
                  d(i,j)=0;
             end
        end
end
D=(1/h)*d

%%-- Function differentiation using Operational matrix D --%%

C_bpf=zeros(1,m);
for i=1:m
     C_bpf(i)=(m/T)*int(f,tt(i),tt(i+1));
end
C_bpf;

C_diff_BPF=C_bpf*D
C_diff_BPF_plot=[C_diff_BPF C_diff_BPF(m)];

hold on
stairs(tt,C_diff_BPF_plot,'-m','LineWidth',2)

%%-- Calculation of Percentage Error --%%

C_direct=Coeff_direct(1:m);
Percentage_Error_BPF=(C_direct-C_diff_BPF)./C_direct*100

%%--- Calculation of MISE for BPF ---%%

for i=1:m
        Fdd_BPF=(fd-C_diff_BPF(i))^2;
        ise_BPF(i)=double(int(Fdd_BPF,tt(i),tt(i+1)));
end

ise_BPF;

ISE_BPF=sum(ise_BPF);   % ISE over m sub-intervals

MISE_BPF=ISE_BPF/T;     % Mean Integral Square Error

%%--- GBPF Based Representation ---%%

C_GBPF=zeros(1,m);

t2(1)=0;
for i=1:m
     t2(i+1)=t2(i)+h1(i);
end
t2
```

```
for i=1:m
    t_Lower(i)=t2(i);
    t_Upper(i)=t2(i+1);
    T1(i)=h1(i);
    C_GBPF(i)=(1/h1(i))*int(f,t,t_Lower(i),t_Upper(i));
end
C_GBPF;

%%--- GBPF Coefficients of Direct Expansion ---%%

Cd_GBPF=zeros(1,m);
for i=1:m
    Cd_GBPF(i)=(1/h1(i))*int(fd,t2(i),t2(i+1));
end
Coeff_direct_GBPF=[Cd_GBPF Cd_GBPF(m)];

%%-- Operational matrix DG in BPF domain --%%

DG=zeros(m,m);

for i=1:m
    for j=1:m
        if j-i==0
            DG(i,j)=2/h1(i);
        elseif j-i>0
            DG(i,j)=(4/h1(j))*(-1)^(j-i);
        else
            DG(i,j)=0;
        end
    end
end

%%-- Differentiation using Operational matrix DG --%%

C_diff_GBPF=C_GBPF*DG;

C_diff_GBPF_plot=[C_diff_GBPF  C_diff_GBPF(m)];

figure
plot(t1,Fd,'--k','LineWidth',2)   % exact plot
hold on
stairs(t2,C_diff_GBPF_plot,'-b','LineWidth',2)

%%--- Calculation of Percentage Error ---%%

C_direct=Coeff_direct(1:m);

Percentage_Error_GBPF=(C_direct-C_diff_GBPF)./C_direct*100
```

```
%%---------- Calculation of MISE for GBPF -----------%%

for i=1:m
        Fdd_GBPF=(fd-C_diff_GBPF(i))^2;
        ise_GBPF(i)=double(int(Fdd_GBPF,tt(i),tt(i+1)));
end

ise_GBPF;

ISE_GBPF=sum(ise_GBPF);   % ISE over m sub-intervals

MISE_GBPF=ISE_GBPF/T   % Mean Integral Square Error (MISE)
```

3.7 **To determine double differentiation of three time functions in BPF domain using (a) first-order operational matrix for differentiation twice and (b) one-shot operational matrix for differentiation once and comparison of the results.**
(Chapter 3, Figures 3.9 to 3.11, pages 83 and 84)

```
clc
clear all
format long
%%-- Number of Sub-intervals and Total Time --%%

m = input('Enter the number of sub-intervals chosen:\n')
T = input('Enter the total time period:\n')
h=T/m;
tt=0:h:T;

%%--- Approximation of a time-function ---%%

syms t
x1=t^2/2;    % function to be approximated

%%--- BPF Coefficients of Direct Expansion ---%%

C_d=zeros(1,m);
for i=1:m
    C_d(i)=(m/T)*int(x1,tt(i),tt(i+1));
end
Coeff_d=[C_d C_d(m)];

%%-- Double differentiation and Expansion in BPF --%%

xd_diff2=diff(diff(x1,t),t);

t1=0:0.01:T;
X1=subs(xd_diff2,t1);
```

```
plot(t1,X1,'--k','LineWidth',2)
hold on

C_direct=zeros(1,m);
for i=1:m
    C_direct(i)=(m/T)*int(xd_diff2,tt(i),tt(i+1));
end
Coeff_direct=[C_direct C_direct(m)];

%%--- Function Plotting ---%%

stairs(tt,Coeff_direct,'-k','LineWidth',2)

%%--- Operational matrix D in BPF domain ---%%

d=zeros(m,m);

for i=1:m
    for j=1:m
        if j-i==0
            d(i,j)=2;
        elseif j-i>0
            d(i,j)=4*(-1)^(j-i);
        end
    end
end
D=(1/h)*d

Coeff_D_FirstOrder=C_d*D^2;

Coeff_D_FirstOrder_plot=[Coeff_D_FirstOrder
                         Coeff_D_FirstOrder(m)];

hold on
stairs(tt,Coeff_D_FirstOrder_plot,'-m','LineWidth',2)

%%-- Formation of One Shot Operational Matrix --%%

n=2;
pn=zeros(m,m);
for i=1:m
    for j=1:m
        if j-i==0
            pn(i,j)=1;
        elseif j-i>0
            pn(i,j)=(j-i+1)^(n+1)-2*(j-i)^(n+1)+(j-i-1)
                                              ^(n+1);
        end
    end
end
```

```
Pn=((h^n)/factorial(n+1))*pn;
Dn=inv(Pn);

%%-- Solution of the Function using One Shot Dn --%%

Coeff_Dn_OneShot=C_d*Dn;

Coeff_Dn_OneShot_plot=[Coeff_Dn_OneShot Coeff_Dn_
                                        OneShot(m)];

hold on
stairs(tt,Coeff_Dn_OneShot_plot,'-b','LineWidth',2)

Coeff_D_FirstOrder

C_direct

Coeff_Dn_OneShot
```

4.1 Analysis of the open loop system of Example 4.1 for a step input using block pulse operational transfer function (BPOTF). (Chapter 4, Figure 4.1, Table 4.1, pages 92–94)

```
clc
clear all
format long

%%--- Number of Sub-intervals and Total Time ---%%

m = input('Enter the number of sub-intervals chosen:\n')
T = input('Enter the total time period:\n')
h=T/m;
tt=0:h:T;

%%---- Exact curve ----%%

syms t
c=1-exp(-t);          % output of the system

t1=0:0.01:T;
Ct=subs(c,t1);
plot(t1,Ct,'--k','LineWidth',2)
hold on

C_exact=subs(c,tt);
```

```
%%-- Direct Expansion of Output in BPF domain --%%

C_direct=zeros(1,m);
for i=1:m
    C_direct(i)=(m/T)*int(c,tt(i),tt(i+1));
end

C_direct

%%-- Operational matrix D in BPF domain --%%

d=zeros(m,m);

for i=1:m
    for j=1:m
        if j-i==0
            d(i,j)=2;
        elseif j-i>0
            d(i,j)=4*(-1)^(j-i);
        else
            d(i,j)=0;
        end
    end
end
D=(1/h)*d

%%-- Function Integration using Operational matrix D --%%

I=eye(m);              % Identity matrix
R=ones(1,m);           % Input in BPF domain

C_BPF=R*inv(D+I);      % Output using BPOTF

C_BPF_Plot=[C_BPF C_BPF(m)];

stairs(tt,C_BPF_Plot,'k-','LineWidth',2)
ylim([0 1.2])

%%--- Calculation of Percentage Error ---%%

Percentage_Error=(C_direct-C_BPF)./C_direct*100
```

4.2 Analysis of the closed loop system for a step input using block pulse operational transfer function (BPOTF).
(Chapter 4, Figure 4.3, Table 4.2, pages 93–96)

```
clc
clear all
format long

%%-- Number of Sub-intervals and Total Time --%%

m = input('Enter the number of sub-intervals chosen:\n')
T = input('Enter the total time period:\n')
h=T/m;
tt=0:h:T;

%%--- Exact curve ---%%

syms t
c=0.5*(1-exp(-2*t));          % output of the system

t1=0:0.01:T;
Ct=subs(c,t1);
plot(t1,Ct,'--k','LineWidth',2)
hold on

C_exact=subs(c,tt);

%%-- Direct Expansion of Output in BPF domain --%%

C_direct=zeros(1,m);
for i=1:m
    C_direct(i)=(m/T)*int(c,tt(i),tt(i+1));
end

C_direct

%%-- Operational matrix D in BPF domain --%%

d=zeros(m,m);

for i=1:m
    for j=1:m
        if j-i==0
            d(i,j)=2;
        elseif j-i>0
            d(i,j)=4*(-1)^(j-i);
        else
            d(i,j)=0;
```

```
            end
        end
end
D=(1/h)*d

%%-- Function Integration using Oprational matrix D --%%

I=eye(m);                    % Identity matrix
R=ones(1,m);                 % Input in BPF domain

C_BPF=R*inv(D+(2*I));        % Output using BPOTF

C_BPF
C_BPF_Plot=[C_BPF C_BPF(m)];

stairs(tt,C_BPF_Plot,'k-','LineWidth',2)
ylim([0 0.6])

%%--- Calculation of Percentage Error ---%%

Percentage_Error=(C_direct-C_BPF)./C_direct*100
```

4.3 Analysis of the open loop system of Example 4.3 for a step input for a typical value of Lambda.
(Chapter 4, Figure 4.4, page 99)

```
clc
clear all
format long

%%--- Number of Sub-intervals and Total Time ---%%

m = input('Enter the number of sub-intervals chosen:\n')
T = input('Enter the total time period:\n')
h=T/m;
tt=0:h:T;

%%---- Exact curve ----%%

syms t
c=(1/5)*(1-exp(-5*t));         % output of the system
t1=0:0.01:T;
Ct=subs(c,t1);
plot(t1,Ct,'--k','LineWidth',2)
hold on

C_exact=subs(c,tt);
```

```
%%-- Direct Expansion of Output in BPF domain --%%

C_direct=zeros(1,m);
for i=1:m
    C_direct(i)=(m/T)*int(c,tt(i),tt(i+1));
End

C_direct

%%-- Operational matrix D in BPF domain --%%

d=zeros(m,m);

for i=1:m
    for j=1:m
        if j-i==0
            d(i,j)=2;
        elseif j-i>0
            d(i,j)=4*(-1)^(j-i);
        else
            d(i,j)=0;
        end
    end
end
D=(1/h)*d

%%-- Function Integration using Operational matrix D --%%

I=eye(m);                          % Identity matrix
R=ones(1,m);                       % Input in BPF domain

Lambda=4;

C_BPF=R*inv((D/Lambda)+(5*I));     % Output for a
                                             typical Lambda

C_BPF

C_BPF_plot=[C_BPF C_BPF(m)];

stairs(tt,C_BPF_plot,'k-','LineWidth',2)

%%--- Calculation of Percentage Error ---%%

Percentage_Error=(C_direct-C_BPF)./C_direct*100
```

4.4 Analysis of the first-order system of Example 4.1 for a step input using modified block pulse operational transfer function (MBPOTF) approach.
(Chapter 4, Table 4.3, page 105)

```
clc
clear all
format long

%%--- Number of Sub-intervals and Total Time ---%%

m = input('Enter the number of sub-intervals chosen:\n')
T = input('Enter the total time period:\n')
h=T/m;
tt=0:h:T;

%%---- Exact curve ----%%

syms t
c=1-exp(-t);                              % output of the system

t1=0:0.01:T;
Ct=subs(c,t1);
% plot(t1,Ct,'--k','LineWidth',2)   % exact plot
% hold on

C_exact=subs(c,tt);

%%--- Direct Expansion of Output in BPF domain ---%%

C_direct=zeros(1,m);
for i=1:m
    C_direct(i)=(m/T)*int(c,tt(i),tt(i+1));
end
C_direct;

%%--- Formation of MBPOTDF Matrix ---%%

q11=(exp(-h)+h-1)/(1-exp(-h))^2;

for i=1:m
    for j=1:m
        if i==j
            q(i,j)=q11;
        elseif j>i
            q(i,j)=exp(-((j-i+1)-2)*h);
        end
    end
end
q;
```

```
MBPOTF1=((1-exp(-h))^2/h)*q;

%%-- Function Integration using MBPOTF --%%

I=eye(m);                  % Identityy matrix
R1=ones(1,m);              % Input in BPF domain

C1_BPF=R1*MBPOTF1;         % Output using MBPOTF approach

C1_BPF

%%--- Calculation of Percentage Error ---%%

Percentage_Error=(C_direct-C1_BPF)./C_direct*100
```

4.5 Analysis of the closed loop system of Example 4.2 for a step input using modified block pulse operational transfer function (MBPOTF) approach.
(Chapter 4, Table 4.4, page 106)

```
clc
clear all
format long

%%-- Number of Sub-intervals and Total Time --%%

m = input('Enter the number of sub-intervals chosen:\n')
T = input('Enter the total time period:\n')
h=T/m;
tt=0:h:T;

%%--- Exact curve ---%%

syms t
c=0.5*(1-exp(-2*t));          % output of the system

t1=0:0.01:T;
Ct=subs(c,t1);

C_exact=subs(c,tt);

%%-- Direct Expansion of Output in BPF domain --%%

C_direct=zeros(1,m);
for i=1:m
    C_direct(i)=(m/T)*int(c,tt(i),tt(i+1));
end
C_direct;

%%-- Formation of MBPOTDF Matrix --%%
```

```
a=2;
r11=(exp(-a*h)+(a*h)-1)/(1-exp(-a*h))^2;

for i=1:m
    for j=1:m
        if i==j
            r(i,j)=r11;
        elseif j>i
            r(i,j)=exp(-((j-i+1)-2)*a*h);
        end
    end
end
r;
MBPOTF2=((1-exp(-a*h))^2/(4*h))*r;

%%-- Function Integration using MBPOTF  --%%

I=eye(m);              % Identity matrix
R1=ones(1,m);          % Input in BPF domain

C2_BPF=R1*MBPOTF2;     % Output using MBPOTF approach

C2_BPF

%%--- Calculation of Percentage Error ---%%

Percentage_Error=(C_direct-C2_BPF)./C_direct*100
```

4.6 Analysis of the second order system with the imaginary roots $\pm j\sqrt{2}$ or a step input using modified block pulse operational transfer function (MBPOTF) approach.
(Chapter 4, Table 4.5, page 108)

```
clc
clear all
format long

%%-- Number of Sub-intervals and Total Time --%%

m = input('Enter the number of sub-intervals chosen:\n')
T = input('Enter the total time period:\n')
h=T/m;
tt=0:h:T;

%%--- Exact curve ---%%

a=sqrt(2);
syms t
c=0.5-(0.5*cos(a*t));        % output of the system
```

```
t1=0:0.01:T;
Ct=subs(c,t1);

C_exact=subs(c,tt);

%%-- Direct Expansion of Output in BPF domain --%%

C_direct=zeros(1,m);
for i=1:m
    C_direct(i)=(m/T)*int(c,tt(i),tt(i+1));
end

C_direct;

%%--- Formation of MBPOTF Matrix ---%%

v11=((a*h)-sin(a*h))/(a^3*h);
for i=1:m
    for j=1:m
        if i==j
            v(i,j)=v11;
        elseif j>i
            v(i,j)=(4/(a^3*h))*((sin(a*h/2)^2)*
                                sin(((j-i+1)-1)*a*h));
        end
    end
end
v;
MBPOTF4=v;

%%-- Function Integration using MBPOTF --%%

I=eye(m);               % Identity matrix
R1=ones(1,m);           % Input in BPF domain

C4_BPF=R1*MBPOTF4;      % Output using MBPOTF approach
C4_BPF

%%--- Calculation of Percentage Error ---%%

Percentage_Error=(C_direct-C4_BPF)./C_direct*100
```

4.7 Analysis of a second-order system with complex roots for a step input using modified block pulse operational transfer function (MBPOTF) approach.
(Chapter 4, Table 4.6, page 110)

```
clc
clear all
```

```
format long

%%-- Number of Sub-intervals and Total Time --%%

m = input('Enter the number of sub-intervals chosen:\n')
T = input('Enter the total time period:\n')
h=T/m;
tt=0:h:T;

%%---- Exact curve -----%%

syms s t
C=1/(s^3+s^2+s);

c=ilaplace(C);      % output of the system

t1=0:0.01:T;
Ct=subs(c,t1);

C_exact=subs(c,tt);

%%-- Direct Expansion of Output in BPF domain --%%

C_direct=zeros(1,m);
for i=1:m
    C_direct(i)=(m/T)*int(c,tt(i),tt(i+1));
End

C_direct

%%-- Formation of MBPOTDF Matrix --%%

a=1; b=1;

alpha=0.5*a;
beta=0.5*sqrt((4*b)-a^2);

  w11=exp(-alpha*h)*((alpha^2-
          beta^2)*sin(beta*h)+2*alpha*beta*cos(beta*h))
              +(beta*h)*(alpha^2+beta^2)-(2*alpha*beta);
for i=1:m
    for j=1:m
        if i==j
            w(i,j)=w11;
        elseif j>i
            A=exp(-(j-i+1)*alpha*h)*exp(2*alpha*h)
                        *((alpha^2-beta^2)*sin(((j-
                                    i+1)-2) *beta*h));
            B=exp(-(j-i+1)*alpha*h)*[exp(2*alpha*h)
                        *(2*alpha*beta*cos((j-i+ 1)-2)
                      *beta*h))-2*exp(alpha*h)* ((alpha
```

```
                              ^2-beta^2) *sin(((j-i+1)
                                          -1)*beta*h))];
              C=exp(-(j-i+1)*alpha*h)*[-(2*exp(alpha*h)
                           *(2*alpha*beta*cos (((j-i+1)-1)
                 *beta*h))) +((alpha^2-beta^2)*sin ((j-i+1)
                                            *beta*h))];
              D=exp(-(j-i+1)*alpha*h)
                      *[2*alpha*beta*cos ((j-i+1) *beta*h)];
              w(i,j)=A+B+C+D;
           end
       end
end
w;
MBPOTF5=(1/(beta*h*(alpha^2+beta^2)^2))*w;

%%--------- Function Integration using MBPOTF ----------%%

I=eye(m);                 % Identity matrix
R1=ones(1,m);             % Input in BPF domain

C5_BPF=R1*MBPOTF5;        % Output using MBPOTF approach

%%---------- Calculation of Percentage Error -----------%%

Percentage_Error=(C_direct-C5_BPF)./C_direct*100;
```

5.1 Analysis of the open loop system of Example 5.1 for a step input using the convolution matrix in block pulse function (BPF) domain.
(Chapter 4, Figure 5.4, Table 5.1, page 120)

```
clc
clear all
format long

%%-- Number of Sub-intervals and Total Time --%%

m = input('Enter the number of sub-intervals chosen:\n')
T = input('Enter the total time period:\n')
h=T/m;
tt=0:h:T;

%%---- Exact curve ----%%

syms t
c=0.25*(1-exp(-4*t));  % exact expression of output

t1=0:0.01:T;
Ct=subs(c,t1);
```

```
plot(t1,Ct,'--k','LineWidth',2)
hold on

%%-- Direct Expansion of Output in BPF domain --%%

C_direct=zeros(1,m);
for i=1:m
    C_direct(i)=(m/T)*int(c,tt(i),tt(i+1));
end

C_direct;

C_direct_plot=[C_direct C_direct(m)];

stairs(tt,C_direct_plot,'m-','LineWidth',2)
hold on

%%-- Impulse response vector of System --%%

g=exp(-4*t);
G=subs(g,tt);

G_direct=zeros(1,m);
for i=1:m
    G_direct(i)=(m/T)*int(g,tt(i),tt(i+1));
end

G_direct;

%%--------------- Convolution matrix Qc ---------------%%

r=ones(1,m);

Cc=zeros(m,m);

for i=1:m
    for j=1:m
        if j-i==0
            Cc(i,j)=r(1);
        elseif j-i>0
            Cc(i,j)=r(j-i)+r((j-i)+1);
        end
    end
end

%%--------- Convolution operation using Qc ----------%%

C_convolution=(h/2)*G_direct*Cc

C_convolution_plot=[C_convolution C_convolution(m)];
```

```
stairs(tt,C_convolution_plot,'b-','LineWidth',2)

%%-- Calculation of Percentage Error --%%

Percentage_Error=(C_direct-C_convolution)./C_direct*100
```

5.2 Identification of the plant of the feedback system of Example 5.3 for a step input using the "deconvolution" matrix in block pulse function (BPF) domain.
(Chapter 5, Figures 5.5 and 5.6, Table 5.3, pages 125 and 126)

```
clc
clear all
format long

%%-- Number of Sub-intervals and Total Time --%%

m = input('Enter the number of sub-intervals chosen:\n')
T = input('Enter the total time period:\n')
h=T/m;
tt=0:h:T;

%%--- Exact curve ---%%

syms t
c=exp(-2*t)*sin(2*t);      % exact expression of output

%%-- Direct Expansion of Output in BPF domain --%%

C_direct=zeros(1,m);
for i=1:m
    C_direct(i)=(1/h)*int(c,tt(i),tt(i+1));
End

C_direct;

%%-- Impulse response vector of System --%%

g=2*exp(-2*t)*(cos(2*t)-sin(2*t));
G=subs(g,tt);

G_direct=zeros(1,m);
for i=1:m
    G_direct(i)=(1/h)*int(g,tt(i),tt(i+1));
end

G_direct;

G_direct_plot=[G_direct G_direct(m)];
```

```
stairs(tt,G_direct_plot,'m-','LineWidth',2)

%%--------------- Convolution matrix Cc ---------------%%

r=ones(1,m);

Cc_r=zeros(m,m);

for i=1:m
    for j=1:m
        if j-i==0
            Cc_r(i,j)=r(1);
        elseif j-i>0
            Cc_r(i,j)=r(j-i)+r((j-i)+1);
        end
    end
end

%%--------- Convolution operation using Qc ----------%%

G_convolution=(2/h)*C_direct*inv(Cc_r)

G_convolution_plot=[G_convolution G_convolution(m)];

hold on
stairs(tt,G_convolution_plot,'b-','LineWidth',2)

hold on
plot(tt,G,'k--','LineWidth',2)

%%-- Calculation of Percentage Error --%%

Percentage_Error=(G_direct-G_convolution)./G_direct*100
```

5.3 Identification of the plant of the closed loop system of Example 5.4 for a step input using the "deconvolution" matrix in block pulse function (BPF) domain.
(Chapter 5, Figure 5.9, Table 5.5, pages 137 and 138)

```
clc
clear all
format long

%%-- Number of Sub-intervals and Total Time --%%

m = input('Enter the number of sub-intervals chosen:\n')
T = input('Enter the total time period:\n')
h=T/m;
tt=0:h:T;
```

```
t1=0:0.01:T;

%%--- Exact curve ---%%

syms t
c=exp(-2*t)*sin(2*t);      % exact expression of output

%%-- Direct Expansion of Output in BPF domain --%%

C_direct=zeros(1,m);
for i=1:m
    C_direct(i)=(1/h)*int(c,tt(i),tt(i+1));
End

C_direct;

%%-- Impulse response vector of System --%%

g=2*exp(-4*t);
G=subs(g,t1);

G_direct=zeros(1,m);
for i=1:m
    G_direct(i)=(1/h)*int(g,tt(i),tt(i+1));
end

G_direct;

G_direct_plot=[G_direct G_direct(m)];

% hold on
stairs(tt,G_direct_plot,'m-','LineWidth',2)

%%--------------- Convolution matrix Cc ---------------%%

R=ones(1,m);    % Input to the system

Cc_output=zeros(m,m);

for i=1:m
    for j=1:m
        if j-i==0
            Cc_output(i,j)=C_direct(1);
        elseif j-i>0
            Cc_output(i,j)=C_direct(j-i)+
            C_direct((j-i)+1);
        end
    end
end
```

```
%%--------- Identication of system ----------%%

H=4*ones(1,m);        % BPF expansion of feedback element

Z=(h/2)*H*Cc_output; % Feedback signal after convolution

E=R-Z;

Cc_E=zeros(m,m);

for i=1:m
    for j=1:m
        if j-i==0
            Cc_E(i,j)=E(1);
        elseif j-i>0
            Cc_E(i,j)=E(j-i)+E((j-i)+1);
        end
    end
end

G_convolution=(2/h)*C_direct*inv(Cc_E)

G_convolution_plot=[G_convolution G_convolution(m)];

hold on
stairs(tt,G_convolution_plot,'b-','LineWidth',2)

hold on
plot(t1,G,'k--','LineWidth',2)

%%-- Calculation of Percentage Error --%%

Percentage_Error=(G_direct-G_convolution)./G_direct*100
```

6.1 **Integration of the function** $f(t) = t^2/2$ **of Example 6.1 in delayed unit step function (DUSF) and block pulse function (BPF) domain, using respective operational matrices.**
(Chapter 6, Table 6.1, page 149)

```
clc
clear all
format long

%%-- Number of Sub-intervals and Total Time --%%

m=10;
T=1;
h=T/m;
tt=0:h:T;
```

```
%%--- Integration of a time-function ---%%

syms t
f=t^2/2;        % function to be integrated

fi=int(f,t);   % integrated function

t1=0:0.01:T;
Fi=subs(fi,t1);

%%-- BPF Coefficients of Direct Expansion --%%

Cd_BPF=zeros(1,m);
for i=1:m
    Cd_BPF(i)=(m/T)*int(fi,tt(i),tt(i+1));
End

%%-- Operational matrix in BPF domain P --%%

P=zeros(m,m);
for i=1:m
    P(i,i)=h/2;
    for j=(i+1):m
        P(i,j)=h;
    end
end
P;

%%-- Function Integration using Oprational matrix P --%%

Coeff_BPF=zeros(1,m);
for i=1:m
    Coeff_BPF(i)=(m/T)*int(f,tt(i),tt(i+1));
end
Coeff_BPF;

C_int_BPF=Coeff_BPF*P

%%--- Calculation of Percentage Error ---%%

Percentage_Error_BPF=(Cd_BPF-C_int_BPF)./Cd_BPF*100

%%--- Direct expansion of the integrated function in DUSF
                                        domain ---%%

Cd_DUSF(1)=Cd_BPF(1);

for i=1:(m-1)
    Cd_DUSF(i+1)=Cd_BPF(i+1)-Cd_BPF(i);
end
Cd_DUSF
```

```
%%-- DUSF Coefficients of the function --%%

Coeff_DUSF(1)=Coeff_BPF(1);

for i=1:(m-1)
    Coeff_DUSF(i+1)=Coeff_BPF(i+1)-Coeff_BPF(i);
end
Coeff_DUSF

%%-- Operational matrix in DUSF domain Pd --%%

Pd=P;

%%-- Function Integration using Operational matrix Pd --%%

C_int_DUSF=Coeff_DUSF*Pd

%%-- Calculation of Percentage Error --%%

Percentage_Error_DUSF=(Cd_DUSF-C_int_DUSF)./Cd_DUSF*100
```

6.2 To determine the function with a stretched argument (t/λ) of the time function $f(t) = \sin(\pi t)$ of Example 6.2 in DUSF and BPF domain.
(Chapter 6, Figures 6.4 and 6.5, pages 165 and 166)

```
clc
clear all
format long

%%-- Number of Sub-intervals and Total Time --%%

m = input('Enter the number of sub-intervals chosen:\n')
T = input('Enter the total time period:\n')

%%-- For Conventional BPF based Approximation --%%

h=T/m;
t=0:h:T;

%%--- Exact Plot of Function ---%%

syms x
f1=sin(pi*x);

j1=0:0.001:1;
j2=1.001:0.001:1.5;
j=[j1 j2];

f1j=subs(f1,j1);
```

```
f2j=zeros(1,length(j2));

fj=[f1j f2j];

plot(j,fj,'--k','LineWidth',2)  % exact plot
hold on

%%--- DUSF Based Approximation ---%%

Coeff_BPF=zeros(1,m);
for i=1:(m-1)
    if i<=8
        Coeff_BPF(i)=(1/h)*int(f1,t(i),t(i+1));
    end
end
Coeff_BPF

Coeff_BPF_plot=[Coeff_BPF 0];

for i=1:m
    for j=1:m
        if j>=i
            N(i,j)=1;
        end
    end
end

Coeff_DUSF=Coeff_BPF*inv(N)

stairs(t,Coeff_BPF_plot,'-k','LineWidth',2)
xlim([0 1.5])
ylim([0 1.01])

%%---- Stretch Matrix S ----%%

Lambda=1.25;
S=zeros(m);

for i=1:m
    ni=fix((i-1)*Lambda);
    alpha_i=((i-1)*Lambda)-ni;
    k=ni;
    if (k+1)<=m
        S(i,(k+1))=(1-alpha_i);
        if (k+2)<=m
            S(i,(k+2))=alpha_i;
        end
    end
end

S;
```

```
Coeff_DUSF_Stretched=Coeff_DUSF*S;

%%-- Equivalent BPF  stretched coefficients --%%

Coeff_BPF_Stretched=Coeff_DUSF_Stretched*N

Coeff_BPF_Stretched_plot=[Coeff_BPF_Stretched 0];

%%--- Function Plotting ---%%

figure
stairs(t,Coeff_BPF_Stretched_plot,'-k','LineWidth',2)
hold on
j=0:0.001:(1.25);
plot(j,sin(pi*j/1.25),'--k','LineWidth'',2)

xlim([0 1.5])
ylim([0 1.01])

%%---- Plot ----%%

j1=0:0.001:1;
j2=1.001:0.001:1.5;
j=[j1 j2];

figure
plot(j,fj,'-k','LineWidth',2)
hold on

j3=0:0.001:(1.25);
plot(j3,sin(pi*j3/1.25),'-m','LineWidth',2)

ylim([0 1.2])

t=0:h:T;
figure
stairs(t,Coeff_BPF_plot,'--k','LineWidth',2)
hold on
stairs(t,Coeff_BPF_Stretched_plot,'-m','LineWidth',2)

ylim([0 1.2])
```

6.3 Solution of the functional differential equation of Example 6.3 both in DUSF and BPF domain.
(Chapter 6, Tables 6.2 and 6.3, Figure 6.6, pages 169 and 170)

```
clc
clear all
format long
```

```
%%-- Number of Sub-intervals and Total Time --%%

m = input('Enter the number of sub-intervals chosen:\n')
T = input('Enter the total time period:\n')

%%-- For Conventional BPF based Approximation --%%

h=T/m;
t=0:h:T;

I=eye(m);    % Identity matrix

Lambda=1.25;

%--- Operational Matrix for Integration ---%%

P=zeros(m);

for i=1:m
    for j=1:m
        if j-i==0
            P(i,j)=(h/2);
        elseif j-i>0
            P(i,j)=h;
        end
    end
end

P;

%%---- Stretch Matrix S ----%%

S=zeros(m);

for i=1:m
    ni=fix((i-1)*Lambda);
    alpha_i=((i-1)*Lambda)-ni;
    k=ni;
    if (k+1)<=m
        S(i,(k+1))=(1-alpha_i);
        if (k+2)<=m
            S(i,(k+2))=alpha_i;
        end
    end
end

S;
```

```
%%---------------- Solution of D matrix -------------%%

% z=[-1 zeros(1,(m-1))];
% c=(transpose(S)*transpose(z))+transpose(z);
% sp=(I+(transpose(S)*transpose(P))+transpose(P));
% d=inv(sp)*c
% X=(transpose(d)*P)+z

z=[1 zeros(1,(m-1))];

c=-(z*S)-z;

sp=I+(P*S)+P;

d=c*inv(sp);

X=(d*P)+z;

%%--- Equivalent BPF coefficients ---%%

X_BPF(1)=X(1);

for k=1:(m-1)
    X_BPF(k+1)=X_BPF(k)+X(k+1);
End

X_BPF

X_BPF_plot=[X_BPF X_BPF(m)];

D_BPF(1)=d(1);

for k=1:(m-1)
    D_BPF(k+1)=D_BPF(k)+d(k+1);
End

D_BPF

%%---- Function Plotting ----%%

% stairs(t,X_BPF_plot,'-k','LineWidth',2)
% ylim([0 1])

%%------------ Result for Table --------------%%

x_transient(1)=1;

for i=1:m
    x_transient(i+1)=x_transient(i)+(h*D_BPF(i));
end
x_transient
```

7.1 Integration of the function $f(t) = t^2/2$ of Example 7.1 in sample-and-hold function (SHF) domain using operational matrix. (Chapter 7, Figure 7.6, page 185)

```
clc
clear all
format long

%%-- Number of Sub-intervals and Total Time --%%

m = input('Enter the number of sub-intervals chosen:\n')
T = input('Enter the total time period:\n')
h=T/m;
tt=0:h:T;

%%--- Integration of a time-function ---%%

syms t
% f=t;
f=t^2/2;       % function to be integrated

fi=int(f,t);   % integrated function

t1=0:0.01:T;
Fi=subs(fi,t1);
plot(t1,Fi,'--k','LineWidth',2)   % exact plot
hold on
Fi_h=subs(fi,tt);
plot(tt,Fi_h,'ko','LineWidth',2,'MarkerSize',7,'MarkerFac
eColor','k')

%%--- Function integration in SHF domain ---%%

C_shf=subs(f,tt);
C_SHF=C_shf(1:m);

P1s=zeros(m);
for i=1:m
    for j=1:m
        if j>i
            P1s(i,j)=1;
        end
    end
end
P1S=h*P1s;
C_int_SHF=C_SHF*P1S
C_int_SHF_plot=[C_int_SHF C_int_SHF(m)];

hold on
stairs(tt,C_int_SHF_plot,'-m','LineWidth',2)
```

7.2 BPF domain solution of the sampled-data system shown in Figure 7.1(a) with the sample-and-hold matrix S.
   (Chapter 7, Table 7.2, page 181)

```
clc
clear all
format long

%%-- Number of Sub-intervals and Total Time --%%

m=40;
T=10;

q=4; n=10;

h=T/m;
tt=0:h:T;

%%-- Direct Expansion of Output in BPF domain --%%

syms t
c=1-exp(-t);

C_direct=zeros(1,m);
for i=1:m
    C_direct(i)=(m/T)*int(c,tt(i),tt(i+1));
                              % Calculating BPF Coefficients
end

C_direct;

%-- Operational matrix D in BPF domain --%%

d=zeros(m,m);

for i=1:m
    for j=1:m
        if j-i==0
            d(i,j)=2;
        elseif j-i>0
            d(i,j)=4*(-1)^(j-i);
        else
            d(i,j)=0;
        end
    end
end
D=(1/h)*d;

%-- Function Integration using Operational matrix D --%%

I=eye(m);              % Identity matrix
```

```
R1=ones(1,m);        % Input in BPF domain

BPOTF1=inv(D+I);

S=zeros(m);
for k=1:n
    i1=1+q*(k-1);
    j=i1:(k*q);
    S(i1,j)=ones(1,length(j));
End

S

C_BPF=R1*BPOTF1;     % Output in BPF domain

C1_new=R1*S*BPOTF1
```

7.3 Solution of the sampled-data system shown in Figure 7.1(a) using sample-and-hold operational transfer function (SHOTF) and comparison of the result with the result obtained through z-transform analysis.
(Chapter 7, Table 7.3, page 190)

```
clc
clear all
format long

%%-- Number of Sub-intervals and Total Time --%%

m = input('Enter the number of sub-intervals chosen:\n')
T = input('Enter the total time period:\n')
h=T/m;
th=0:h:T;

%%-- Output using conventional z-transform --%%

num=1-exp(-1);
den=[1 -(1+exp(-1)) exp(-1)];

x=[1 zeros(1,(m-1))];
y=filter(num,den,x)

%%-- Formation of SHOTF1 Matrix --%%

SH=zeros(m);

for i=1:m
    for j=1:m
        if j-i==1
```

```
            SH(i,j)=1;
        elseif j-i>=2
            SH(i,j)=exp(-(j-i-1)*h);
        end
    end
end

SHOTF1=(1-exp(-h))*SH;

%%-- Output of system using SHOTF1 --%%

R1=ones(1,m);

C1d=R1*SHOTF1
```

8.1 Integration of the function $f(t) = t^2/2$ of Example 8.1 in delta function (DF) domain using operational matrix for integration. (Chapter 8, Figure 8.3, page 207)

```
format long
clc
clear all

%%-- Number of Sub-intervals and Total Time --%%

m = input('Enter the number of sub-intervals chosen:\n')
T = input('Enter the total time period:\n')
h=T/m;
tt=0:h:T;

%%-- Integration of a time-function --%%

syms t
f=t^2/2;      % function to be integrated
fi=int(f,t);  % integrated function

t1=0:0.01:T;
Fi=subs(fi,t1);
plot(t1,Fi,'k','LineWidth',2)   % exact plot
hold on

%%--- Function integration in DF domain ---%%

C_df=subs(f,tt);
C_DF=C_df(1:m);

P1d=zeros(m);
for i=1:m
    for j=1:m
```

```
            if j>=i
                P1d(i,j)=1;
            end
        end
end
P1d;

C_int_DF=C_DF*P1d

C_int_DFm1=sum(C_df,2);
C_int_DF_plot=[C_int_DF C_int_DFm1]

stem(tt,C_int_DF_plot,'-k','LineWidth',2)
hold on

%%-- Integrated function in DF domain comparision --%%

C_DF_integrated=zeros(1,m);

for i=1:m
    C_DF_integrated(i)=sum(C_DF(1:i));
end

C_DF_integrated
C_DF_integrated_m1=sum(C_df,2);
C_DF_integrated_plot=[C_DF_integrated C_DF_integrated_m1]

stem(tt,C_DF_integrated_plot,'--m','LineWidth',2)
```

8.2 Solution of the discrete control system shown in Figure 8.5, with delta domain operational transfer function (DOTF). (Chapter 8, Table 8.1, page 213)

```
clc
clear all
format long

%%-- Number of Sub-intervals and Total Time --%%

m = input('Enter the number of sub-intervals chosen:\n')
T = input('Enter the total time period:\n')
h=T/m;
tt=0:h:T;

%%-- Output using conventional z-transform --%%

num=[1 0 0];
den=[1 -(1+exp(-1)) exp(-1)];

x=[1 zeros(1,9)];
```

```
y=filter(num,den,x)

%%--- Result obtained using DOTF1 ---%%

R1=ones(1,m);      % Input in BPF domain

a=1;

DOTF1=eye(m);
for i=1:m
    for j=1:m
        if j>i
            DOTF1(i,j)=exp(-(j-i)*a*h);
        end
    end
end

DOTF1;

C1_DOTF=R1*DOTF1
```

9.1 **Integration of the function** $f(t) = \sin(\pi t)$ **in non-optimal block pulse function (NOBPF) domain using operational matrix for integration and comparison of the result with BPF domain result. (Chapter 9, Table 9.2, page 230)**

```
clc
clear all
format long

%%-- Number of Sub-intervals and Total Time --%%

m = input('Enter the number of sub-intervals chosen:\n')
T = input('Enter the total time period:\n')
h=T/m;
t=0:h:T;
%%-- Functions for Approximated --%%

syms x
f=sin(pi*x);

%%-- BPF Based Representation --%%

C_BPF=zeros(1,m);
for i=1:m
    C_BPF(i)=(m/T)*int(f,t(i),t(i+1));   % Coefficients
end
C_BPF
```

```
%%-- NOBPF Based Representation --%%

Samples=subs(f,t);
C_NOBPF=zeros(1,m);

for i=1:m
    C_NOBPF(i)=(Samples(i)+Samples(i+1))/2;
end
C_NOBPF

%%--- Percentage Error ---%%

Percentage_Error=(C_BPF-C_NOBPF)./C_BPF*100
```

9.2 Analysis of the open loop first-order system having a plant impulse response of exp(-*t*), for a unit ramp input, using the convolution matrix in NOBPF domain.
(Chapter 9, Table 9.5, page 235)

```
clc
clear all
format long

%%-- Number of Sub-intervals and Total Time --%%

m = input('Enter the number of sub-intervals chosen:\n')
T = input('Enter the total time period:\n')
h=T/m;
tt=0:h:T;

%%--- Exact curve ---%%

syms t
c=t-1+exp(-t);

t1=0:0.01:T;
Ct=subs(c,t1);
plot(t1,Ct,'--k','LineWidth',2)  % exact plot
  hold on

%%-- Direct Expansion of Output in optimal BPF domain --%%

C_direct=zeros(1,m);
samples_c=subs(c,tt);

for i=1:m
    C_direct(i)=(samples_c(i)+samples_c(i+1))/2;
end
```

```
C_direct;
C_direct_plot=[C_direct C_direct(m)];
stairs(tt,C_direct_plot,'m-','LineWidth',2)
hold on

%%-- Convolution matrix Qc in NOBPF domain --%%

r=t;
samples_r=subs(r,tt);
R=zeros(1,m);
for i=1:m
    R(i)=(samples_r(i)+samples_r(i+1))/2;
end

gt=exp(-t);
samples_gt=subs(gt,tt);

E=zeros(1,m);
for i=1:m
    G(i)=(samples_gt(i)+samples_gt(i+1))/2;
end

Cc=zeros(m,m);

for i=1:m
    for j=1:m
        if j-i==0
            Cc(i,j)=G(1);
        elseif j-i>0
            Cc(i,j)=G(j-i)+G((j-i)+1);
        end
    end
end

%%--------- Convolution operation using Qc ----------%%

C_convolution=(h/2)*R*Cc

C_convolution_plot=[C_convolution C_convolution(m)];
stairs(tt,C_convolution_plot,'b-','LineWidth',2)

%%-- Calculation of Percentage Error --%%

Percentage_Error=(C_direct-C_convolution)./C_direct*100

AMP_ramp_NOBPF=sum(abs(Percentage_Error),2)/m
```

9.3 Identification of the plant of a second-order underdamped system having the impulse response $g(t) = \sin(t)$, for a unit step input using the "deconvolution" matrix in NOBPF domain. (Chapter 9, Table 9.20, page 248)

```
clc
clear all
format long

%%-- Number of Sub-intervals and Total Time --%%

m = input('Enter the number of sub-intervals chosen:\n')
T = input('Enter the total time period:\n')
h=T/m;
tt=0:h:T;

%%--- Exact curve ---%%

syms t
g=sin(t);

t1=0:0.01:T;
Gt=subs(g,t1);
plot(t1,Gt,'--k','LineWidth',2)   % exact plot
hold on

%%-- Direct Expansion of Impulse response in optimal BPF
                                             domain --%%

G_direct=zeros(1,m);
samples_g=subs(g,tt);

for i=1:m
    G_direct(i)=(samples_g(i)+samples_g(i+1))/2;
end
G_direct;
G_direct_plot=[G_direct G_direct(m)];

stairs(tt,G_direct_plot,'m-','LineWidth',2)
hold on

%%-- Convolution matrix Qc in NOBPF domain --%%

r=ones(1,m);

ct=1-cos(t);
```

```
samples_ct=subs(ct,tt);

C=zeros(1,m);
for i=1:m
    C(i)=(samples_ct(i)+samples_ct(i+1))/2;
end

Cc=zeros(m,m);

for i=1:m
    for j=1:m
        if j-i==0
            Cc(i,j)=r(1);
        elseif j-i>0
            Cc(i,j)=r(j-i)+r((j-i)+1);
        end
    end
end

%%-- Convolution operation using Qc ---%%

G_convolution=(2/h)*C*inv(Cc)

G_convolution_plot=[G_convolution G_convolution(m)];

stairs(tt,G_convolution_plot,'b-','LineWidth',2)

%%--- Calculation of Percentage Error ---%%

Percentage_Error=(G_direct-G_convolution)./G_direct*100

AMP_step_NOBPF=sum(abs(Percentage_Error),2)/m
```

10.1 Convolution of two time functions $f_1(t) = u(-t)$ and $f_2(t) = \exp(-t)$ for $0 < t < 1$ s, of Example 10.1, using the convolution matrix in NOBPF domain.
(Chapter 10, Figure 10.8, page 279)

```
clc
clear all
format long

%%-- Number of Sub-intervals and Total Time --%%

m=32;
T1=1; T2=1;
T=T1+T2;
h=T/m;
th=0:h:T;
```

```
%%--- Exact curve ---%%

syms t
c1=1-exp(-t);               % for 0 to 1 sec
c2=exp(-(t-1))-exp(-1);    % for 1 to 2 sec

t1=0:0.01:1;
Ct1=subs(c1,t1);

t2=1.01:0.01:2;
Ct2=subs(c2,t2);

tt=[t1 t2];
CT12=[Ct1 Ct2];

plot(tt,CT12,'-k','LineWidth',2)   % exact plot
hold on

%%-- Direct Expansion of Output in BPF domain --%%

m1=m/2;
C_direct1=zeros(1,m1);
C_direct2=zeros(1,m1);

for i=1:m1
    C_direct1(i)=(m1/T1)*int(c1,th(i),th(i+1));
end

th2=th((m1+1):(m+1));
for i=1:m1
    C_direct2(i)=(m1/T2)*int(c2,th2(i),th2(i+1));
end

C_direct=[C_direct1 C_direct2]

C_direct_plot=[C_direct C_direct(m)];

stairs(th,C_direct_plot,'m-','LineWidth',2)

%%-- Impulse response vector of System --%%

g=exp(-t);

G_direct=zeros(1,m1);
for i=1:m1
    G_direct(i)=(m1/T1)*int(g,th(i),th(i+1));
end

G_direct;
G_direct_new=[G_direct zeros(1,m1)];
```

```
%%-- Convolution matrix Qc --%%

r=[ones(1,m1)  zeros(1,m1)];
Cc=zeros(m,m);

for i=1:m
    for j=1:m
        if j-i==0
            Cc(i,j)=r(1);
        elseif j-i>0
            Cc(i,j)=r(j-i)+r((j-i)+1);
        end
    end
end

%%-- Convolution operation using Qc --%%

C_convolution=(h/2)*G_direct_new*Cc

  C_convolution_plot=[C_convolution C_convolution(m)];

hold on
stairs(th,C_convolution_plot,'b-','LineWidth',2)

%%-- Calculation of Percentage Error --%%

Percentage_Error=(C_direct-C_convolution)./C_direct*100;
```

Index